RECIPROCATING PUMPS

RECIPROCATING PUMPS

Terry L. Henshaw

VNR VAN NOSTRAND REINHOLD COMPANY
New York

Library of Congress Catalog Card Number 86-32577
ISBN 0-442-23251-9

Printed in the United States of America

Van Nostrand Reinhold Company Inc.
115 Fifth Avenue
New York, New York 10003

Van Nostrand Reinhold Company Limited
Molly Millars Lane
Wokingham, Berkshire RG11 2PY, England

Van Nostrand Reinhold
480 La Trobe Street
Melbourne, Victoria 3000, Australia

Macmillan of Canada
Division of Canada Publishing Corporation
164 Commander Boulevard
Agincourt, Ontario M1S 3C7, Canada

16 15 14 13 12 11 10 9 8 7 6 5 4 3 2 1

Library of Congress Cataloging-in-Publication Data

Henshaw, Terry L., 1935-
Reciprocating pumps.

Includes index.
1. Reciprocating pumps. I. Title.
TJ915.H46 1987 621.6′5 86-32577
ISBN 0-442-23251-9

Dedicated to the guys who keep 'em running.

FOREWORD

Engineers, operators, and mechanics have found themselves extremely frustrated for a number of years because they were unable to find any modern book dealing comprehensively with reciprocating pumps. Terry Henshaw's book has now filled a major gap in the technical literature, and he has made available to users of reciprocating pumps the up-to-date accumulated knowledge and technical data necessary to select, operate, and maintain this type of equipment most effectively.

Such a lack of information has not existed in the case of centrifugal pumps. One naturally asks oneself why this should be so. To understand the underlying circumstances, one needs to examine the historical record of pumping equipment in general. In 1840, Henry R. Worthington invented the first direct acting reciprocating pump and thereby opened the door for a completely new and effective means of moving liquids from one place to another, or from one level of energy to a higher level. His invention was, of course, followed very quickly by many improvements over that first pump, and reciprocating pumps became most willing and capable servants of our technological civilization. Parallel to this, many books became available for the user to consult and use as a guide in his search for a better understanding of this type of equipment.

As always, technological progress kept up with the need for larger capacities and a more economical utilization of resources. Very shortly after the commercialization of the centrifugal pump at the turn of this century, it began to rapidly displace the reciprocating pump. More and more manufacturers began to abandon what they perceived as a dwindling market. Indeed, when I was starting out on my own career a little over fifty years ago, there were those who were predicting the complete demise of the older design—quite mistakenly, as it turned out to be. The reciprocating pump is alive and well and shows every promise to continue to be so, for the simple reason that there is a need for certain conditions of service which can never be met as effectively by any other type of pumping equipment.

Thus, we can say that reciprocating pump technology is certainly older than that of centrifugal pumps, but it is not necessarily wiser. This is to some extent due to the fact that the smaller market available for reciprocating pumps and the concomitant smaller number of manufacturers of these pumps had failed to attract the critical size of engineering personnel to sustain the chain reaction so necessary to the continued growth of technological progress.

Fortunately, there are always exceptions to all rules and this book presents an excellent illustration of this observation. Not only, as I have said, did Terry Henshaw fill a glaring gap in the technical literature on his subject, but he has done so with considerable flair and distinction. The material he has selected to present cannot be found in such detail and with such clarity anywhere else.

Not accidentally, Terry Henshaw has addressed himself to the user rather than to the designer of reciprocating pumps, for it is the user who stands most in need of guidance and advice, and who merits our sympathy the most. In this, he shares my own predilection, for it is to the user that I have always addressed my own contributions to the technical literature on centrifugal pumps. In his task, he has succeeded most brilliantly and I am proud to be given the opportunity to attest to this publicly in signing this foreword,

IGOR J. KARASSIK

PREFACE

Recent years have seen a plethora of articles and books on centrifugal pumps, but little on reciprocating units. The reciprocating pump is thought by some to be obsolete, since the centrifugal has become predominant, but a sizable portion of pumping services are still performed by "recips."

Since the last general book on reciprocating pumps appeared, the industry has undergone an extensive evolution. New and improved reciprocating pumps, primarily power pumps, are being introduced each year. The newer designs allow operation at two to three times the speeds of 1940. These higher speeds allow utilization of smaller units, but impose more stringent requirements on system design, and on operating and maintenance procedures. This book attempts to update the literature in this area, and, because of their greater popularity, will concentrate on power pumps; while still providing a significant amount of information on direct-acting (steam) pumps.

We will explore the reasons that recips are used in many services; their designs, characteristics, and problems; and how to avoid these problems in their selection, application, and maintenance. Because it is often a high maintenance area, two chapters are devoted to packing and stuffing boxes. Because of the impact of NPSH on the operation of recips, two chapters are also devoted to that subject (the first of which is on *acceleration head*).

This book is not intended as a pump design guide, although pump designers will find portions of value. It *is* intended as a guide to system designers, purchasers, installers, maintainers, and operators of reciprocating pumps; with its major goal being satisfactory installations. It is actually a compilation of answers to many of the questions I faced as I worked my way through the processes of selection, application, installation, maintenance, and operation of these units.

Special thanks needs to be expressed to all of those who willingly contributed to this effort. Numerous pump companies, auxiliary equipment suppliers, and individuals have helped to make this book more complete.

It is my hope that you find that this book provides a logical progression of thought on each subject covered, and that it is a valuable tool in the performance of your job.

TERRY L. HENSHAW
Battle Creek, Michigan, U.S.A.

ACKNOWLEDGMENTS

A special thanks is due to each of the following individuals and companies which provided material and/or information for this book:

Contributor	Contribution
John C. Baumann, P.E. Consulting Chemical Engineer Lafayette, LA	Process Information
Barry W. Brown, LCDR USN (Retired) District Manager Union Pump Co. Burlingame, CA	Appendix E
Samuel L. Collier, P.E. Consulting Engineer Houston, TX	Information on Drilling Mud & Cementing Services
Durabla Manufacturing Company Paoli, PA	Figures 2-6, 2-8, 2-9
The Duriron Company Pressure Products Industries Division Warminster, PA	Figures 10-12, 10-13
Evans Pump Equipment Company Long Beach, CA	Figure 2-10
Fluid Kinetics Corporation Ventura, CA	Figures 10-15, 10-20, 10-22
Gary L. Fuller, Drafter Battle Creek, MI	Most of the drawings

Contributor	Contribution
Gaso Pumps, Inc. A Goulds Pumps Company Tulsa, OK	Figures 1-18, 2-40, 10-3, 10-11, 10-25, 11-7
Gaulin Corporation An APV Group Company Everett, MA	Figures 2-30, 3-15
Greer Hydraulics A VSI Company City of Commerce, CA	Figures 10-18, 10-23, 10-24
Gulf Publishing Company Book Publishing Division Houston, TX	Figure 9-8
Ingersoll-Rand Company Allentown, PA	Figures 1-1, 1-2, 1-13, 1-17, 2-12, 3-1
Kobe, Inc. Huntington Park, CA	Figures 2-4, 2-11, 2-31, 3-7, 3-13, 3-14, 4-25, 4-26, 10-7
Mission Drilling Products Division TRW Energy Products Group Houston, TX	Figures 1-15, 2-19, 2-26, 2-27, 5-13, 5-23
National Museum of American History Smithsonian Institution Washington, DC	Figure 1-3
NLB Corporation Wixom, MI	Figures 3-12, 8-1
Oil-Rite Corporation Manitowoc, WI	Figure 10-14
The Timken Company General Offices Canton, OH	Figure 2-35

Contributor	Contribution
Union Pump Co. Battle Creek, MI	Figures 1-5, 1-6, 1-10, 1-11, 1-14, 1-16, 1-19, 1-21, 1-22, 2-1, 2-2, 2-3, 2-5, 2-7, 2-13, 2-14, 2-15, 2-17, 2-21, 2-22, 2-23, 2-24, 2-25, 2-28, 2-29, 2-32, 2-41, 3-2, 3-3, 3-4, 3-5, 3-6, 3-8, 3-10, 4-14, 4-17, 4-18, 4-19, 4-21, 4-22, 4-23, 4-27, 4-28, 4-29, 8-4, 8-5, 9-1, 10-4, 10-6, 10-8, 10-9, 10-10, 10-26, 12-2, 12-3
Utex Industries, Inc. Houston, TX	Figures 4-15, 5-4, 5-7, 5-8, 5-11, 5-12, 5-21, 5-24, 5-25
Worthington Pump Division Dresser Industries, Inc. Harrison, NJ	Figures 1-20, 3-9, 3-11, 4-20

NOTICE

Readers familiar with reciprocating pumps will recognize that some statements in this book are not *always* true. As is the case with any subject, for just about any rule, an exception can be found. Recognizing that most of the statements contained herein do have one or more exceptions, but preferring to keep the text as clear and clean as possible, the author has tried to avoid the over-use of such words as "usually," "generally," and "typically." Statements are therefore intended to state the general condition. The reader is therefore cautioned that exceptions can be found to many of the statements in this book.

It should also be pointed out that, although both the author and publisher have been diligent in attempting to avoid all errors, it is possible that one or more have crept into the text, or more likely, into one or more equations, considering the number of hands that have processed the material to produce this book.

In light of the above, it is therefore necessary that the author and publisher supply this material without representation or warranty, and assume no liability, consequential or otherwise, which may arise from its use.

CONTENTS

Chapter 1
INTRODUCTION TO RECIPROCATING PUMPS

1.1. A BRIEF HISTORY AND OVERVIEW

The pump may have been man's fourth invention, following the wedge, the lever, and the wheel,[1] although one source[2] places its development even before the lever. Reciprocating pumps have been in use for over 2000 years. The development of the very early power-driven reciprocating pump shown in Fig. 1.1 is attributed to Ctesibius of Alexandria, Egypt about 150 B.C.,[2] about 100 years after Archimedes invented his famous screw pump. The Ctesibius unit consisted of a water wheel which drove a cam shaft, which, through levers, caused a pair of pistons to alternately reciprocate. These same basic principles are found in power pumps of current design.

One of the first applications of steam power to drive a pump was the Newcomen Pumping Engine, shown in Fig. 1.2, which was introduced about 1711.[2] A conventional reciprocating water well pump was connected through a rocking beam to a steam-driven piston. After steam from the boiler would push the piston up, water would condense the steam, and atmospheric pressure would push the piston back down. The reference does not say that the process was automatic, and the drawing indicates manual valves for the steam and water. A simpler predecessor of this machine did use manual valves, so the Newcomen Engine was probably also cycled manually.

The *direct-acting* steam pump was invented by Henry Worthington in 1840 while engaged in experiments with a steam-powered canal boat.[3] Figure 1.3 is a nineteenth century woodcut of the pump, a working replica of which is on display at the National Museum of American History in Washington, DC.

The direct-acting steam pump experienced its maximum popularity between 1850 and 1950. The power pump has claimed a sizable share of the pump market from about 1900 to the present. The relative popularity of both types of reciprocating pumps has waned with the development of the centrifugal pump and low cost energy sources. The centrifugal pump gradually encroached on those areas which were previously the domain of reciprocating units, although recips have concurrently expanded into new fields, and therefore still command a significant share of today's pump

FIG. 1.1. The Ctesibius Pump, circa 150 B.C. (*Courtesy Ingersoll-Rand Co.*)

FIG. 1.2. The Newcomen Pumping Engine, about 1711. (*Courtesy Ingersoll-Rand Co.*)

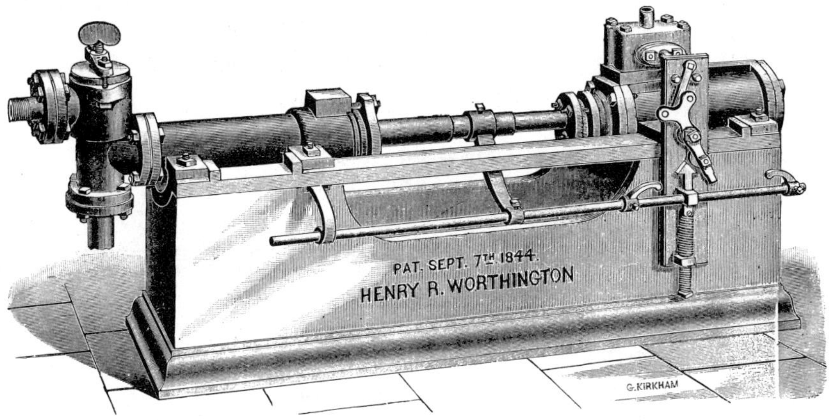

FIG. 1.3. The Worthington Direct-Acting Pump, 1840. (*Courtesy National Museum of American History—Smithsonian Institution*)

market. With the cost of energy now generally increasing, they may regain some of the market share recently taken by centrifugals.

1.2. DEFINITIONS

A *reciprocating pump* is a positive-displacement (''P-D'' or just ''displacement'') pump. A positive-displacement pump is one which traps a fixed volume of liquid at suction (inlet) conditions, compresses it to discharge pressure, and pushes it out the discharge nozzle. You will note that this definition is equally applicable to a rotary type pump, which is another type of P-D pump. In a recip pump, though, this pumping action is accomplished by the reciprocating motion of a piston, plunger, or diaphragm. This principle is illustrated in Fig. 1.4. The reciprocating pump is also distinguished from the rotary pump by a check valve on the discharge side of each pumping element, and usually one on the suction side also, to prevent the back-flow of the fluid from the discharge line to the suction line.

The reciprocating pump is not a kinetic machine, as is a centrifugal pump, because it does not require velocity to achieve pressure. High pressures are achieved at low velocities, and therein lies one of the advantages of the recip, particularly for pumping abrasive slurries and high-viscosity liquids.

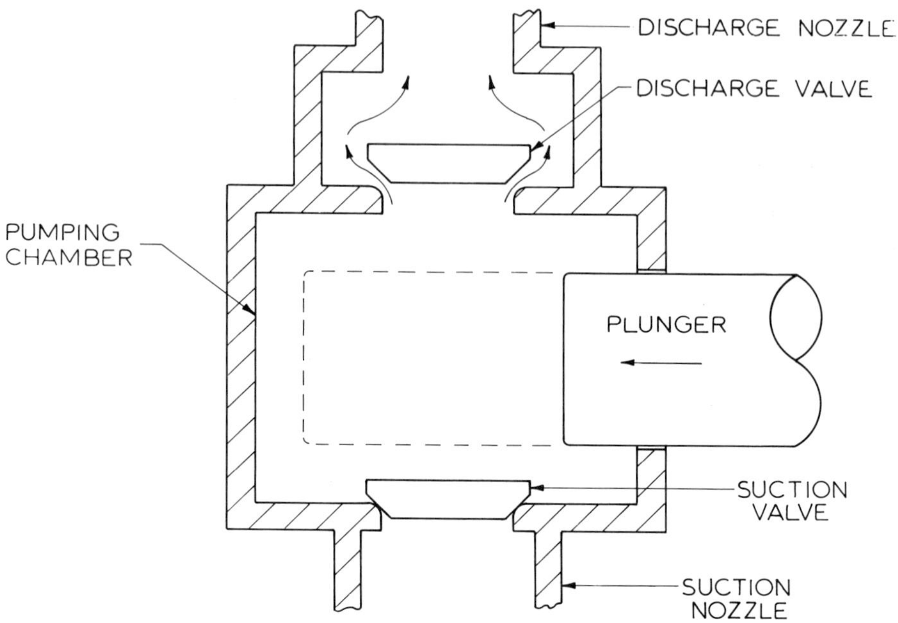

FIG. 1.4. A reciprocating pump is characterized by the pumping element compressing the trapped fluid and pushing it past the discharge valve.

A *power pump,* as shown in Fig. 1.5, is a reciprocating pump which drives the pumping element(s) with a crankshaft or cam shaft. It requires a driver with a rotating shaft (such as a motor, engine, or turbine).

A *direct-acting pump,* as shown in Fig. 1.6, is a reciprocating pump driven directly by fluid (gas or liquid). The drive end is similar to the steam end of a steam engine, but the unit contains no crankshaft. The force from the drive end is transmitted *directly* to the liquid end, thence the term "direct-acting." Since the direct-acting pump was originally designed to be driven by steam (and initially most were), it was known for decades as a "steam" pump—not by the fluid pumped, but by the motive fluid. As other gases and liquids have become more generally used as motive fluids, the term "direct-acting," being more descriptive, although more awkward, has been adopted.

The *liquid end* of a reciprocating pump, as shown in Figs. 1.5 and 1.6, is that portion of the pump which surrounds, and is in contact with, the pumpage.

The *drive end* of a reciprocating pump is that portion of the pump which provides power to the liquid end. The drive end of a power pump, called a "*power end*," is illustrated in Fig. 1.5. It normally contains a power

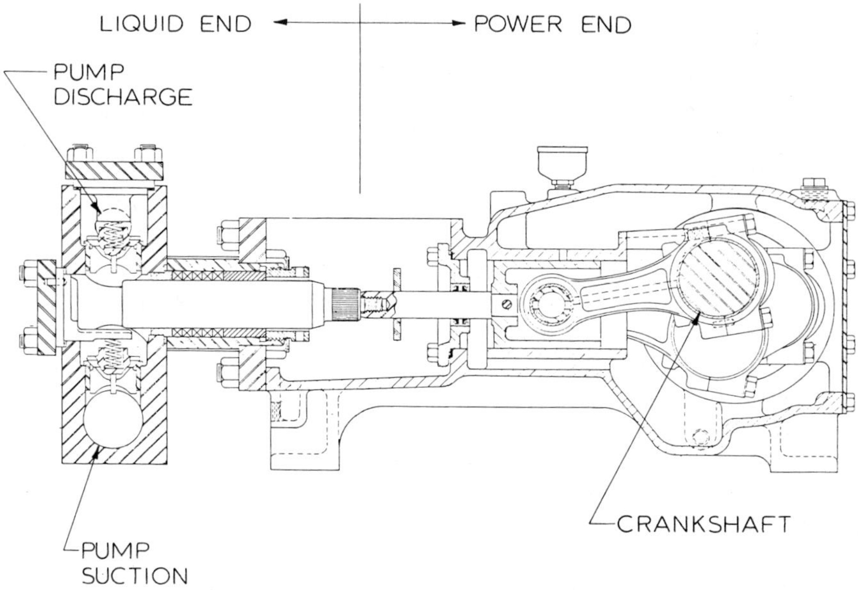

FIG. 1.5. Power pump. (*Courtesy Union Pump Co.*)

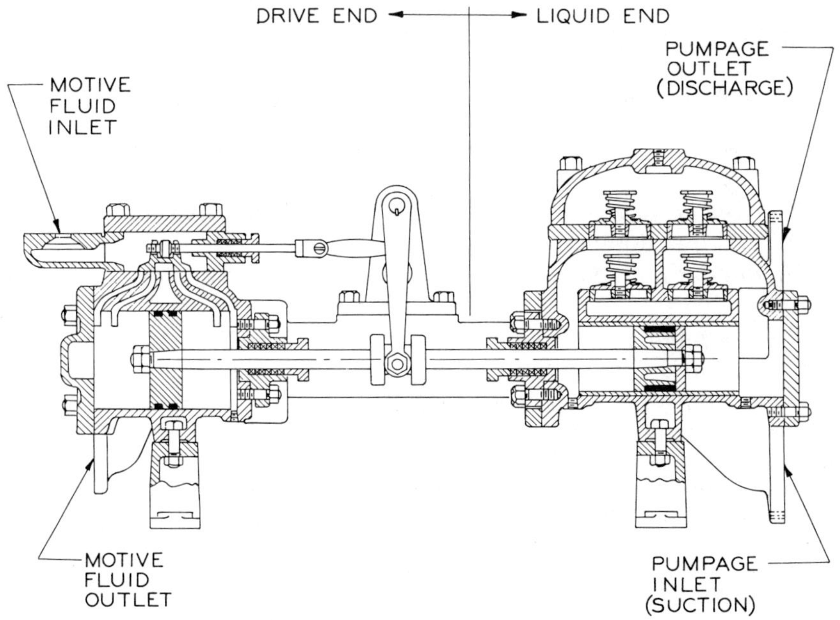

FIG. 1.6. Direct-acting pump. (*Courtesy Union Pump Co.*)

frame, crankshaft, connecting rod(s), and crosshead(s). The drive end of a direct-acting pump, as shown in Fig. 1.6, normally contains a drive cylinder, drive piston(s), piston rod(s), and control valving.

The word *stroke* has different meanings to different pump people. Some use "stroke" to mean *one* unidirectional motion of a piston (or plunger or diaphragm), i.e. equivalent to half of a revolution of the crankshaft. Others use the word to mean a complete cycle or revolution of the shaft (*two* motions of the piston). Since one of these values is twice the other, confusion often arises when, for example, one person says a pump is running 80 strokes per minute (meaning 40 rpm), while a second individual describes the same pump as running 40 strokes per minute. The author prefers the first definition, which agrees with the dictionary and with terminology used for internal-combustion engines, i.e.: *stroke*–one unidirectional motion.

Because of the different meanings, caution is recommended in the use of such terms as "strokes per minute" or "stroking rate." Confusion will be minimized if we use "revolutions per minute" (rpm) or "cycles per minute."

Pumpage is the term used to describe the fluid (sometimes containing solids and/or gases) which is pumped. "Product" is another popular term for pumpage.

1.3. CLASSIFICATION

Reciprocating pumps are usually classified by the following features:

1. *Orientation* of the centerline of pumping element—"horizontal" or "vertical."
2. *Number of discharge strokes per cycle* of each drive rod—"single-acting" or "double-acting."
3. *Number of noncoaxial drive rods*—"simplex," "duplex," or "multiplex."*
4. *Type of pumping element*—"piston," "plunger," or "diaphragm."
5. *Type of drive end*—"power" or "direct-acting."
6. *Adjustability of stroke length*—"fixed-stroke length" or "adjustable-stroke length."

Figure 1.7 illustrates this classification in a diagram. A typical description would be a "horizontal, single-acting, triplex, plunger, power pump."

* The "triplex" is a "multiplex" pump—and it's pronounced with a long "i," as in "tricycle."

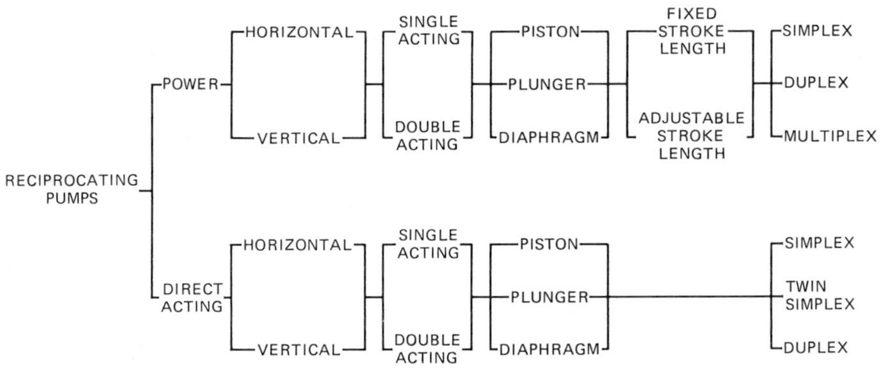

FIG. 1.7. Classification diagram for reciprocating pumps.

1.3.1. Horizontal vs Vertical

A reciprocating pump is classified as either horizontal or vertical, based on the orientation of the axis of its pumping element. For example, if the plunger(s) reciprocates vertically (an element must reciprocate along its axis), it is a vertical pump.

1.3.2. Single-Acting vs Double-Acting

Figure 1.4 illustrates a single-acting pump. As the plunger moves to the left, liquid is pushed out of the pumping chamber around the discharge valve. As the plunger moves to the right, liquid flows around the suction valve and into the pumping chamber.

Figure 1.8 illustrates a double-acting pump. As the piston moves to the left, liquid is pushed out of the left pumping chamber, around the left discharge valve, through the discharge manifold, and into the discharge line. Simultaneously, liquid flows from the suction line through the suction manifold, around the right suction valve, and into the right pumping chamber. When the piston moves to the right, liquid is pushed out of the right pumping chamber, and simultaneously flows into the left pumping chamber.

Although these figures show the typical arrangements of a plunger pump as single-acting and a piston pump as double-acting, such is not always the case. Plunger pumps are sometimes double-acting and piston pumps are sometimes single-acting. A plunger pump can be made double-acting by tying two coaxial plungers together with side-rods and cross-

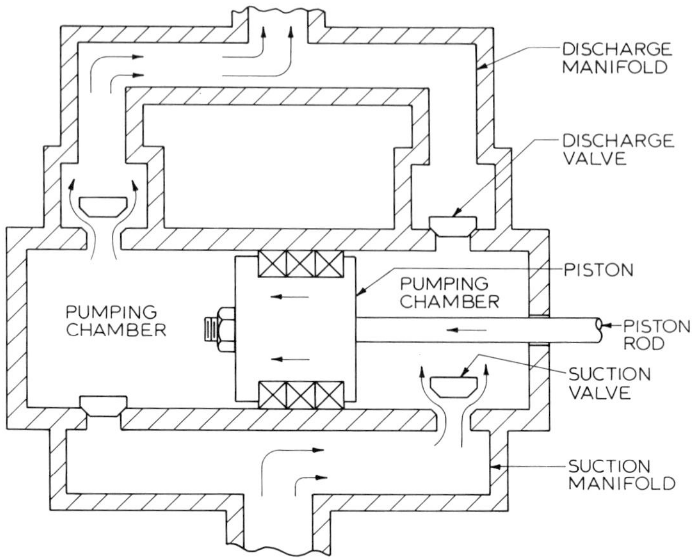

Fig. 1.8. Liquid end of a double-acting piston pump.

heads, as illustrated in Fig. 1.9. A piston pump can be made single-acting by exposing one end of the piston to the atmosphere, as seen in Fig. 1.15.

1.3.3. Simplex vs Duplex vs Triplex vs ____

Table 1.1 lists the names used by the industry[4] to designate the number of drive rods in a pump. The number of pumping elements or pumping chambers cannot always be used to determine the pump classification because a simplex pump, for example, may contain two or more plungers and pumping chambers, or two or more drive end pistons or liquid end pistons. Some direct-acting simplex pumps also contain two drive rods (one on each end of the drive cylinder). Therefore, the column heading reads "Number of *Noncoaxial* Drive Rods."

1.3.4. Piston vs Plunger vs Diaphragm

A question common among students of reciprocating pumps is the difference between a "*piston*" and a "*plunger*." In fact, these terms are often used interchangeably by users, particularly maintenance personnel, a fact which occasionally leads to confusion in communications.

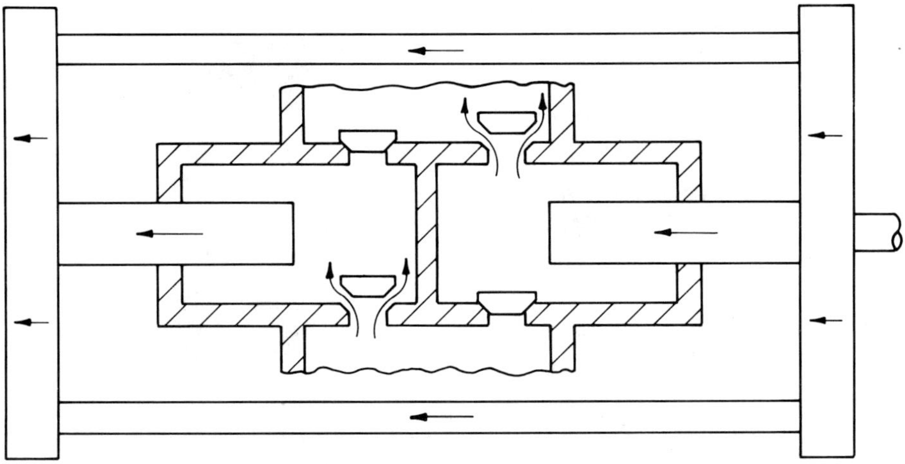

Fig. 1.9. Liquid end of a double-acting plunger pump.

A *piston,* as shown in Fig. 1.10, is a cylindrical disc, mounted on a smaller-diameter rod and usually fitted with some type of sealing rings. The rings move with the piston.

A *plunger,* as shown in Fig. 1.11, is a smooth rod, similar to a piston rod. The sealing rings are stationary, and the plunger slides through the rings.

Table 1.1. Types of Reciprocating Pumps as Determined by Number of Noncoaxial Drive Rods.

Number of Noncoaxial Drive Rods	Type of Pump
1	Simplex
2	Duplex
3	Triplex
4	Quadruplex
5	Quintuplex
6	Sextuplex
7	Septuplex
8	Octuplex
9	Nonuplex
10	Decaplex

FIG. 1.10. A piston mounted on a rod. (*Courtesy Union Pump Co.*)

It is interesting to note that rotary pump terminology does not follow these definitions. A rotary-type axial plunger pump is called an axial "piston" pump. The plungers are called "pistons."

A *diaphragm* is a flexible disc or tube, as shown in Fig. 1.12, which serves to isolate the pumpage from the piston, plunger, or atmosphere. A diaphragm may be mechanically-actuated (connected directly to a reciprocating rod) or hydraulically-actuated (driven by liquid which is oscillated by a piston or plunger).

1.3.5. Fixed Stroke Length vs Adjustable Stroke Length

Most of the larger industrial power pumps (stroke lengths above 3 inches) are manufactured with a fixed stroke length. Many of the smaller units are also made with fixed stroke lengths, but many of the smaller units intended for precise injection services are made with a variable stroke length. A large power pump made with a variable stroke length is shown in Fig. 1.13.

FIG. 1.11. A plunger. (*Courtesy Union Pump Co.*)

Fig. 1.12. Two types of diaphragms used in Reciprocating pumps.

1.4. CONFIGURATIONS

Power pumps are provided in a variety of configurations. Power ends are available both open and enclosed, although most current models are enclosed. Vertical units are provided with power ends above or below the liquid ends, the latter being more common. Some vertical plunger pumps *push* the plunger into the liquid cylinder, placing the connecting rods in compression, and the power frame and liquid end tie studs in tension. Other vertical pumps *pull* the plunger into the liquid cylinder, placing the power end in compression, and the connecting rods, connecting rod bolts, and side rods in tension. Horizontal plunger pumps normally *push* the plunger into the liquid cylinder.

On vertical units which have the liquid end above the power end, the power end supports the liquid end.

On most horizontal units, the power end supports the liquid end, although with some large horizontal units, the liquid end is supported independently.

Valve assemblies are normally integral with the fluid cylinder, but in some special applications, such as for slurry and/or hot services, the

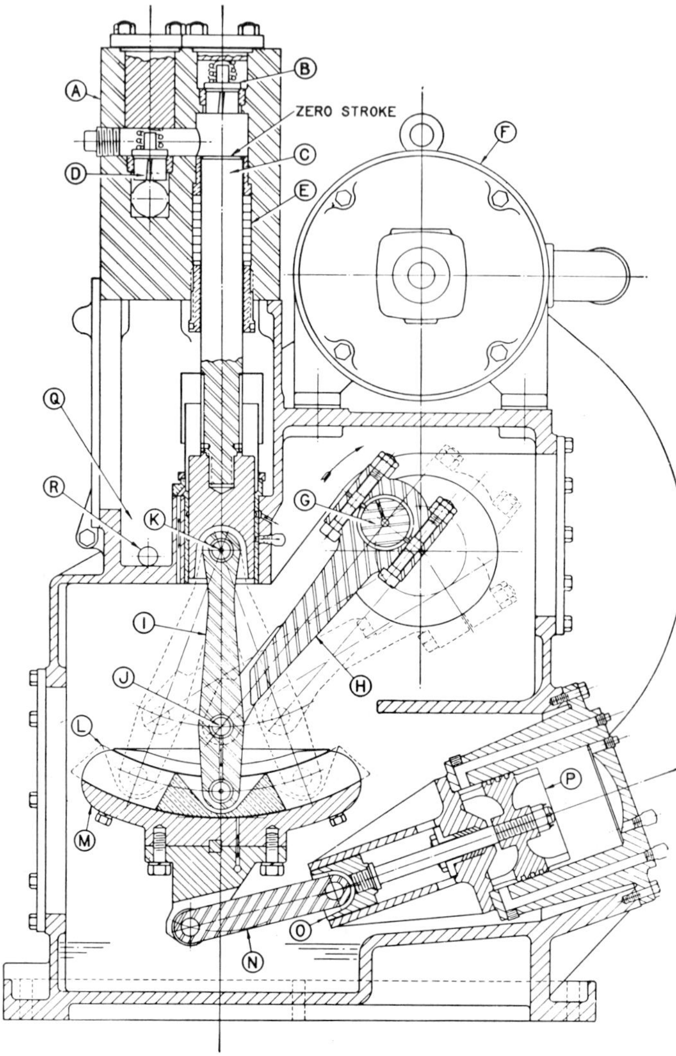

Pump adjusted for ZERO STROKE

FIG. 1.13. Large variable-stroke-length power pump. (*Courtesy Ingersoll-Rand Co.*)

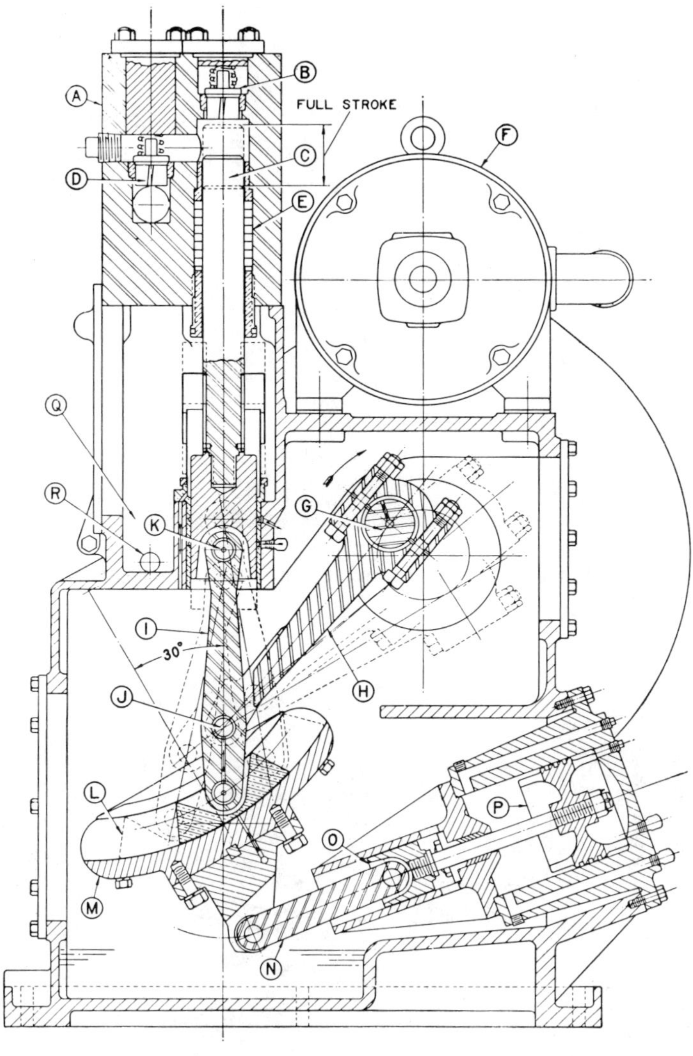

Pump adjusted for FULL STROKE

FIG. 1.13 (continued)

FIG. 1.14. Horizontal, triplex, plunger, power pump with top-mounted motor. (*Courtesy Union Pump Co.*)

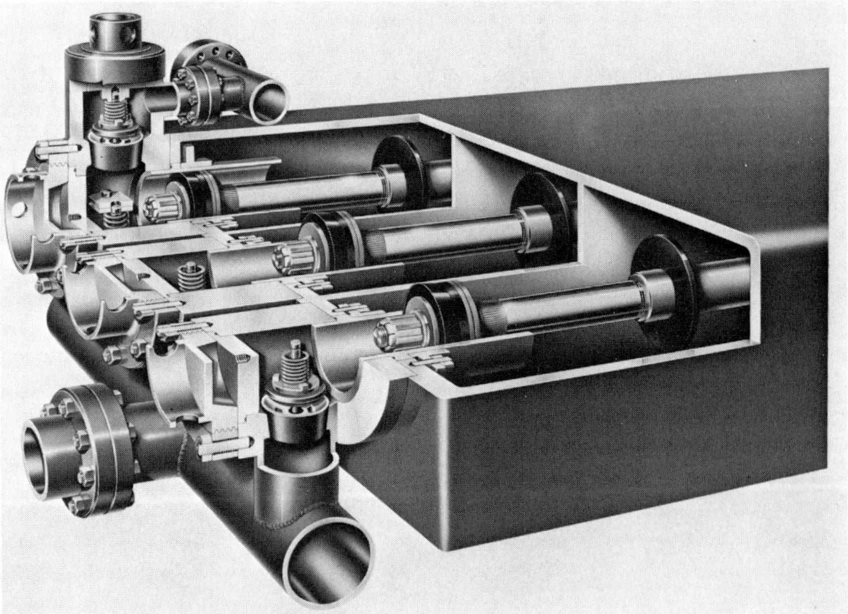

FIG. 1.15. Liquid end of a horizontal, triplex, piston, power pump. (*Courtesy TRW/Mission Drilling Products Div.*)

FIG. 1.16. Horizontal, quintuplex, plunger, power pump with top-mounted motor. (*Courtesy Union Pump Co.*)

valves are mounted remotely so that the plunger (or piston or diaphragm) and stuffing box are isolated from the pumpage.

Figures 1.14 through 1.22 illustrates some of the classifications and configurations in which reciprocating pumps are provided.

1.5. SIZE DESIGNATIONS

1.5.1. Power Pumps

The size of a power pump is normally designated by stating first the diameter of the plunger (or piston) and second the length of the stroke, as illustrated in Fig. 1.23. In the U.S., the units are inches.

EXAMPLE:

2 x 3

Plunger diameter ↗ ↖ Stroke length

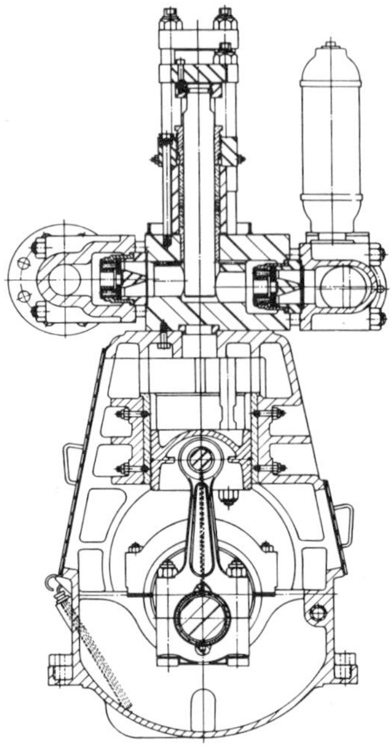

FIG. 1.17. Vertical, plunger, power pump. (*Courtesy Ingersoll-Rand Co.*)

FIG. 1.18. Horizontal, duplex, piston, power pump. (*Courtesy of GASO Pumps, Inc., a Goulds Pumps Company*)

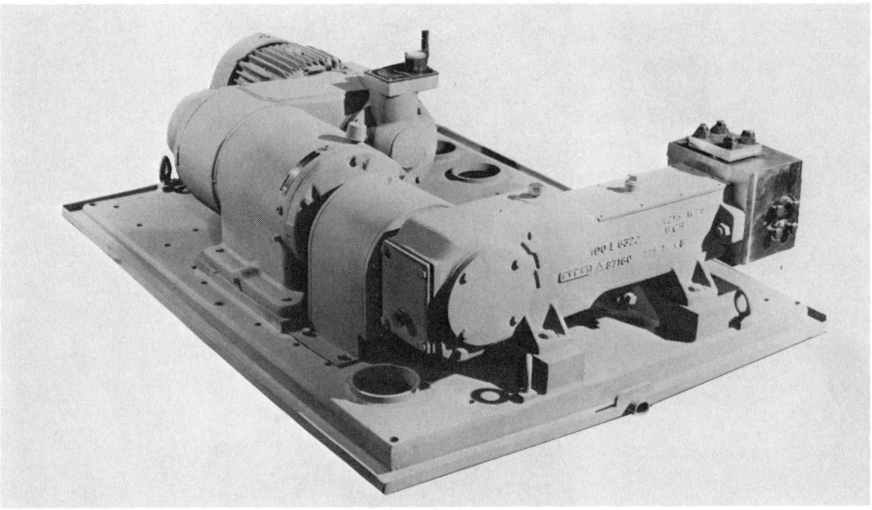

FIG. 1.19. Horizontal, simplex, plunger, power pump with side-mounted variable-speed driver. (*Courtesy Union Pump Co.*)

The size of a power pump is also sometimes designated by the size of the driver required at maximum continuous speed and load, such as "60 horsepower." This size driver is seldom used for continuous duty, though, because actual operating speed and load are usually less than the maximum, and suction pressure is often above atmospheric, all of which reduce the required power. The pump may, though, be operated on inter-

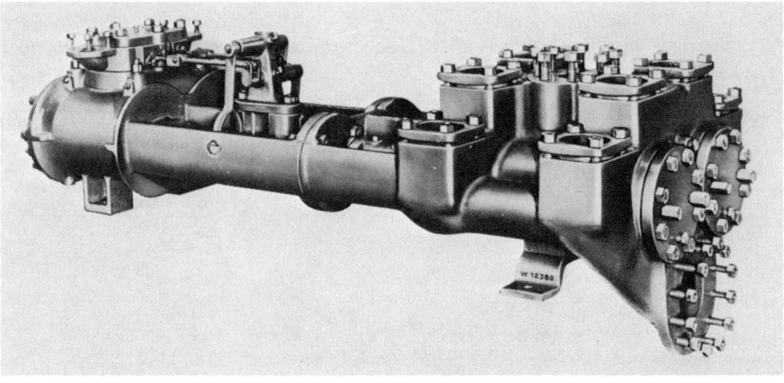

FIG. 1.20. Horizontal, duplex, piston, direct-acting pump. (*Courtesy Worthington Pump Div., Dresser Industries, Inc.*)

FIG. 1.21. Vertical, simplex, piston, direct- acting pump. (*Courtesy Union Pump Co.*)

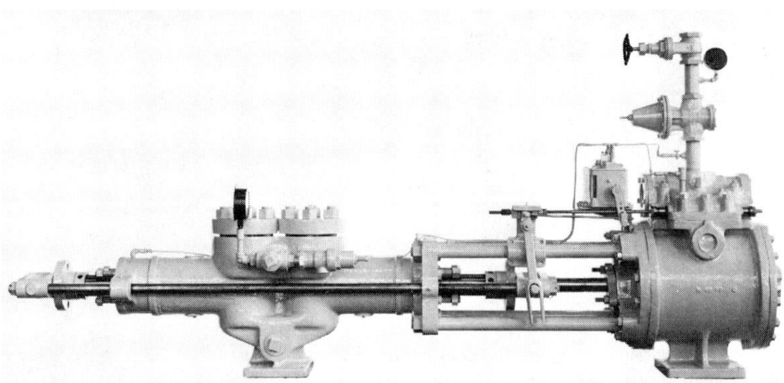

FIG. 1.22. Horizontal, simplex, plunger, direct-acting pump. (*Courtesy Union Pump Co.*)

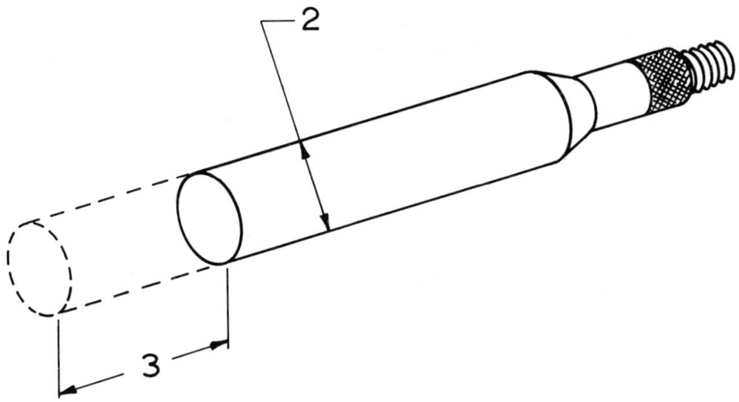

Fig. 1.23. Size designation for a power pump.

mittent duty at speeds and loads above the maximum continuous ratings, requiring a driver larger than indicated by its designation.

1.5.2. Direct-Acting Pumps

The size of a direct-acting pump is designated similarly to that of a power pump, except that the diameter of the drive piston preceeds the liquid element diameter.

Example:

6 x 4 x 6

Drive piston diameter → 6; Liquid piston diameter → 4; Stroke length → 6

This example is illustrated in Fig. 1.24.

1.6. REASONS FOR USE

A large field of hydraulic applications is best served by reciprocating pumps. Some of these services are beyond the capabilities of currently-produced centrifugal and rotary pumps. Others *could* be performed by centrifugals or rotaries, but not without an increase in power requirements and/or maintenance costs. Descriptions of some of the more common services performed by recips are contained in Chapter 3.

The justification for selecting a reciprocating pump, instead of a centrifugal or rotary, must be cost—not just initial cost, but total cost—includ-

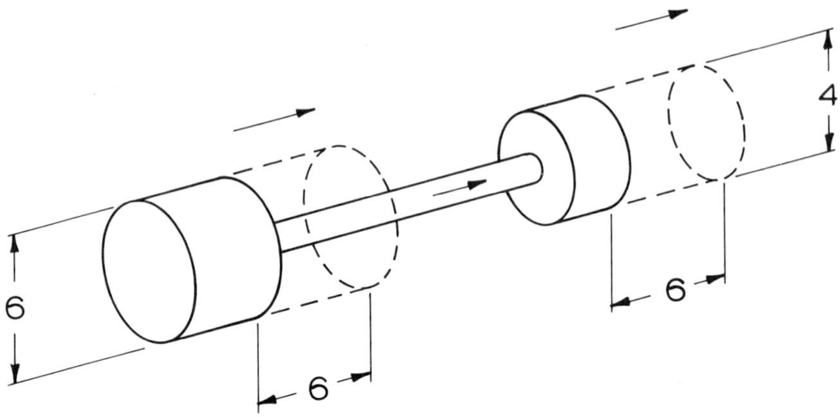

FIG. 1.24. Size designation for a direct-acting pump.

ing costs for energy and maintenance. Some applications, particularly in the high-pressure/low-flow-rate areas, are inherently best suited for reciprocating units. Typical services for power pumps include capacities up to 500 gpm and pressures above 1000 psi.

Direct-acting pumps are sometimes selected because of the availability of driving fluid. Fuel gas, which would otherwise be throttled through a pressure regulator, is often piped through a direct-acting pump to provide "free" pumping. The exhaust is piped into the fuel header for plant use. Compressed air is sometimes used to drive small units in such services as hydrostatic testing and chemical metering.

1.7. ADVANTAGES

1.7.1. Power Pumps

Possibly the best characteristic of the power pump is its inherently high mechanical efficiency. Overall efficiencies normally range from 85% to 95%. The approximate 10% loss includes all losses of belts, gears, bearings, packing, and valves. This means that approximately 90% of the power delivered by the driver is converted into liquid power.

Another (sometimes desirable) characteristic of the reciprocating pump is that its capacity is proportional to speed, and is relatively independent of discharge pressure. Therefore, a constant-speed power pump which pumps, for example, 100 gpm at 500 psig discharge pressure, will pump very nearly 100 gpm at 3000 psig discharge pressure. It is this feature which allows a controlled-volume pump to be used as a precise metering device.

Power pumps, with special fittings and operating at low speeds, are successfully applied to abrasive slurry services.

Under the right conditions, a power pump may be self-priming.

1.7.2. Direct-Acting Pumps

The direct-acting pump has some of the same advantages as the power pump plus some additional ones. These units also are well-suited for high-pressure, low-flow applications. Discharge pressures normally range from 300 to 5000 psig, but may exceed 10,000 psig. Capacity is proportional to speed from "stall" to maximum speed, regardless of discharge pressure. Speed is controlled by throttling the motive fluid. The unit is sometimes self-priming, especially if it is the low-clearance-volume type.

A variety of gases, such as steam, air, or natural gas, can be used to drive the direct-acting pump.

Because there is no bearing housing, crankcase, or oil reservoir (except when a lubricator is furnished); these units are affected negligibly by hostile environments such as corrosive fumes. Some have been inadvertently inundated by floodwater, and continued to pump, with no adverse effects.

Since they require no electricity, the direct-acting pump is inherently explosion-proof. They are quiet, simple to maintain, and their low speeds and rugged construction result in a very long life. It is not unusual to find a direct-acting pump still operating satisfactorily after 50 years of service.

Direct-acting pumps, with special fittings and operating at low speeds, are successfully applied to abrasive slurry services.

The low thermal efficiency of the direct-acting pump, normally a disadvantage, is sometimes used to advantage. Little heat is lost from the motive fluid from inlet to exhaust. The exhaust temperature is the same as would be obtained by throttling. Therefore, in those cases where steam is being throttled to a lower pressure for heating (such as for deaeration of boiler feed water), the steam can be used to drive a direct-acting pump, and the exhaust steam used for heating, with no loss of heating value. In this circumstance, the drive end (piston rings, valves, etc.) is made to operate without lubrication so that the exhaust steam will not be contaminated with oil.

1.8. DISADVANTAGES

Reciprocating pumps do come, though, with their own set of disadvantages, the most common being pulsations. Because of the pulsing flow, special considerations must be given to system design. Guidelines for system design are provided in Chapter 11.

For most pumping applications, the initial cost and maintenance cost of a recip will be more than for a centrifugal or rotary pump. Packing in a power pump typically lasts only about 2500 hours, considerably less than a typical mechanical seal on a rotating shaft (although some packing, in high-pressure, moderately high-speed, continuous 24 hr/day operation, has lasted more than 2 years (18,000 hours)).

The direct-acting pump, as mentioned earlier, has a low thermal efficiency when driven by a gas, such as steam. The "mechanical efficiency" (force output/force input) is high, but because the unit has no way to store energy, such as with a flywheel, the gas must remain at full inlet pressure in the drive cylinder throughout the entire stroke. At the end of the stroke, the gas is expanded to the exhaust line, but no work is performed during this expansion. Therefore, most of the available thermal energy in the gas is lost to friction, resulting in a low thermal efficiency. Steam consumption of these units is approximately 100 lb/hr/hhp, i.e. for each hydraulic horsepower (hhp) developed on the liquid end, about 100 lb/hr of steam is required. When the motive fluid is natural gas or air, the consumption is approximately 3000 SCFH/hhp, i.e. for each hydraulic horsepower developed on the liquid end, about 3000 standard cubic feet are required each hour.

1.9. AVOIDING PROBLEMS

To make the reciprocating pump acceptable to the user, its characteristics, particularly at higher speeds, must be fully understood. Most problems with reciprocating pumps can be avoided by selecting the pumps at moderate speeds, by carefully designing the pumping systems, by properly applying auxiliary equipment (such as dampeners, relief valves, and lubricators), and by proper maintenance practices (particularly preserving the alignment between the plunger and stuffing box).

The remainder of this book will attempt to provide material to better enable suppliers and users to achieve these goals.

REFERENCES

1. Settles, Gary S., Pumps, in *Collier's Encyclopedia,* vol. 19, E. Friedman, ed. in chief, pp. 518–521, New York: Macmillan Educational Corp. (London and New York: P. F. Collier, Inc.), 1980.
2. *Ingenious Mechanisms,* New York: Ingersoll-Rand, Hydraulic Division, 1960.
3. Hirschfeld, Fritz, Our glorious past: An ASME centennial profile—Henry R. Worthington—Inventor/industrialist and founding father, *Mechanical Engineering Magazine* (ASME), vol. 102, p. 24, March 1980.
4. *Hydraulic Institute Standards, 14th ed.,* Hydraulic Institute, 712 Lakewood Center North, 14600 Detroit Ave., Cleveland, OH 44107, 1983.

Chapter 2
COMPONENTS AND FUNCTIONS

Figure 2.1 identifies the major components of a typical power pump. The liquid end is to the left, the power end to the right.

Figure 2.2 identifies the major components of a typical direct-acting pump, with the drive end to the left, liquid end to the right.

2.1. LIQUID END COMPONENTS AND FUNCTIONS

All reciprocating pumps contain one or more pumping elements (pistons, plungers, or diaphragms) which reciprocate into and out of pumping chambers to produce the pumping action. Each pumping chamber contains at least one discharge (outlet) valve and, usually, one suction (inlet) valve. Typical power pump liquid ends are shown in Figs. 2.3 and 2.4.

The liquid end is that portion of the pump which contains and retains the pumpage. Elements common to the liquid ends of all reciprocating pumps are the liquid cylinder, the pumping element(s), and valve(s).

2.1.1. Liquid Cylinder

The liquid cylinder is the major component of the liquid end. (It often is not cylindrical in shape, particularly in a power pump, where it tends to be more rectangular.) It is the major pressure-retaining part, forms the major portion of the pumping chamber, and usually contains or supports all other liquid end components.

2.1.2. Liquid-end Valves

The valves in the liquid end of a reciprocating pump are simply check valves. They are opened by liquid differential pressure only (not by any mechanical device). Most of these valves are spring-loaded, a feature necessary to get the valve back to its seating surface quickly. (More information on this is given in Chapter 6.)

Disc Valves. Valves come in a variety of shapes, the most common probably being a disc. Figures 2.5, 2.6, and 2.7 illustrate disc valves which are guided by bolts or studs in the center of the discs. The valve in Fig. 2.6

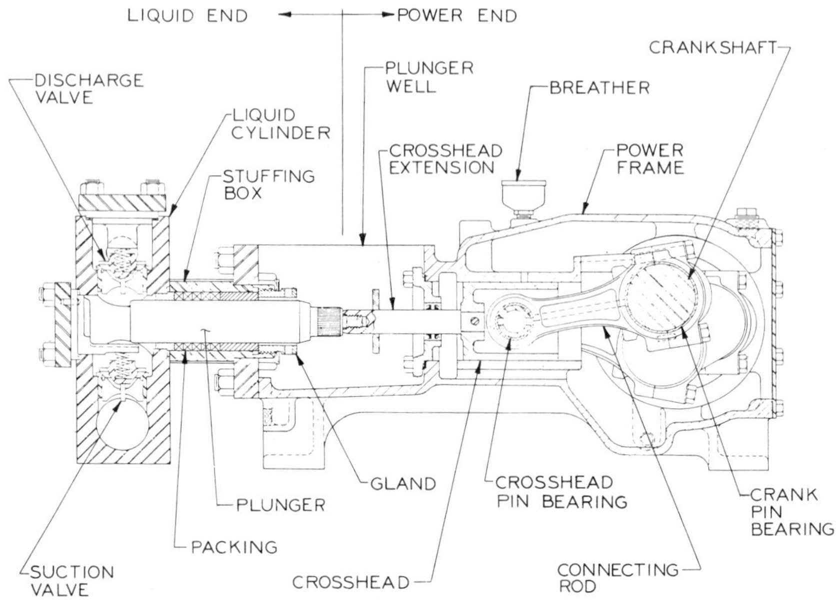

FIG. 2.1. Components of a power pump. (*Courtesy Union Pump Co.*)

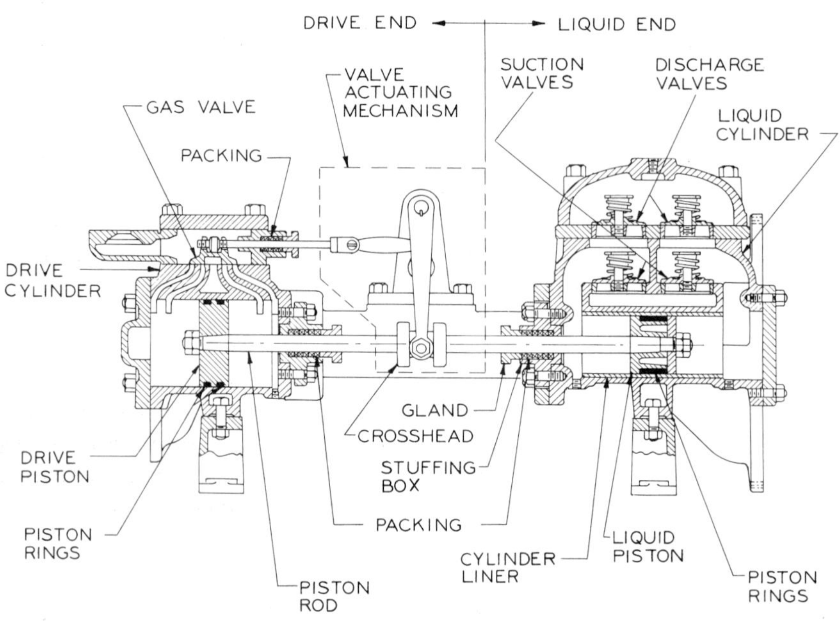

FIG. 2.2. Components of a direct-acting pump. (*Courtesy Union Pump Co.*)

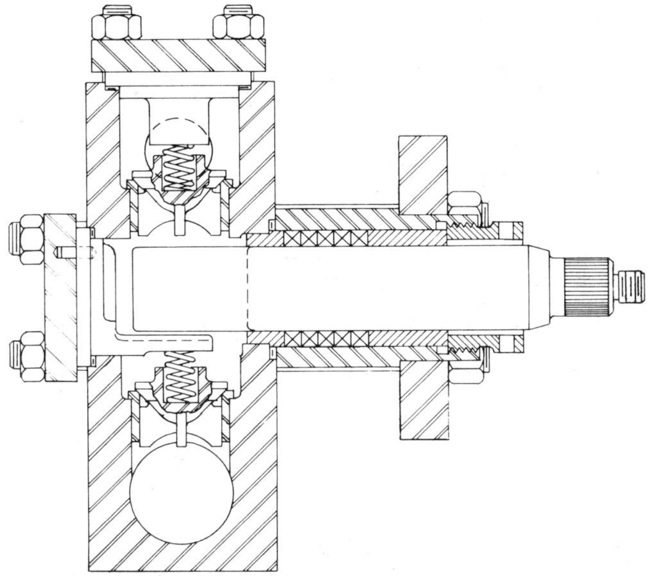

FIG. 2.3. The liquid end of a horizontal power pump. (*Courtesy Union Pump Co.*)

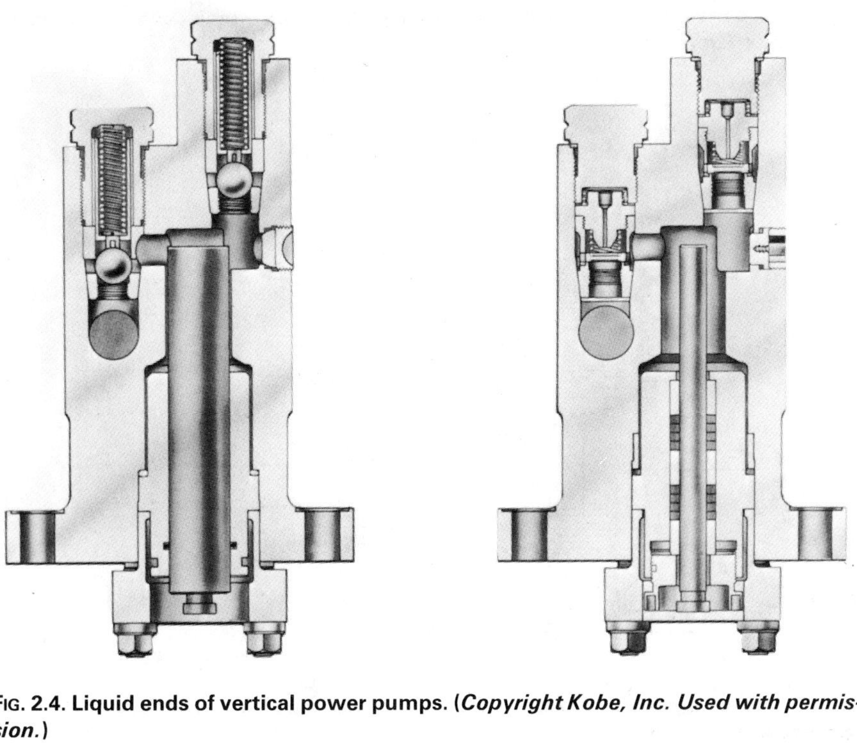

FIG. 2.4. Liquid ends of vertical power pumps. (*Copyright Kobe, Inc. Used with permission.*)

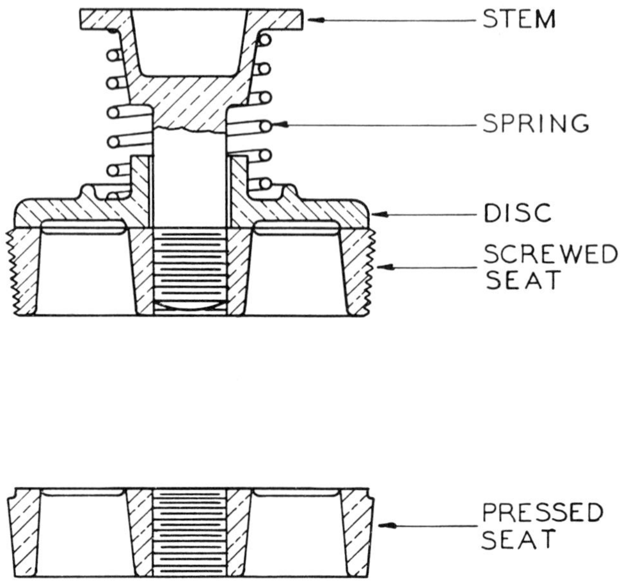

FIG. 2.5. A conventional disc-type valve assembly—Guided by a center stem. (*Courtesy Union Pump Co.*)

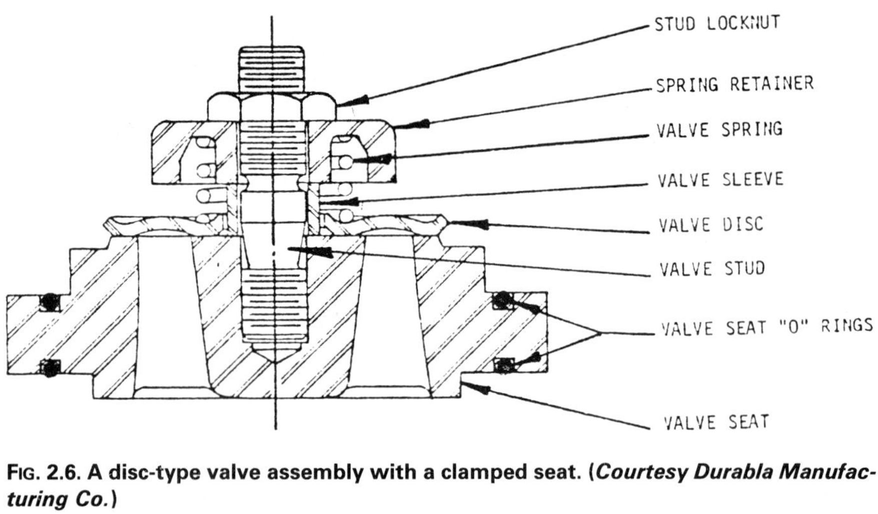

FIG. 2.6. A disc-type valve assembly with a clamped seat. (*Courtesy Durabla Manufacturing Co.*)

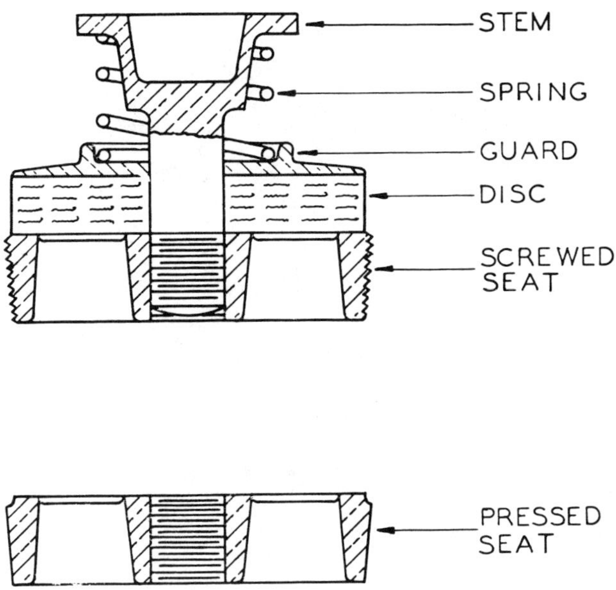

FIG. 2.7. A disc-type valve assembly with a composition disc. (*Courtesy Union Pump Co.*)

is also shown in the photographs of Figs. 2.8 and 2.9. This valve requires two seating surfaces for the disc—one near the outside diameter (O.D.) and one near the center guide. The valve assemblies in Figs. 2.10 and the lower portion of 2.11 guide the disc with the cage at the O.D. of the disc. With no center guide, only an outer seating surface is required.

With each of the above valve units, liquid flows only around the outside of the disc. Figure 2.12 illustrates a disc valve which allows liquid flow

FIG. 2.8. Top view of a disc valve assembly. (*Courtesy Durabla Manufacturing Co.*)

Fig. 2.9. Bottom view of a disc valve assembly. (*Courtesy Durabla Manufacturing Co.*)

through the large center hole, as well as around the disc O.D. The extra flow area allows a higher capacity for the same size valve.

Wing-Guided Valves. Another popular liquid valve, often used for higher pressure applications, is the bevel-seat wing-guided valve as shown in Figs. 2.13 and 2.14. The valve is guided by wings integral with,

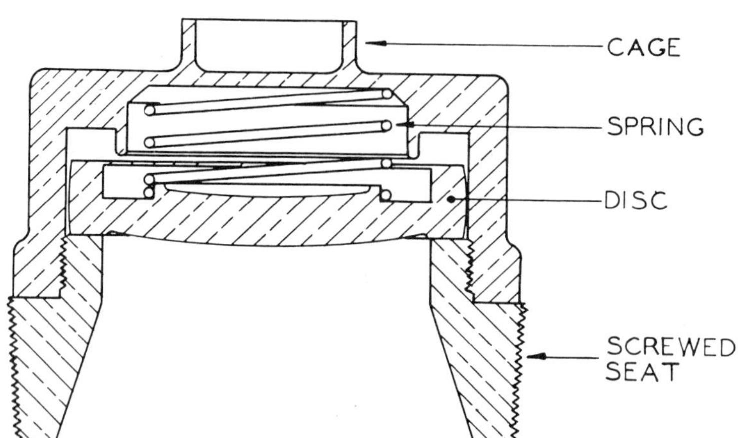

Fig. 2.10. A disc valve assembly with the disc guided on the outside diameter. (*Courtesy Evans Pump Equipment Co.*)

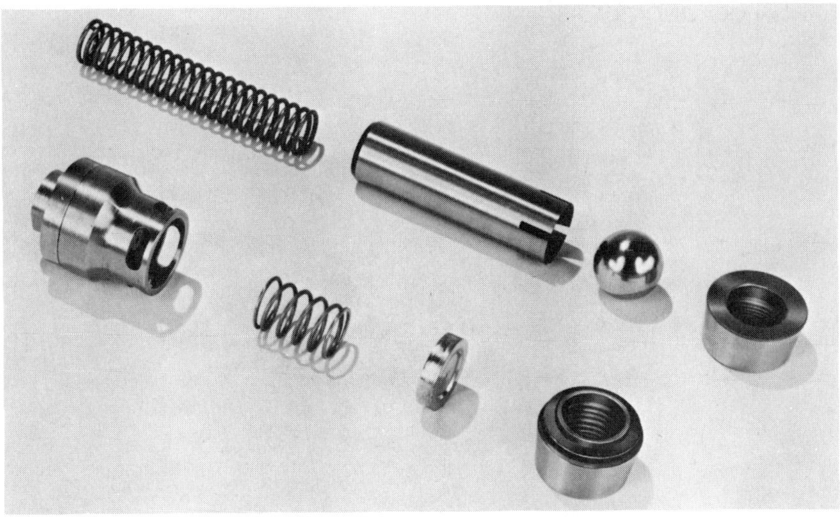

FIG. 2.11. Disc and ball valve assemblies. (*Copyright Kobe, Inc. Used with permission.*)

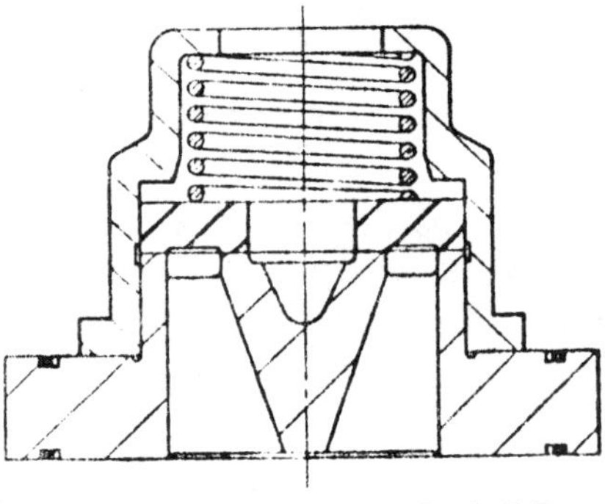

FIG. 2.12. A disc valve assembly with flow area at both O.D. and I.D. (double-ported). (*Courtesy Ingersoll-Rand Co.*)

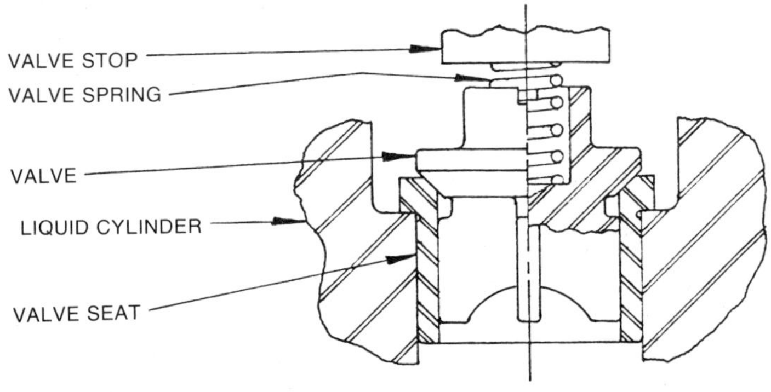

FIG. 2.13. A wing-guided valve assembly. (*Courtesy Union Pump Co.*)

FIG. 2.14. A wing-guided valve and seat. ***(Courtesy Union Pump Co.)***

and on the upstream side of, the disc. The wings have a small radial clearance with the hole in the valve seat, to accurately guide the disc back to its seat upon closing. The seating surfaces on both the valve and the seat are beveled typically at an angle of 45°.

Ball Valves. A valve finding application in a variety of services is the ball, as shown in Figs. 2.15, 2.16, and the top portion of 2.11. Its greatest use is in smaller pumps, such as chemical-injection metering (controlled-volume) units. Ball valves also find use in larger recips, especially for slurry applications. Balls are often used for pumpage containing stringy matter such as sewage sludge, because there are no wings or webs to catch these solids. The balls are sometimes spring-loaded. Because the weight of the ball becomes a significant factor in the larger sizes, they are often made hollow when the diameter exceeds about $3\frac{1}{2}$ inches. Even then, the inertia of the ball usually limits the pump speed to a value below that allowed by lighter-weight valves. Ball valves are usually made of steel or ceramic, but for some low-pressure applications, rubber is used.

Bell Valves. An adaptation of the ball valve is the bell valve (Fig. 2.17). The bottom portion of the valve is spherical (therefore having many of the desirable characteristics of the ball), with a stem attached to the top. The

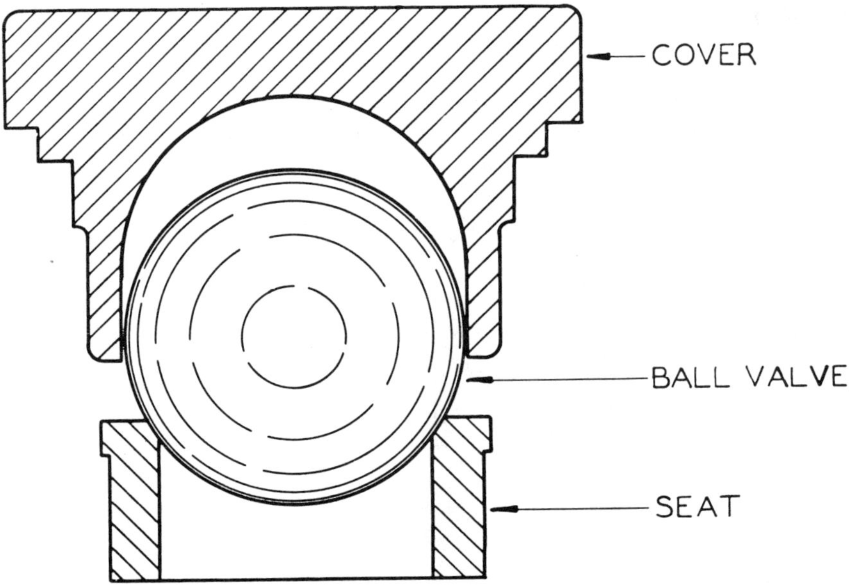

FIG. 2.15. A ball valve assembly. (*Courtesy Union Pump Co.*)

FIG. 2.16. A ball valve and seat.

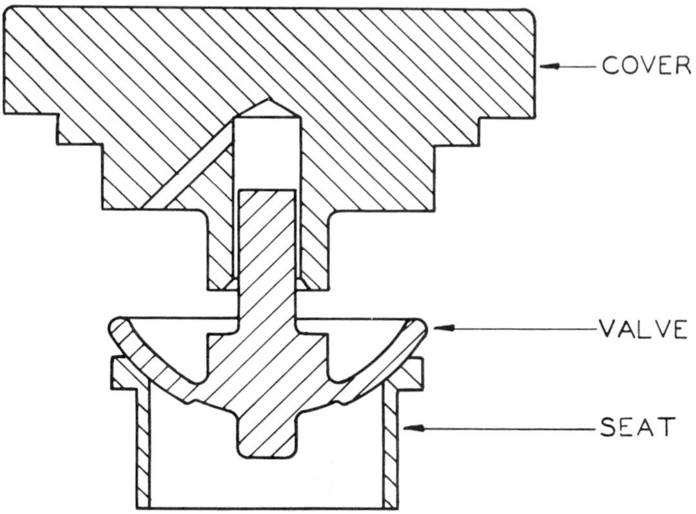

FIG. 2.17. A bell valve assembly. (*Courtesy Union Pump Co.*)

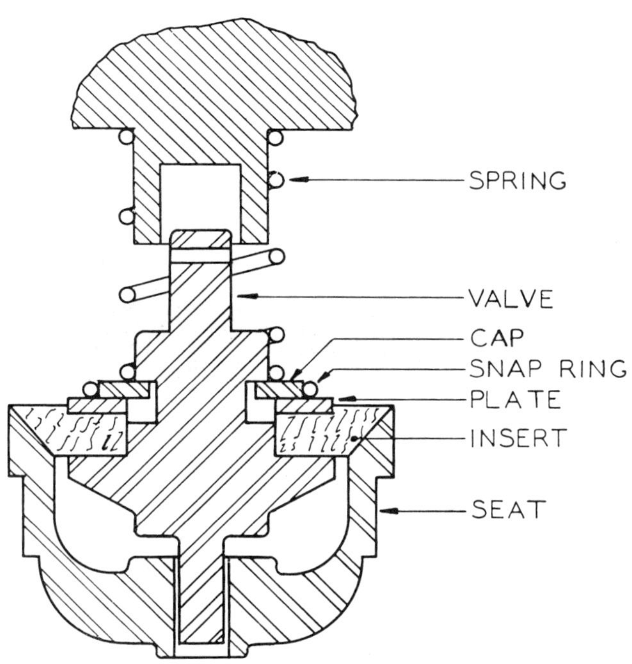

FIG. 2.18. A center-guided valve with an elastomeric sealing element.

stem guides the valve, and allows the valve to be more readily spring-loaded. The lighter weight allows higher cycling frequencies, and the smaller size allows smaller pump dimensions, which results in a lower clearance volume.

Elastomeric Valves. Pumps handling slurries at lower temperatures (below about 250° F) are often equipped with a valve containing some type of rubber seating surface. Figures 2.18 and 2.19 illustrate such a valve used in pumps handling drilling mud, powdered coal, and other such slurries. The rubber will temporarily absorb and seal around small particles that become trapped between the valve and seat when the valve closes. To prevent extrusion, the rubber in the valve in Fig. 2.19 is backed by a steel disc, which presses against the lower portion of the seating surface of the seat when pressure is applied to the top of the valve.

FIG. 2.19. A center-guided valve with an elastomeric sealing element. (*Courtesy TRW/ Mission Drilling Products Div.*)

2.1.3. Valve Seats

Valves sometimes seat directly on the fluid cylinder, but most industrial pumps contain replaceable valve seats. These seats are held to the cylinder by bolting, clamping, threads, or a press fit. In smaller valves the press is sometimes straight. In larger valve units the press is often a taper. Previous illustrations show various interfaces between the cylinder and seat.

2.1.4. Stuffing Box Assemblies

Stuffing boxes (packing boxes) contain the rings of packing which seal around the reciprocating rod or plunger. The box may be a separate piece or integral with the cylinder. The stuffing box assembly may contain a gland (packing nut), throat bushing, gland follower bushing, and lantern ring. The gland may be held to the box with bolting, or may thread into (or onto) the box. A simple stuffing box assembly is shown in Fig. 2.20. Stuffing box assemblies are covered in greater detail in Chapter 4.

2.1.5. Piston-Type Liquid Ends

Piston. A piston is a cylindrical disc, mounted on a rod, and normally containing some type of sealing rings such as piston rings or packing. Figure 2.21 illustrates a one-piece piston with rings. The rings may be one

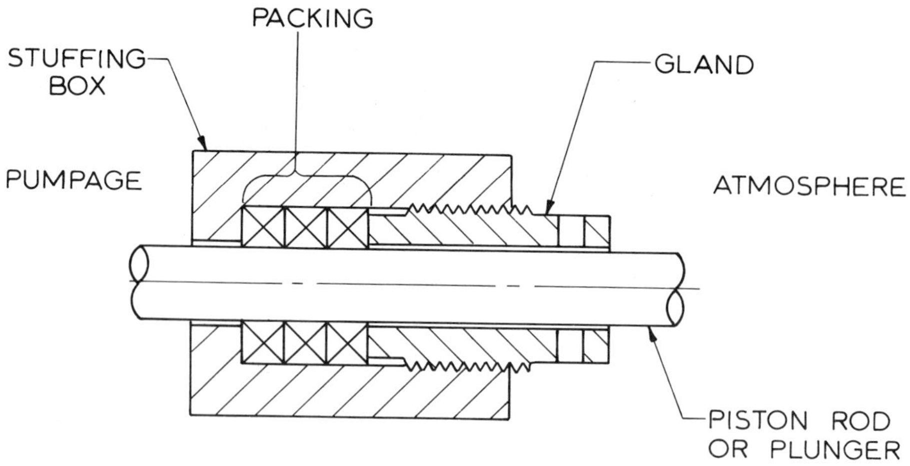

FIG. 2.20. A simple stuffing box assembly.

FIG. 2.21. A one-piece piston with rings and rod. (*Courtesy Union Pump Co.*)

piece, as shown, or segmented (in which case expanders are required), and may be made of a metal, plastic, or fiber material. Figure 2.22 illustrates a two-piece "body-and-follower" type piston made to accept square rings of packing or a three-piece ring arrangement, as shown in Fig. 2.23.

Another type of piston is the cup type, as shown in Fig. 2.24. This cup-type piston is a stacked assembly held together by the nut on the end of the piston rod. The cup is usually made of some type of rubber, cloth, or plastic, or a combination of two or all three of these materials.

It is the function of the piston to transmit the force from the piston rod to the liquid. The rings or cups serve to minimize leakage from the high pressure side of the piston to the low pressure side.

In double-acting pumps, the differential pressure across the piston is the difference between discharge and suction pressures (plus pressure drops across suction and discharge valves). Any leakage around the piston is contained within the pump, not escaping to the atmosphere, but liquid through-put (capacity) is reduced by this leakage.

When a piston is used in a single-acting pump, as seen in Fig. 1.15, leakage past the piston goes to the atmosphere.

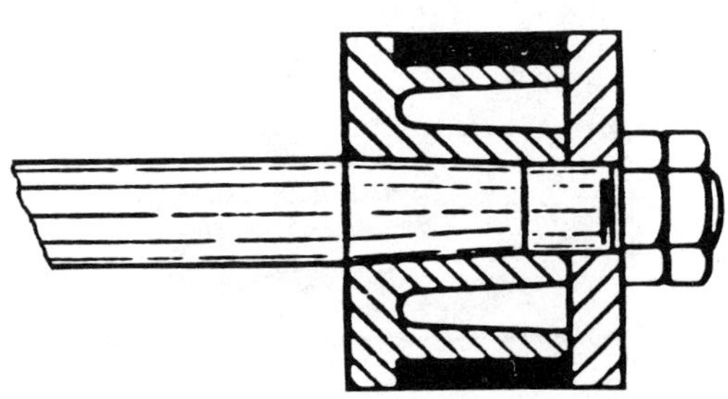

FIG. 2.22. A two-piece body-and-follower piston. (*Courtesy Union Pump Co.*)

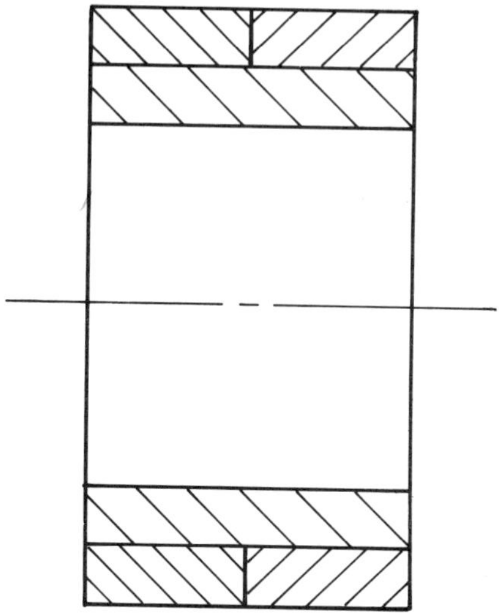

FIG. 2.23. A three-piece piston ring assembly for the body-and-follower piston. (*Courtesy Union Pump Co.*)

Liner. To absorb wear from the sealing elements, most piston pump liquid cylinders are equipped with a replaceable liner. This is a cylindrical sleeve which is fastened to the cylinder by an interference fit or by bolting (see Fig. 2.25). The liner is often made of iron or bronze, but in the more severe services, is made of a harder material for extended life.

If the liner is bolted to the cylinder, a gasket (or other sealing element) is provided to prevent leakage, from one end of the cylinder to the other, between the liner and cylinder. Because leakage at this point would erode

FIG. 2.24. A cup-type piston. (*Courtesy Union Pump Co.*)

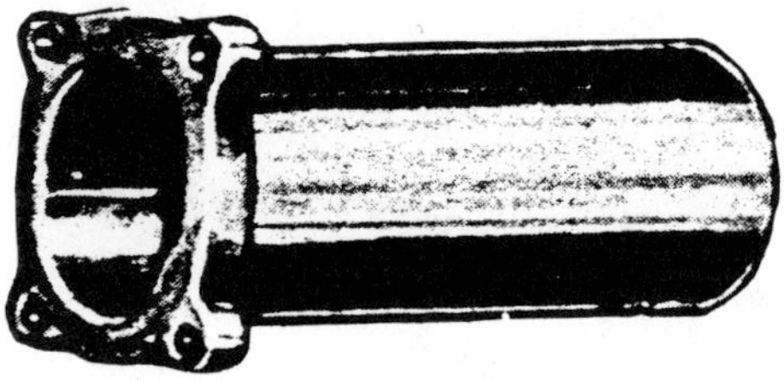

FIG. 2.25. A liner for a liquid cylinder. (*Courtesy Union Pump Co.*)

the cylinder, some pumps are built with a system to indicate such leakage, so that corrective maintenance may be performed before damage occurs. Figures 2.26 and 2.27 show a set of high-pressure liner packings with a lantern ring for indicating leakage through either sealing ring.

Piston Rod. The piston rod is the rod which drives the piston. The piston usually mounts on a tapered section of the rod or against a shoulder on the rod, and is held on the rod with a nut. For a double-acting unit, the shoulder and the nut are made suitable to transmit the required force to the piston. The rod is alternately in compression and tension. To extend the life of the rod, it is often hardened or coated with a hard material in the area which passes through the packing. A typical liquid end piston rod is shown in Fig. 2.21.

FIG. 2.26. A set of high-pressure liner packing. (*Courtesy TRW/Mission Drilling Products Div.*)

FIG. 2.27. Individual pieces of the liner packing assembly in Fig. 2.26. (*Courtesy TRW/ Mission Drilling Products Div.*)

2.1.6. Plunger-Type Liquid Ends

Plunger pumps are simpler than piston pumps, containing fewer components. A plunger pump can be thought of as a piston pump with the piston removed. A plunger, as shown in Fig. 2.28, is simply a smooth rod which pumps by displacing the fluid in the pumping chamber. Because the plunger contacts only stuffing box components, a cylinder liner is not required. Also, since there is no piston, the piston rings and a piston fastener are eliminated.

FIG. 2.28. A plunger is just a smooth rod. (*Courtesy Union Pump Co.*)

Parts common to both piston and plunger pump liquid ends a
cylinder, valve assemblies, and stuffing box assemblies.

2.2. DRIVE END COMPONENTS AND FUNCTIONS

2.2.1. Power Pumps

The drive end of a power pump is called the "power end" (see Fig. 2.29). It bears a close resemblance to an internal combustion engine, containing a crankshaft, connecting rods, wrist pins, and bearings, with the cross-heads being similar to the pistons in an engine. (The direction of the flow of power, though, is opposite.) The major function of the power end is to convert rotary power to reciprocating power, and to supply that reciprocating motion to the pumping elements in the liquid end. Figure 2.30 shows the major internal parts of a horizontal power end. This unit contains gears to reduce driver speed to crankshaft ("eccentric shaft") speed. Figure 2.31 shows the major components of a vertical power end. This unit contains a double-reduction gear and a pressure-lubrication system.

Power Frame. The *power frame* (see Fig. 2.32) is the major part of a power pump. It supports all other parts of the power end, and also often supports the liquid end. It supports the main bearings either directly or through bearing carriers, and supports the crossheads either directly or through liners.

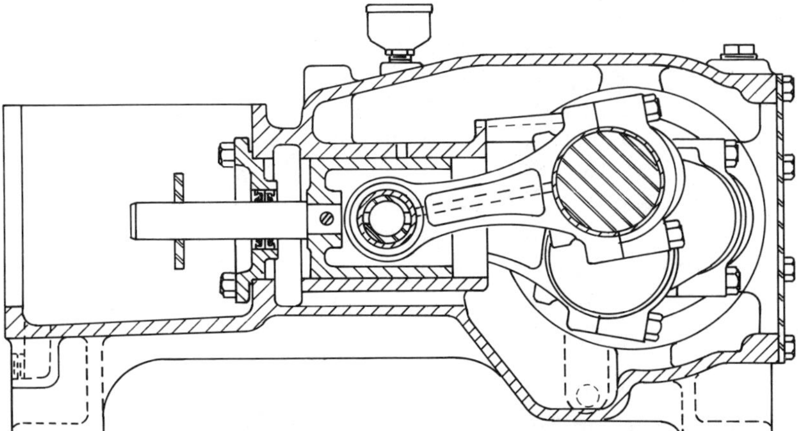

FIG. 2.29. A power pump power end. (*Courtesy Union Pump Co.*)

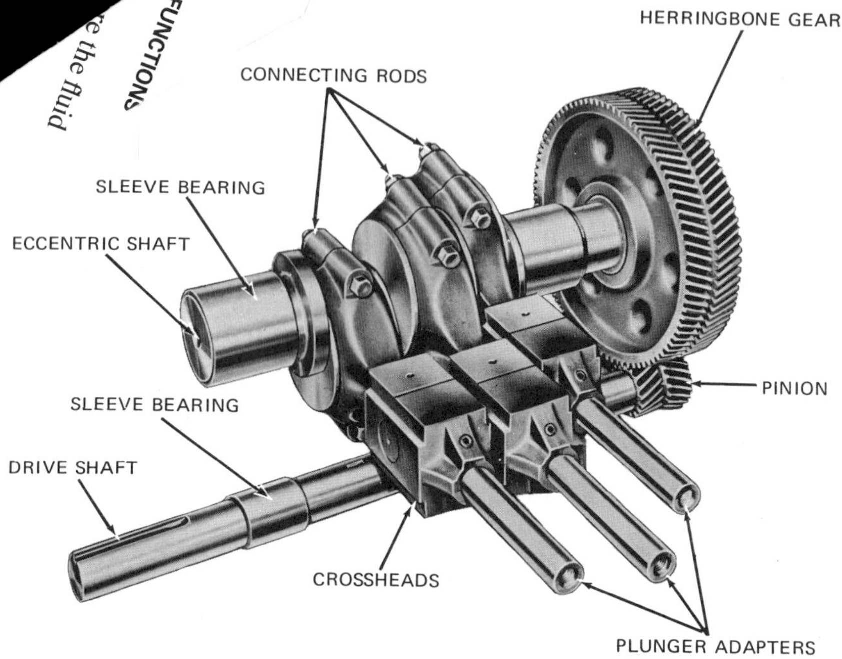

FIG. 2.30. Internals of a horizontal power pump power end. (*Courtesy APV Gaulin, Inc.*)

Power frames are made of various grades of iron, steel, aluminum, and magnesium.

Crankshaft (or Cam Shaft or Eccentric Shaft). The *crankshaft* (Fig. 2.33) drives the connecting rods, and absorbs the loads from the rods and drive system (belts, gears, or chain). Some pumps use eccentrics mounted on a straight shaft (Fig. 2.34). A few power pumps are designed with a cam shaft and followers instead of a crankshaft and connecting rods.

Crankshafts are made of ductile iron, cast steel, and forged steel.

Main Bearings. The *main bearings* support the crankshaft, and absorb all of the loads imposed on the crankshaft. These bearings are of the sleeve type (as seen in Fig. 2.30), roller type (Fig. 2.35), or, in smaller pumps, ball type.

Connecting Rods and Bearings. *Connecting rods* transmit the power from the crankshaft to the crossheads. The crankshaft moves in pure rotary motion, the crosshead in pure reciprocating motion. The connect-

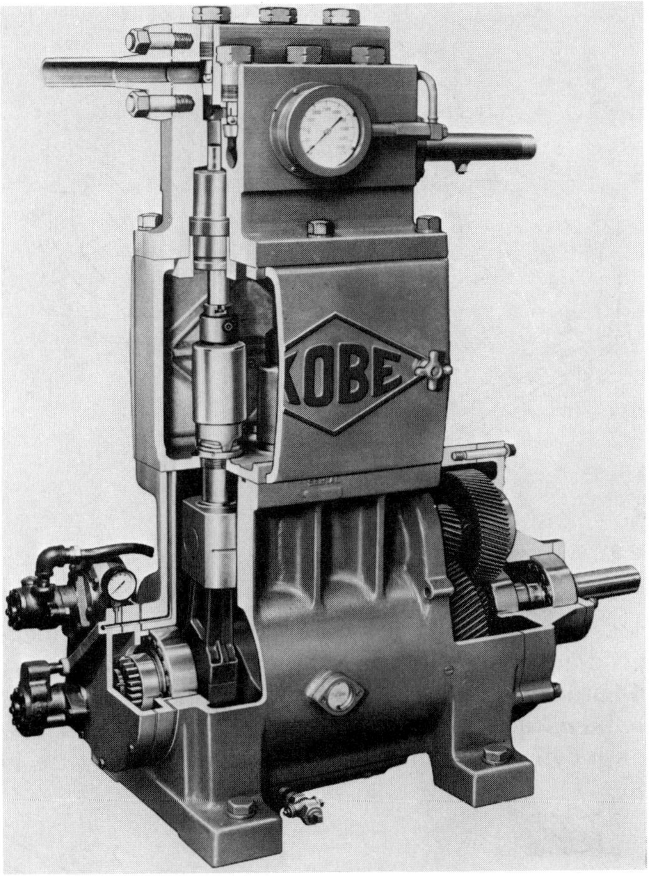

FIG. 2.31. Cutaway vertical power pump power end. (*Copyright Kobe, Inc. Used with permission.*)

ing rod is the link between the two, with one end in pure rotary motion, and the other in pure reciprocating motion. The connecting rod is usually equipped with a replaceable bearing at each end, one which rides on the journal of the crankshaft, the other which oscillates on the wrist pin. The bearings are usually of the sleeve type, but may be rolling-element type. During operation, the rod oscillates above and below the centerline of the crosshead about 11°, swinging through a total arc of about 22°.

Connecting rods are made of iron, steel, and aluminum.

Crossheads and Pins. The *wrist pin* (crosshead pin) is driven by the connecting rod, transmitting the force to the crosshead. On some pumps,

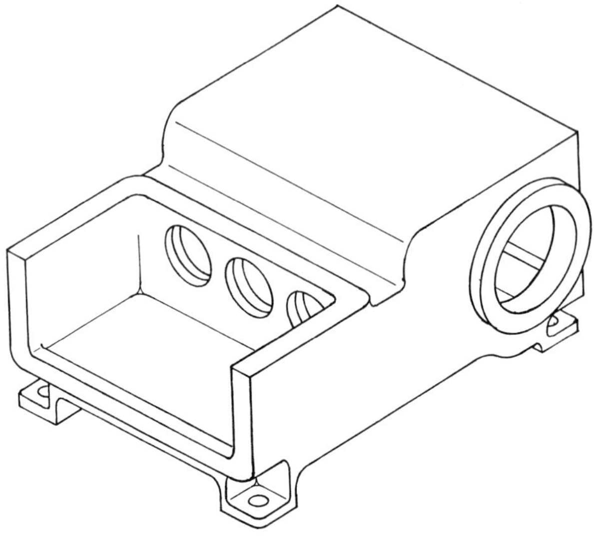

FIG. 2.32. A power pump power frame. (*Courtesy Union Pump Co.*)

the pin is locked in the crosshead. On others, the pin is locked in the connecting rod and rotates in the crosshead. The wrist pin is a tube or solid rod of hardened polished steel. (See Fig. 2.36.)

Some *crossheads* look much like pistons in an internal-combustion engine. They are cylindrical tubes, enclosed on one end. (See Fig. 2.37.)

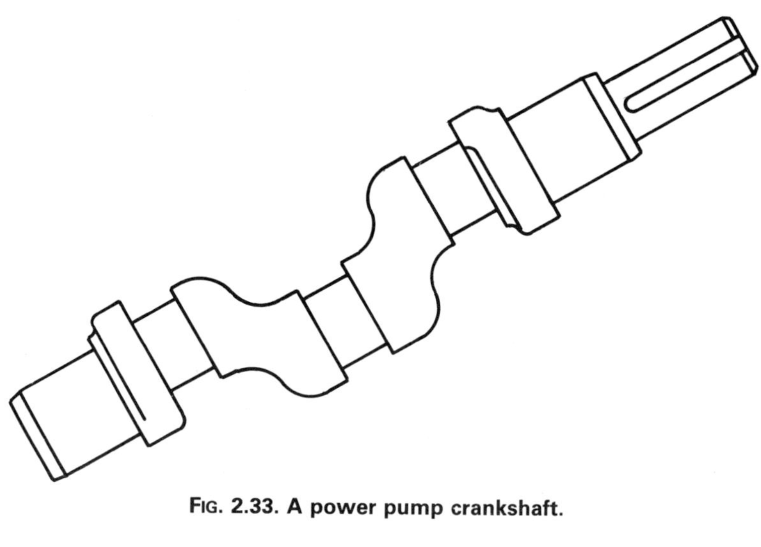

FIG. 2.33. A power pump crankshaft.

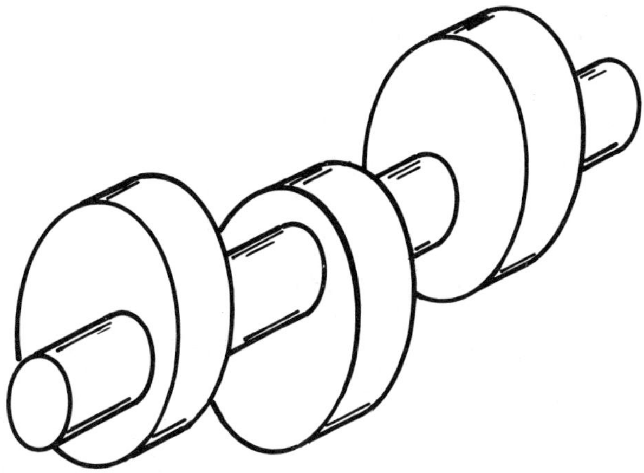

FIG. 2.34. An eccentric-type power pump crankshaft.

Some crossheads have rectangular cross-sections, as seen in Fig. 2.30. A cross-bored hole in the crosshead accepts the wrist pin. The face of the crosshead is drilled and reamed or threaded to accept the rod called a "crosshead extension," "crosshead stub," "pony rod," or "plunger adapter." The crosshead absorbs the force from the wrist pin and transmits it to the extension. In addition to absorbing the total axial force being transmitted from the connecting rod to the liquid end, the crosshead must also absorb the side load from the connecting rod, as shown in Fig. 2.38.

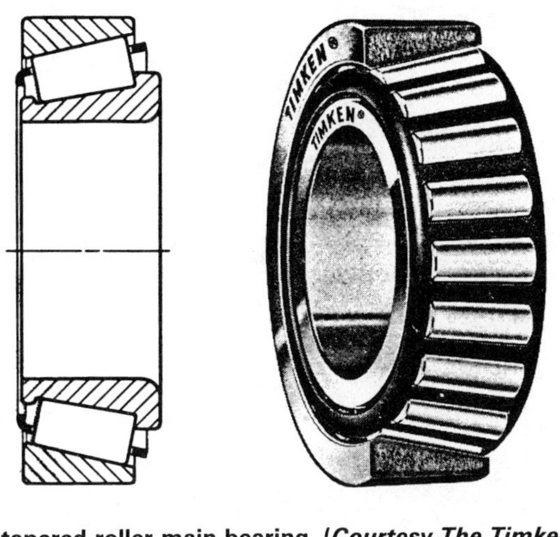

FIG. 2.35. A tapered roller main bearing. (*Courtesy The Timken Company*)

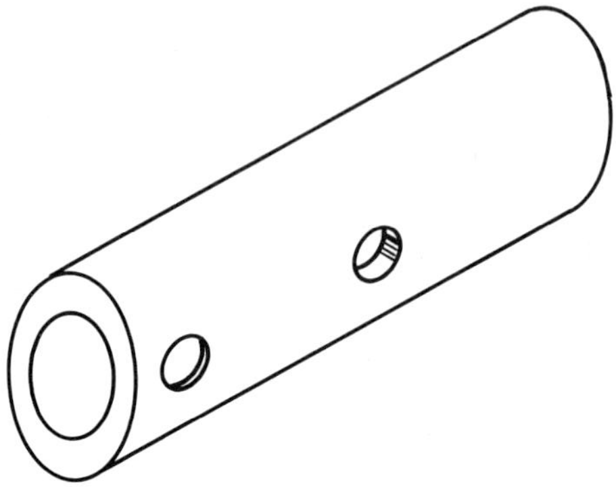

FIG. 2.36. A wrist pin (crosshead pin).

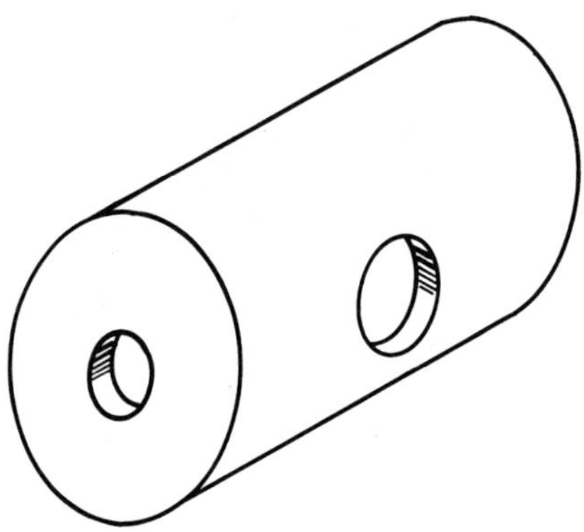

FIG. 2.37. A power pump crosshead.

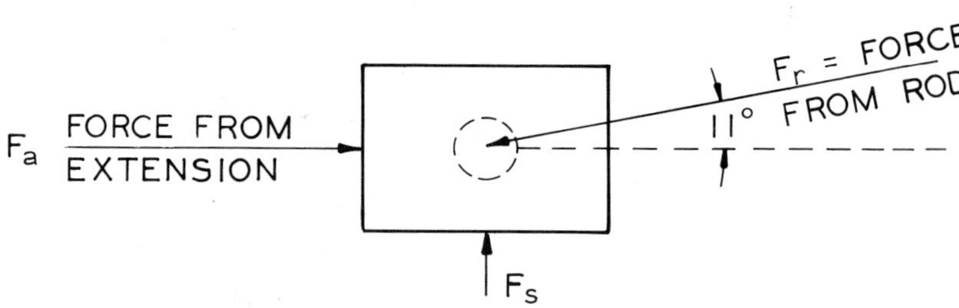

$$F_s = F_a \tan 11^\circ = 0.19\ F_a$$

FIG. 2.38. Force diagram on a crosshead.

When the connecting rod is at an 11° angle from the centerline of the crosshead, the side load on the crosshead is 19% of the axial load.

The crosshead also provides support to the plunger through the crosshead extension.

Crossheads are made of iron, steel, and aluminum.

An additional crosshead is utilized on some vertical power pumps. Those units which "pull" the plunger down into the liquid end use a heavy bar at the top of the pump to connect the plunger to the side rods. This bar is called an "upper crosshead." The crosshead in the power end is then called a "power crosshead."

Crosshead Extensions. The *crosshead extension* (or "crosshead stub," or "pony rod," or "plunger adapter") is a solid rod which normally is pressed or threaded into the end of the crosshead. It not only transmits the axial force to the plunger, but often provides radial support to the plunger, guiding it through the center of the stuffing box. The plunger may be fastened to the extension with a nut or clamp, or may thread into the extension. Crosshead extensions are usually steel or stainless steel, but occasionally are integral with the crosshead, and therefore of the same material as the crosshead.

Seals and Wiper Boxes (Seal Caps). Since most current power ends are enclosed, with oil splashing around inside during operation, some type of sealing devices are required around the crankshaft (or pinion shaft on internally-geared units), and each crosshead extension, to keep the oil from leaking out. This sealing is accomplished on some units by rings of standard packing in a small stuffing box. Other units are equipped with rubber lip seals. To keep the oil in, the lips face inward. Some pumps have extra seals on the crosshead extensions, with lips facing outward, to minimize the migration of foreign matter (particularly pumpage) into the power end. The "plate" which contains the extension seals is called a *wiper box* or *seal cap*.

Lubrication Systems. Many power end components require lubrication to function properly. Main bearings, connecting rod bearings, wrist pin bearings, crossheads, and any internal gears, require oil to provide lubrication and cooling of rolling and sliding surfaces. The simplest of systems is the "splash" system. Oil is maintained at a level high enough to submerge or be splashed onto the lubrication points.

The most complex system is the "pressure" system, utilizing a pump (usually a rotary gear pump) to distribute oil to all, or most, of the lubrication points. The oil pump may be mounted inside or outside of the power end, and is driven directly, or indirectly, by the crankshaft. An alternative arrangement is to drive the pump with an electric motor. Some units utilize a shaft-driven pump as the primary oil pump, and an electric-driven pump as an auxiliary. The auxiliary unit can be used to provide prelubrication to the power end, prior to operation, and as a back-up to the primary pump. A pressure system provides the opportunity to include an oil cooler, filter, and low-pressure switch. The switch may be wired into the driver, to provide an automatic shut-down if oil pressure should drop, or to an alarm to warn of low oil pressure.

A power end lubrication system, intermediate between the simple splash system and the complex pressure system, is the "lift-gravity" system, as shown in Fig. 2.39. One or more discs, mounted on or integral with the crankshaft, lift oil from the sump and deposit it, through scrap-

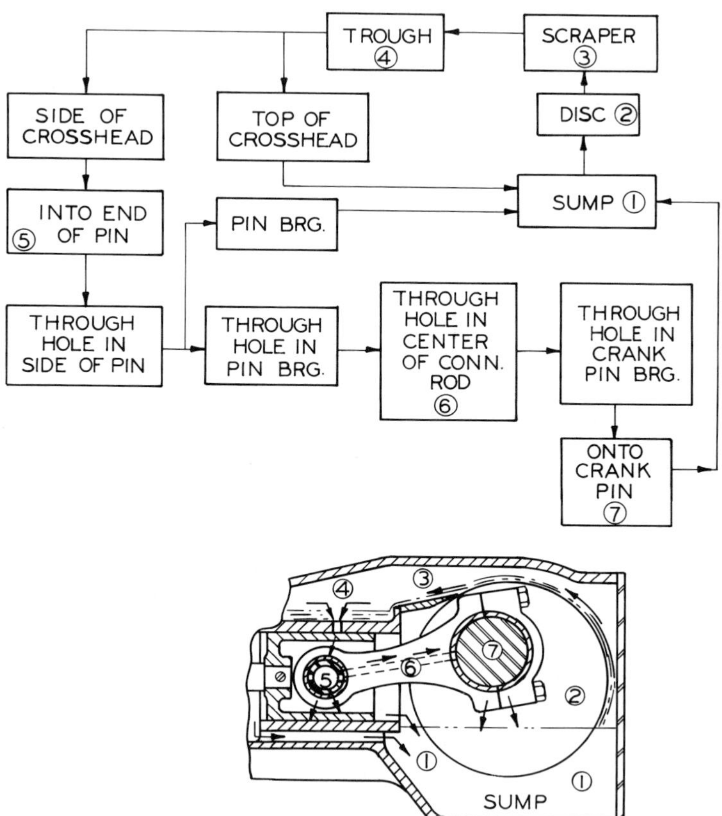

FIG. 2.39. Flow of oil in a typical lift-gravity lubrication system.

ers, into a trough located immediately above the crosshead(s). The oil then flows, pushed by the small elevation head, into the center of the hollow wrist pin, through a hole in the pin into the pin bearing, and also through a hole in the pin bearing and connecting rod into the connecting-rod bearing on the crankshaft. This system provides lubrication at speeds below those allowed by the simpler splash system, and requires the oil level to be no higher than the bottom of the crosshead bore.

Breathers. The power end is enclosed to keep oil in and contaminants out, but must be allowed to "breathe" to prevent a build-up of internal pressure, which could damage oil seals or cause a loss of oil through the seals or sight glass. It is the function of the *breather* to allow air to flow into or out of the power end while minimizing the entrance of solid or liquid contaminants, or the loss of oil. The power pump shown in Fig. 2.40 has a breather mounted on the power end.

These devices often contain a filter to limit the entry of dust and sand into the power end. They also sometimes contain a simple check valve

FIG. 2.40. A power pump with breather mounted on power end. (*Courtesy of GASO Pumps, Inc., a Goulds Pumps Company*)

which allows the air to flow freely from the power end, but restricts the flow of air back into the pump. During operation, this creates a slight vacuum in the power end, which reduces oil loss through the seals. Unfortunately, it also encourages the entry of atmospheric air through the seals, which can be a problem if the air contains a gaseous contaminant, such as lean oil fumes or ammonia, which dissolves in the oil.

If the air around a pump contains vapors which would damage power end bearings or contaminate the oil, the breather may be replaced with a line that feeds dry air or nitrogen into the power end to maintain a very low (approximately 1 inch of water) positive pressure inside the power end. Special seals, oil level gauges, and other components may be required to tolerate the internal pressure. Another way to reduce power end contamination is to mount the breather in a cleaner environment (such as outside of the building containing the pump), and run a pipe from the breather connection on the power end to the remotely-mounted breather.

Deflectors. Many crosshead extensions and piston rods are fitted with a *deflector*. This is a disc which serves as a partial barrier between the liquid end and power end to reduce the migration of pumpage into the power end. Any liquid leaking from the stuffing box will tend to flow along the rod to the power end. Upon reaching the deflector, the pumpage is stopped and forced to run down the face of the deflector and drip from its edge.

Plunger Wells. Many horizontal power pumps are constructed with a *plunger well,* as shown in Fig. 2.1, an integral part of the power frame that forms a pot or collection basin between the liquid end and the enclosed portion of the power end. The plunger well provides access to the stuffing boxes, glands, crosshead extensions, and extension seals. It also collects leakage of pumpage from the liquid end and oil from the power end. The large drain connection in the bottom of the well should be connected with full-size pipe to a sewer or collection system. It has been noted that these connections are often bushed to a smaller pipe, which is adequate when new and clean, but which becomes clogged with bits of packing, wire, straw, bugs, and other foreign matter, causing liquid to accumulate in the well. Upon reaching the level of the bottom of the deflectors, the deflectors begin to splash the pumpage onto the crosshead extensions, whereupon some migrates into the power end. It is ironic that the deflector, which is intended to help protect the power end, in this case helps to contaminate it.

In some cases, the plunger well has been used as a trap for vapors emitted from the liquid end stuffing box(es). For a pump handling a toxic or volatile liquid, the well has been sealed, and a gasket provided between the cover and the well, with a vent connection into this cavity to pipe away vapors. Such an arrangement practically ensures that the power end will be filled with the same vapor, which then escapes through the breather, but also often contaminates the oil and damages bearings. The plunger well should not be used as a trap for vapors. If it is necessary to stop pumpage vapors from entering the environment, they should be stopped with a special tandem stuffing box and/or a pot seal, as described in Chapter 4.

Plunger Covers. To reduce the entrance of extraneous solids, liquids, and hands into the plunger well, a solid cover is usually provided. This plate is not only a fairly good barrier against the entrance of foreign matter, but, even though not sealed as described above, is also a barrier against the escape of vapors (from the pumpage leaking from the liquid end). If, for example, the pump is pumping hot water, some of the water leaking from the stuffing box will vaporize. The humidity inside the plunger well soon reaches 100%, and water droplets form on metal surfaces. This water drips from the underside of the cover and other surfaces onto the crosshead extension, then migrates into the power end, contaminating the oil and reducing bearing life. A similar phenomenon has been observed with vapors of absorption oil (lean oil) migrating into the power end and being absorbed by the oil, again leading to short bearing life. The solution to this problem is to allow air to circulate freely in the plunger well, either by elevating the solid cover, or replacing it with a non-solid cover, such as one made from expanded metal. The plunger well should be adequately ventilated to avoid power end problems.

2.2.2. Direct-Acting Pumps

The drive end of a direct-acting pump, often called a ''steam end'' on gas-driven units, serves the same function as the power end of a power pump—to provide reciprocating motion to the pumping element(s) of the liquid end. But, unlike the power end which converts rotary power to reciprocating power, the drive end of a direct-acting pump converts fluid (usually gas) pressure into reciprocating power.

The drive end of a direct-acting pump looks much like the liquid end of a direct-acting piston pump, consisting of a cylinder, pistons, piston rings, piston rods, stuffing box assemblies, and a valve mechanism. But unlike the check valves on the liquid end, which are opened by differential pressure, the drive-end valves are mechanically operated by a system of levers and rods.

Drive Cylinder. The drive cylinder forms the major portion of the pressure boundary, and supports the other drive end parts. Unlike the power frame in a power pump, the drive cylinder in a direct-acting pump does not support the liquid end. The cylinder is normally made of cast iron. Because it is iron (a good bearing material) and usually lubricated, wear of the cylinder bore is very slow, and a liner is normally not provided.

Piston and Rod. The piston converts the differential pressure of the motive gas to a force which it imparts to the piston rod. The piston is usually cast iron.

Rings. Piston rings provide the seal between high pressure gas on one side of the piston and low pressure gas on the other. For lubricated operation, the rings are made of cast iron. For nonlubricated operation, the rings are a special combination cast iron and bronze or PTFE (polytetrafluoroethylene) plastic.

Main Gas Valve. The function of the *main gas valve* is to direct high pressure gas, first to one side, then the other, of the drive piston, while simultaneously directing exhaust gas from the opposite side of the piston into the exhaust manifold.

The most common type of main valve is the "D" slide valve, shaped somewhat like the letter "D," or to some eyes, more like a "C," as shown in Fig. 2.41. The valve seals against a flat surface of the cylinder or a wear plate. The pressure above the valve is inlet pressure or "chest" pressure. The pressure below the valve is exhaust or outlet pressure. The differential gas pressure presses the valve against the sealing surface. Because the resulting lateral force is relatively high, this surface requires lubrication, and the "D" slide valve is limited to a differential gas pressure of about 300 psi.

For a higher differential gas pressure, and for nonlubricated operation, the piston-type valve is used (see Fig. 2.42). This type of valve has inlet gas pressure only at the ends, and exhaust gas pressure completely surrounding its midsection. This arrangement provides complete radial balance, eliminating the side thrust inherent in the "D" slide valve. The piston valve is usually fitted with rings to seal the high pressure gas from the low pressure (exhaust) area. Main gas valves and rings are normally cast iron.

On a duplex pump, the main gas valves are positioned by valve rods, which are positioned by a system of levers, which in turn are driven by crossheads mounted on the piston rods (as seen in Fig. 2.2). The operation of this system is described in Chapter 6.

FIG. 2.41. Main gas "D" slide valve. (*Courtesy Union Pump Co.*)

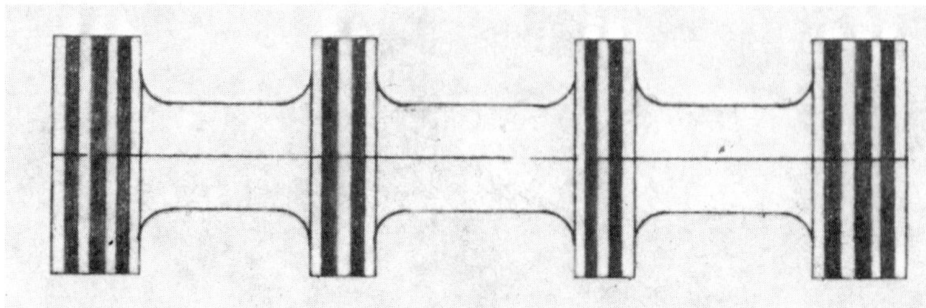

FIG. 2.42. Main gas piston-type valve.

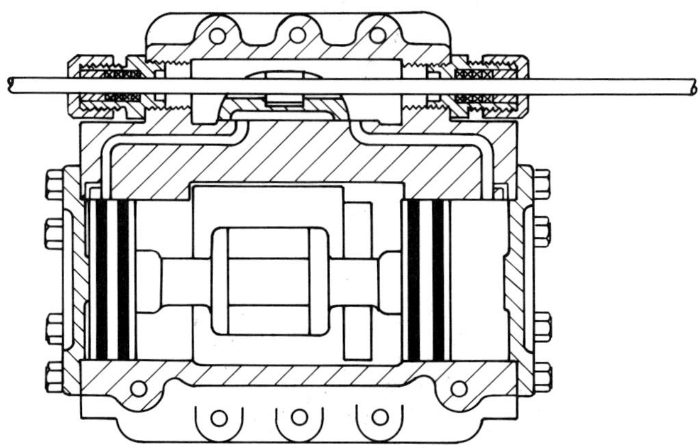

FIG. 2.43. Auxiliary valve and piston which control the main valve of a direct-acting simplex pump.

Auxiliary Valve. Also as explained in Chapter 6, a *simplex* direct-acting pump requires a small *auxiliary valve* to actuate the main gas valve. Its construction and function is similar to that of the main gas valve. It directs high pressure gas, first to one end, then the other, of an auxiliary piston, which drives the main gas valve from one end of its stroke to the other. The auxiliary piston is fitted with rings at both ends to seal the ends of the piston from chest pressure. (See Fig. 2.43.)

Stuffing Box Assemblies. Stuffing box assemblies on the drive end of a direct-acting pump restrict leakage of motive fluid to the atmosphere. They are similar to the stuffing box assemblies on the liquid end.

Chapter 3
TYPICAL SERVICES

Reciprocating pumps perform a plethora of services, from the small unit which provides the pulsing jets of water to clean our teeth, to the large units used to push a powdered coal slurry along a pipeline. This chapter will discuss some of the services performed by reciprocating pumps in the industrial sector of our economy.

3.1. PETROLEUM INDUSTRY

Reciprocating pumps perform services in the petroleum industry which range from the pumping of drilling mud during the drilling of an oil or gas well, to the pumping of the gasoline into the carburetor in our automobile engine.

3.1.1. Drilling Mud

In the drilling of oil and gas wells, drilling mud is pumped by both piston and plunger pumps at pressures up to about 5000 psig. The mud is pumped through the drill bit, where it assists in breaking the rock, then flushes the broken chips away from the bit and up to the surface. The mud flows into a pit where most of the rock settles out, then is pumped back into the well. Figure 3.1 shows a horizontal triplex power pump in mud service.

3.1.2. Cementing

To lock the well casing into the drilled hole in the earth, a mixture of cement and water is pumped to the bottom of the well and into the annular space between the casing and the drilled hole. Engine-driven power pumps are used to pump this slurry into the well.

3.1.3. Blow-Out Preventers

During the drilling operation, heavy hydraulically-driven rams stand ready to slam against the drill pipe and seal the well if excessive pressure threatens to blow the well out. These rams are driven by a hydraulic oil at

FIG. 3.1. A triplex power pump in drilling mud service. (*Courtesy Ingersoll-Rand Co.*)

pressures of 1000 to 3000 psi, which comes from charged accumulators. At the heart of the hydraulic system is either a motor-driven power pump or a gas-driven direct-acting pump which fills the accumulators and maintains the required operating pressure.

3.1.4. Oil Well Pumps

When the hydrocarbon formation pressure is insufficient to force the oil to the surface, it must be pumped to the surface. Reciprocating pumps are often used for this service, as evidenced by the nodding pumping units often seen at producing oil wells. Subsurface reciprocating pumps used in some wells are driven hydraulically by power pumps installed at ground-level near the well.

3.1.5. Utility Services

Duplex piston power pumps are used to gather and transfer crude oil and hydrocarbon condensate.

3.1.6. Secondary Recovery

Some oil formations require externally-applied pressure and/or heat to push the oil to a producing well. Power pumps are sometimes used to inject salt water into peripheral wells to maintain or increase formation pressure. This process is called "waterflooding." Steam is sometimes used to provide pressure and heat to the formation. The steam generator is fed water by a power pump. Liquid carbon dioxide (CO_2) is sometimes injected into the oil formation with power or centrifugal pumps to flush out more oil.

3.1.7. Salt Water Disposal

Salt water, produced with crude oil, is often disposed of by pumping it, with a power pump, into a nonproducing well.

Materials common for salt water service include an aluminum-bronze liquid cylinder, ceramic plungers, and stainless steel or aluminum-bronze fittings.

3.1.8. Well Servicing

Sometimes mother nature is stingy with her black treasure, holding the oil in multiple tight pockets which prevent it from flowing to the well. Such a formation must be "fractured" to create crevices through which the oil will flow. A slurry of small glass beads in water is pumped into the well at a pressure of approximately 15,000 psig. The high pressure fractures the rock in the formation. The glass beads hold the rock apart after the pressure is reduced. This operation also sometimes involves pumping an acid into the formation to partially dissolve the rock. These services are performed by engine-driven power pumps.

3.1.9. Glycol Circulation

Prior to being fed into a pipeline, or processed by a gas plant, natural gas must be dried. Ethylene glycol, as is used in antifreeze solutions, is pumped into an absorber (at about 1000 psig) and flows counter to the moist gas. The glycol absorbs the water, is throttled to atmospheric pressure, is heated to drive out the water, then is cooled and pumped back into the absorber. Both motor-driven power and direct-acting reciprocating pumps are used for this service.

Problems with this application are few, although occasionally the glycol will be considerably hotter than the anticipated 125–150°F. This results when the pump gets installed on the wrong side of the glycol cooler. The higher temperature causes a reduction in the life of packing and plungers.

3.1.10. Amine Circulation

If natural gas contains unacceptable quantities of hydrogen sulfide (H_2S) or carbon dioxide (CO_2), an absorbant such as monoethanol amine, diglycol amine, potassium carbonate, or any of a number of proprietary liquids is pumped into an absorber (at about 1000 psig) to absorb these impurities. The liquid follows a process similar to that of glycol as discussed above. Although larger gas processing plants usually use centrifugal pumps, most smaller plants find the motor-driven power pump best suited for this service.

Amine circulation is usually the most troublesome service for a recip pump in a gas plant. The amine often gets loaded with rust, which becomes embedded in the packing and scores plunger coatings, even those with a hardness of 60 Rockwell C. Special ceramics are utilized as plunger coatings in such cases. The heater also often fails to strip enough CO_2 from the amine to prevent flashing of this gas as it enters the pump. The gas causes problems with valves and valve springs, particularly causing breakage of discharge springs.

3.1.11. Lean Oil Circulation

In an LPG (liquified petroleum gas) extraction plant, an absorption oil, similar to kerosene, is used in a manner very similar to glycol and amine. The oil absorbs heavier hydrocarbons (such as natural gasoline, butane, propane, and ethane) from the natural gas, but instead of being throttled immediately back to atmospheric pressure, it is throttled in stages to lower pressures, the more volatile liquids flashing out at the higher pressures. The desired liquified gases are subsequently recondensed.

As with amine systems, larger gas plants usually use centrifugal pumps for lean oil charge. Smaller plants use motor or engine-driven power pumps.

Problems with power pumps in lean oil service are minimal, if packing is properly selected, installed, and lubricated. On rare occasions, lean oil vapors have been known to migrate into the power end, contaminating the lubricating oil, and causing bearing problems. (See the expanded discussion of this problem in Section 4.15.)

3.1.12. Caustic Charge

Sodium hydroxide (NaOH) is sometimes pumped into liquified petroleum gas to remove sulfur-bearing compounds such as hydrogen sulfide (H_2S). Small power pumps are usually used for this service.

3.1.13. Wash Water Injection

In lower pressure amine absorption systems, water is sometimes pumped by a small power pump into the top of the absorber to absorb, and prevent the loss of, amine vapors.

3.1.14. Rerun

Every gas processing plant will occasionally produce an off-spec product, i.e., it will contain too many light ends or heavy ends. This product must be "metered" (about 15 gpm) back into the process stream for reprocessing. Because the tank of bad product must be emptied, and the product is at its bubble-point, only about 4 feet of NPSH is available when the tank is near empty. With a product like propane (0.5 SG), this amounts to less than 1 psi of NPSH available. Small power pumps are available which perform admirably in this service. They usually have their suction valves oriented vertically, have no springs on the suction valves, and operate at less than 200 rpm.

Problems with power pumps in this service are negligible if proper allowance has been made for any significantly-high suction pressure.

3.1.15. Knockout Drum Pump-Out

Each refinery contains a knockout drum. This vessel separates liquids from the gases destined for the flare stack. These fluids come from various points in the plant where pumps and other equipment are being vented or drained for maintenance. The liquid can therefore range from light hydrocarbons with minimal NPSHA to heavy, viscous oil. The minimum-clearance simplex direct-acting pump is particularly well suited for this application. It can readily be slowed down for the viscous and low-NPSHA liquids, and the low-clearance-volume feature allows the pump to clear itself of any ingested gas.

3.1.16. Hot Oil

Refineries occasionally require pumps to handle hydrocarbons at relatively low flow rates (often below 100 gpm), at pressures up to about 4000 psi, and at temperatures up to 700°F. One such application is in "white oil'' units which produce pharmaceutical grade mineral oil. Both power and direct-acting steam pumps are used in these services.

Power pump liquid ends are segmented and supported on their centerlines so as to maintain alignment with the power end at the elevated temperature. To prevent the expansion of the manifolds from imposing a strain on the liquid end segments, the manifolds are separated from the segments, and connected to them by looping ''C''-shaped piping. Stuffing boxes are water cooled to reduce the temperature inside the boxes, and to prevent the heat from reaching the plate, at the face of the power end, which maintains the alignment of the boxes. Figures 3.2 and 3.3 show the special construction of such units.

Figures 3.4 and 3.5 show the special construction used for direct-acting pumps in hot oil service. The stuffing boxes are water jacketed, and the liquid cylinder supports (below the stuffing boxes) are cooled to maintain

FIG. 3.2. A triplex power pump in hot oil service. (*Courtesy Union Pump Co.*)

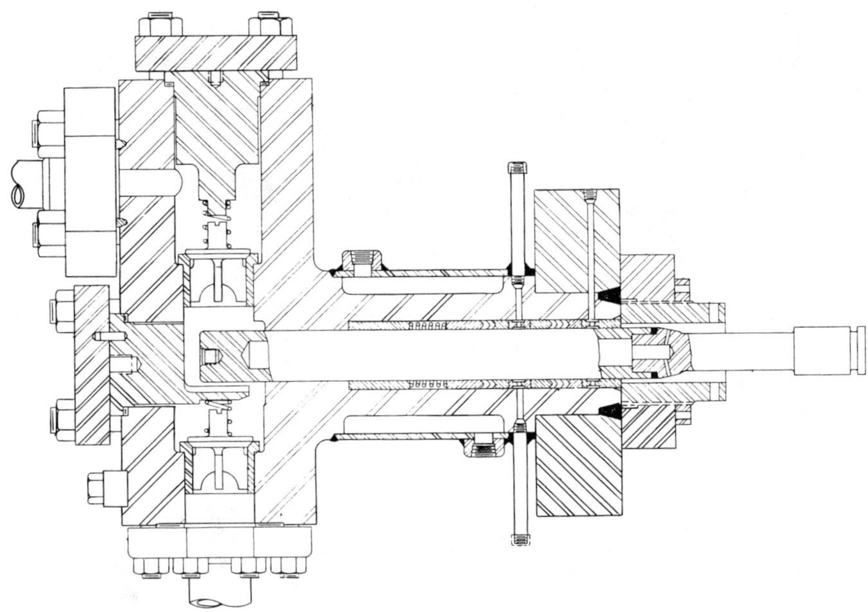

FIG. 3.3. A triplex power pump liquid end for hot oil service. (*Courtesy Union Pump Co.*)

cylinder alignment. Because of the comparative ease of maintaining alignment on such hot oil steam pumps, liquid end design features are less elaborate than those required for multiplex power pumps. The inherent low speed of these units also generally leads to less maintenance than power pumps in similar hot services.

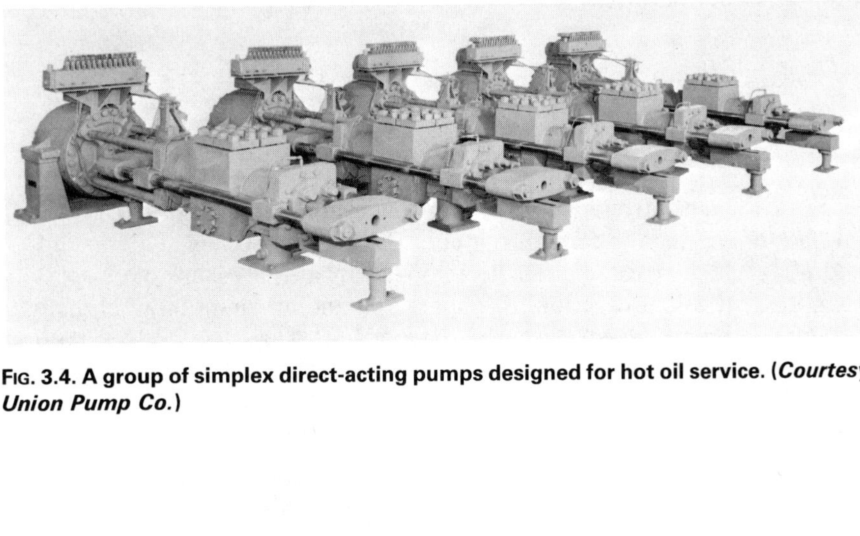

FIG. 3.4. A group of simplex direct-acting pumps designed for hot oil service. (*Courtesy Union Pump Co.*)

FIG. 3.5. The jacketed stuffing box and cylinder support are two of the special features provided for hot oil service. (*Courtesy Union Pump Co.*)

3.2. CHEMICAL INDUSTRY

3.2.1. Fertilizer Plants

Urea is a high-nitrogen content fertilizer produced in *prill* form by a number of chemical plants throughout the world. Urea is made from CO_2 and NH_3 which react chemically in a reactor at a pressure of about 3000 psig. The chemical reaction produces urea and an ammonium carbamate solution, which consists of carbamate, water, CO_2, and NH_3. The urea and carbamate solution are subsequently separated, and the solution pumped back into the reactor. Reciprocating power pumps are used to inject the liquid ammonia and carbamate solution into the reactor.

Carbon steel components are used for the liquid ammonia. The stuffing boxes are often of the double-gland or tandem spring-loaded design. *Bronze* bushings are often used inside the stuffing boxes to support the packing. Because the ammonia is anhydrous (contains no water), it will not corrode the bronze. The outer bushing is often made of cast iron because the moisture from the air will combine with the ammonia to corrode bronze.

Because the carbamate solution is very corrosive, a special stainless steel is used for all wetted parts of the pump liquid end. Valve seats are often clamped in to avoid the galling which may result from a pressed fit. Stuffing boxes are usually the double-gland design, packed with square braided interlocking PTFE filament packing impregnated with a PTFE suspensoid. Because of the corrosiveness, sometimes abrasivity, and dissolved gases, pumps in this service usually require more maintenance than units handling less aggressive liquids. Figure 3.6 shows a triplex pump designed to pump the carbamate solution.

FIG. 3.6. A power pump designed for carbamate. (*Courtesy Union Pump Co.*)

3.2.2. Detergent Production

The granules (prills) of detergent used to wash our clothes are produced in a prill tower, by droplets of detergent slurry falling in the tower against a counter flow of air. The air evaporates the water from the droplets, forming dry granules before they reach the bottom of the tower. The slurry has some air mixed with it before being pumped to the tower.

Low-speed power pumps, usually driven by variable-speed drivers, are used to pump this mixture of detergent, water, and air to the top of the tower. Discharge pressure is usually about 900 psig. The detergent tends to cling to the plunger surface, and when exposed to the air on the atmospheric side of the packing, it forms a hard coating on the plunger. As this hard crust gets carried back into the packing, on the discharge stroke, the packing is damaged. To prevent the solids from accumulating on the plunger, the pump is either fitted with a *pot seal* (as discussed in Chapter 4) which is filled with water, or water is run over the plunger just outboard of the packing. With this latter arrangement, because splashing water is in close proximity with the oil seals of the power end, there is a good possibility that water will migrate through the seals and into the power end. The power end oil must therefore be periodically checked for water, and changed if it becomes contaminated.

Because of the air content, it is necessary to boost the suction pressure on these pumps to about 100 psig to obtain satisfactory volumetric efficiency. The pumps used to boost the suction pressure are often direct-acting units, and often of the low-clearance-volume design because of their suitability for gas-laden liquids.

3.2.3. Plastic Slurries

In the production of plastics such as polyethylene, polypropylene, and polybutylene, a plastic slurry is pumped at a pressure of about 800 psi. Both power and direct-acting plunger pumps are used for this service. Because of the need for variable capacity and low speed, the direct-acting unit has been used more frequently. Pump features include special valves and stuffing boxes.

3.3. METALS INDUSTRIES

3.3.1. Ore Slurry Pipeline

In some parts of the world, where rough terrain limits surface transportation, metal ore produced from mines is ground and transported in pipelines in a water slurry. Plunger type power pumps are used to push the slurry through the pipe.

3.3.2. Aluminum and Steel Production

Aluminum Production. In the production of aluminum, a slurry of bauxite (aluminum ore) and a hot concentrated caustic solution is pumped into a digester at about 800 psig. Both motor-driven plunger power pumps and steam-driven direct-acting piston pumps are used for this service. Some units use remotely-mounted valves to isolate the pumping elements and stuffing boxes from the abrasive ore.

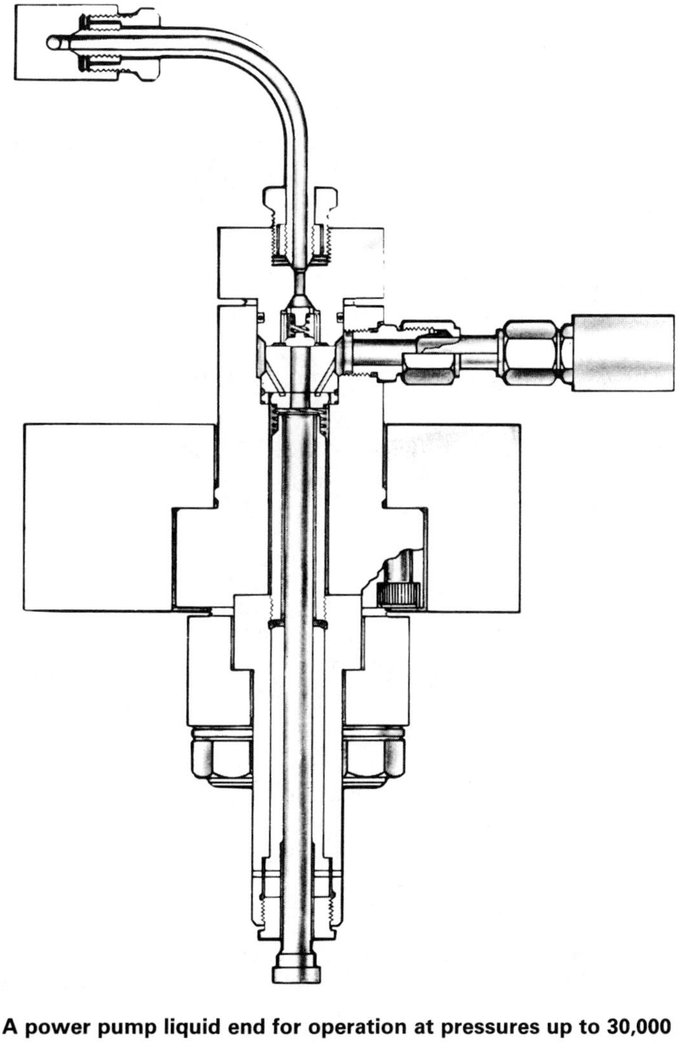

FIG. 3.7. A power pump liquid end for operation at pressures up to 30,000 psi. (*Copyright Kobe, Inc. Used with permission.*)

The abrasivity of this service reduces the life of plungers, pistons, liners, packing, and valve components.

Hydraulic Systems. Both steel and aluminum mills utilize large hydraulic systems to provide power to equipment in various parts of the plants. In steel mills the hydraulic fluid is usually a soluble oil and water. Because the oil-water solution stains aluminum, aluminum mills usually use kerosene or a similar liquid. Both types of mills commonly use motor-driven power pumps to charge the hydraulic systems. System pressures are usually between 1000 and 5000 psig.

Descaling. Smaller steel mills use power pumps to provide the high-pressure water used to blast scale from billets and sheets of hot steel.

Coke Oven Tar. Steel mills will sometimes utilize steam-driven direct-acting piston pumps to transfer the viscous hot tar from the coke ovens.

Hydrostatic Lubrication Systems. Rolling mills sometimes use a hydrostatic lubrication system to support the large rollers. These systems operate at pressures up to 30,000 psig. Figure 3.7 shows the cross-section of a power pump liquid end used for such services.

3.4. NUCLEAR POWER PLANTS

Nuclear power plants use power pumps in two major critical services.

3.4.1. Charging

Plants with pressurized-water reactors are designed to bleed liquid (a weak nitric acid solution called "boronated water") from the pressurized primary coolant loop. This "loss" is from mechanical seals in the large centrifugal circulating pumps, and through a throttling valve. This fluid is cleansed and returned to the high-pressure loop (at approximately 2300 psig) with power pumps or centrifugal units.

Problems with the pumps in this service have been serious. The pumpage is deliberately saturated with hydrogen gas in the volume control tank on the suction side of the pump, creating a condition of marginal NPSH for the pump. Standard operating procedures call for the routine reduction in tank pressure, causing the gas to flash out of solution in the suction line. The ingestion of these bubbles by the power pump results in rough operation and shock conditions which are damaging to pumping equipment and the system. A number of plants have reported complete or partial "vapor-locking" of these charging pumps. Continued operation of the pumps, with some of the pumping chambers vapor bound, has resulted in surging conditions in both suction and discharge piping. Because many of these systems are not designed to allow the discharge pressure on the pump to be reduced below system pressure, it is necessary to stop the pump and resort to an undesirable procedure, such as partial disassembly of the liquid end, to reprime each pumping chamber. Operation of pumps under these cavitating conditions has resulted in pump problems ranging from short packing life to fractured liquid cylinders.

One user resorted to drilling a small hole through each discharge valve to allow each pumping chamber to reprime itself when it became vapor-locked. Although resulting in lower volumetric and mechanical efficiencies, this clever procedure did succeed in allowing the pump to reprime itself while operating. Although the long-range effects are unknown at this writing, it is expected that the hole will enlarge due to the continual high-pressure "leak," requiring frequent changing of the valves, and that power end bearing life may be reduced because of the extra load imposed on the connecting rod during the repriming sequence. (It should be noted that a leaky discharge valve would accomplish the same results.)

3.4.2. Standby Liquid Control

Nuclear power plants which utilize boiling water reactors use power pumps in a service called "Standby Liquid Control." These "SLC" units are a part of the emergency shut-down system. If normal methods of controlling a nuclear reactor fail, these units pump sodium pentaborate into the reactor to shut the unit down. (The boron absorbs the neutrons.) At this writing, the author knows of no case of this back-up system being required to flood a reactor. These pumps undergo periodic testing to assure their readiness to serve. No significant problems have been noted in any of these systems.

3.4.3. Control-Rod Drive

Power pumps have also been used for control-rod drive service.

3.5. PIPELINE SERVICES

A liquid ammonia pipeline runs north-south through the central part of the United States. Power pumps are used to pump the ammonia into the pipeline.

A number of pipelines carry various liquid hydrocarbons along the U.S. Gulf Coast and into other parts of the U.S. When the flow rate is too low to be suitable for a centrifugal, a power pump is used to inject the hydrocarbon into the pipeline.

3.6. SLURRIES

"Slurry" is the term used to describe any mixture of solid particles and a liquid. Pipelines transport slurries of powdered coal and water, and various pulverized ores and water. Power pumps are often used to push these slurries through the pipe.

Because of the abrasiveness of the slurry, liquid end parts such as piston packing, liners, valve components, stuffing box packing, rods, and plungers require replacement more frequently than in clean services.

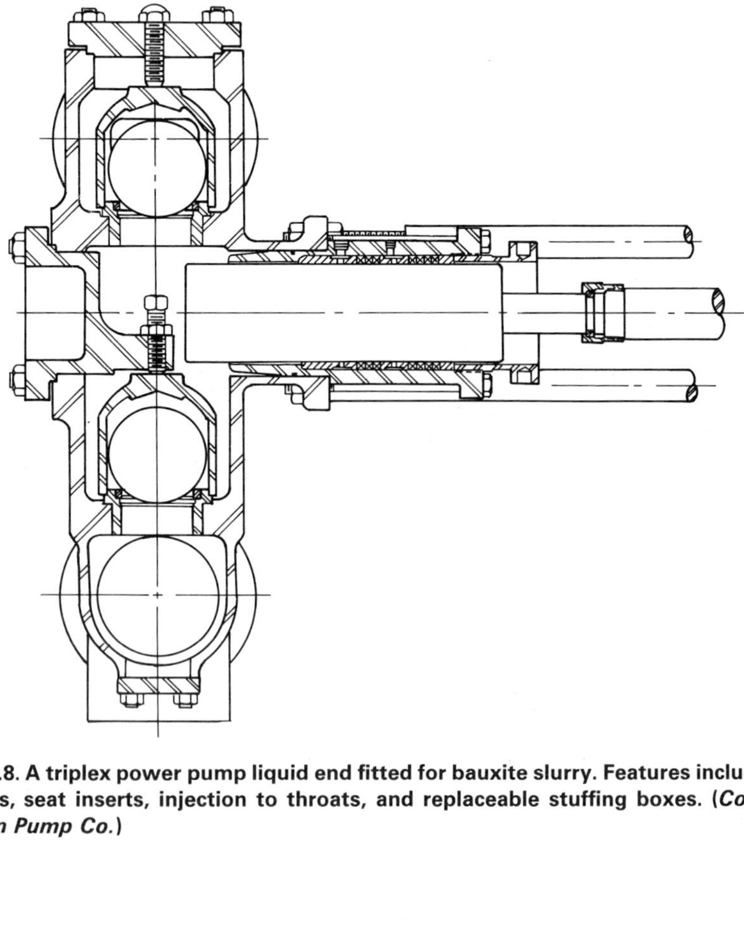

FIG. 3.8. A triplex power pump liquid end fitted for bauxite slurry. Features include ball valves, seat inserts, injection to throats, and replaceable stuffing boxes. (*Courtesy Union Pump Co.*)

The standard reciprocating pump is not designed to handle slurries. Modifications to standard designs, and in some cases special designs, are required to achieve satisfactory operation and component life.

To achieve satisfactory packing and plunger life, abrasive slurry must be prevented from entering the packing. Methods to achieve this (as discussed in Chapter 4) include a wiper ring between the pumpage and packing, a long throat bushing, injection of a clean liquid into the throat area, insertion of a diaphragm or floating piston between the plunger and pumpage, and complete removal of the valve assemblies from the stuffing box area. This last arrangement requires a liquid column between the valve assembly and stuffing box, which increases clearance volume and acceleration head within the pump.

If the slurry has little or no abrasiveness, such as detergent, it can be allowed to enter the packing without significantly reducing the life of the packing or plunger, but detergent must be washed from the surface of the plunger on the atmospheric side of the box, as previously discussed.

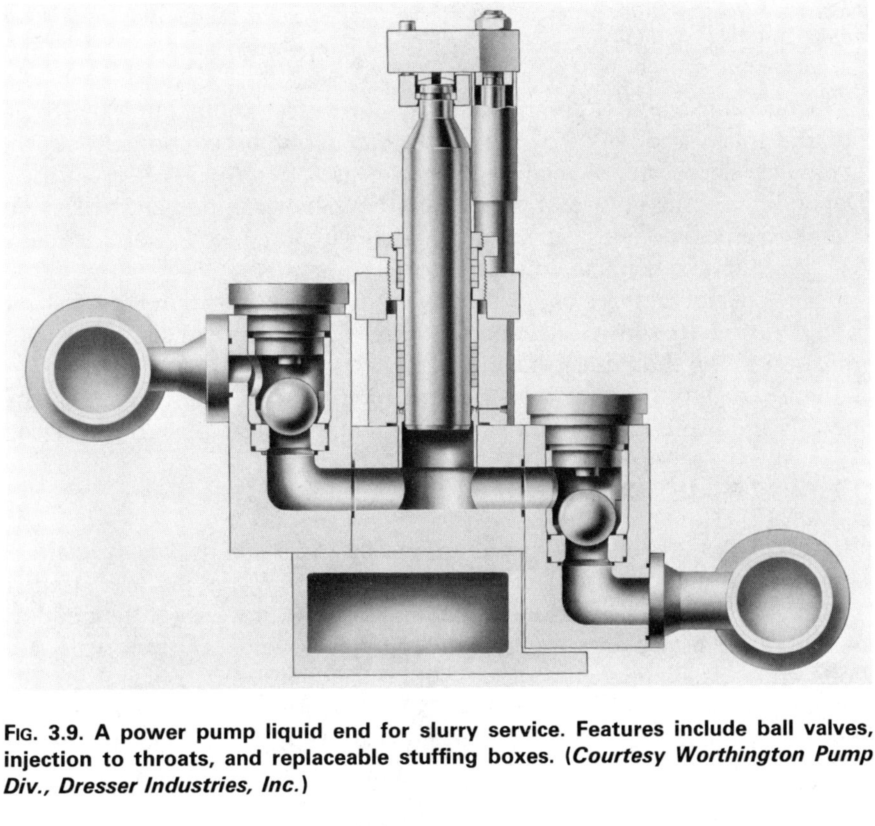

FIG. 3.9. A power pump liquid end for slurry service. Features include ball valves, injection to throats, and replaceable stuffing boxes. (*Courtesy Worthington Pump Div., Dresser Industries, Inc.*)

FIG. 3.10. A bauxite slurry pump and its stuffing box injection pump. (*Courtesy Union Pump Co.*)

Special construction can prevent the slurry from contacting the packing, but there's no way that the valves can avoid direct contact with the pumpage. Therefore, special pump valves are often required for slurries. Depending on the nature of the solids, the valves can be ball, bell, bevel seat with an elastomer insert, wing-guided with a reduced seating area, or disc type with special seating surfaces.

Some special features of slurry power pumps are shown in Figs. 3.8 and 3.9. Figure 3.10 shows the pump of Fig. 3.8 with a small stuffing box injection pump mounted adjacent to it.

Figure 3.11 is a pump constructed for high temperature coal slurry, thereby requiring features for both high temperature and slurry services.

3.7. WATER-JET CLEANING

Water is used at pressures from about 800 to 20,000 psi in a variety of commercial and industrial cleaning operations. Water jets are used in such services as cleaning floors, cleaning airport runways (of the build-up of rubber from airplane tires), removing paint overspray from walls and floor grates, cleaning masonary walls of accumulated soil and graffiti, cleaning ships' hulls, cleaning heat exchanger tube bundles, cleaning the interior of commercial boilers, and cutting plastic which has coated the inside of reactors used in the production of the plastic. Power pumps,

FIG. 3.11. A triplex power pump designed for high temperature coal slurry. (Note the small injection pump with individual discharge openings.) (*Courtesy Worthington Pump Div., Dresser Industries, Inc.*)

from about 1 to 400 horsepower, driven by electric motors and engines, both trailer-mounted and permanently-installed, are used for these services.

Water-jet cleaning has replaced sand blasting and chemical cleaning in some areas because the operation is dust-free, the need for clean-up is minimal, chemical disposal is eliminated, and water is usually readily available.

Figure 3.12 demonstrates the cleaning function of a high-pressure hand-held lance with the portable pumping unit in the foreground.

FIG. 3.12. An engine-driven triplex provides high-pressure water to the cleaning lance. (*Courtesy NLB Corp.*)

Maintenance of the high-pressure units is required more frequently than for the lower-pressure units. At pressures above approximately 7000 psi, packing life is shorter; and, with the high-stress cyclic loading and oft-inadequate suction conditions contributing to hydraulic shock, the probability of a failure of a liquid cylinder or other fluid end component increases.

3.8. WATER-JET CUTTING

In a process similar to that of water cleaning, high pressure water jets are used as cutting tools. At pressures up to 100,000 psi, water flows through nozzles with openings as small as 0.002 inch. These jets cut material as diverse as logs (in the production of lumber) and leather (in the production of shoes). To reduce the tendency of the water jet to expand as it leaves the nozzle, a small amount of polymer is sometimes added to the water.[1,2]

The pump used to produce this high pressure is often a low-speed direct-acting unit (often called an "intensifier") driven by water or oil at a pressure of 3,000 to 10,000 psi.

3.9. BOILER FEED

The feeding of water into boilers is normally a job performed by centrifugal pumps, but smaller boilers often use reciprocating units because the combination of capacity and pressure are not suited for the efficient operation of a centrifugal.

The direct-acting steam pump is used at pressures up to about 1000 psi. It is driven by steam from the boiler, and the exhaust, because it experiences negligible heat loss driving the pump, may be fed into the deaerator. To avoid contamination of the exhaust steam with lubricating oil, the steam end may be designed to operate nonlubricated.

Larger boilers will sometimes use a power pump for start-up or standby service, when the capacity is below the range for efficient operation of the main (centrifugal) feed pump. A number of merchant ships use motor-driven power pumps to feed their boilers when tied up at dockside. In some cases, with the recent increase in fuel costs, these ships have found it economical to reduce fuel consumption by reducing their cruising speed. The resulting reduction in boiler feed rate has made it attractive to operate the power pumps, in lieu of the centrifugal, while under way.

The major problem with reciprocating pumps in boiler feed service is short packing life. The water is usually hot (200+ °F). Because hot water is a very poor lubricant, it provides almost no lubrication for the packing. Lubricating the packing with a good oil such as steam cylinder oil would solve the packing problem, but any oil which found its way into the water would cause problems in the boiler. Therefore, oil is not used. Steps taken to improve the packing environment include water-jacketed stuffing boxes and cool water injected into the packing or run over the plungers outboard of the packing. The recent use of a square braided packing made of fibers of PTFE and aramid has significantly increased packing life in boiler feed service.

3.10. REVERSE OSMOSIS

Contaminated, brackish, and salt water can be made potable by a process called reverse osmosis. A membrane, when exposed to the contaminated water under a pressure of 400 to 800 psi, allows only clean water to pass through. Power pumps are often used to pump the contaminated water into the system. An aluminum-bronze liquid end, as described in Section 3.1.7, Salt Water Disposal, is commonly used for this service.

3.11. CRYOGENIC SERVICES

Pumping of low temperature liquids such as carbon dioxide, ethylene, hydrogen, and nitrogen are accomplished at pressures up to 10,000 psi with specially-fitted power pumps, as shown in Figs. 3.13 and 3.14. Because these liquids are often near their boiling points at the pump inlet, it is necessary that the system be designed to minimize the gain of heat from the environment. Otherwise the liquid will boil before reaching the pump, or upon entering the pump, possibly causing shock operation or vapor-locking of one or more pumping chambers. (For a more complete discussion, see Chapter 8–NPSH.)

FIG. 3.13. The heavy ice coating on this power pump is indicative of the low-temperature capabilities of specially-built units. (*Copyright Kobe, Inc. Used with permission.*)

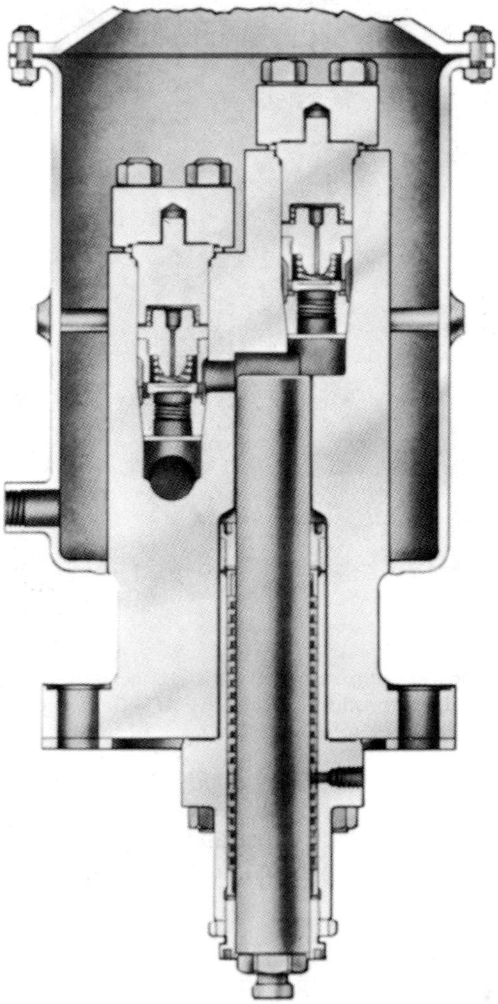

FIG. 3.14. This cryogenic fluid end is jacketed to allow circulation of the pumpage to provide the required cooling. (*Copyright Kobe, Inc. Used with permission.*)

3.12. FOOD SERVICES

Both power and direct-acting pumps are used to pump a variety of foods, from peanut butter to ketchup. Power pumps, fitted with special throttling valves, are used to homogenize milk and other food and industrial prod-

FIG. 3.15. A special power pump with fittings and special throttling valves for homogenizing food and industrial products. (*Courtesy APV Gaulin, Inc.*)

ucts. Power pumps are also used to meter milk through pasteurization processes. These units are usually made of stainless steel and include features which meet strict cleaning and sanitation requirements. Figure 3.15 illustrates a special power pump used for homogenizing service.

3.13. HYDROSTATIC TESTING

Both power and direct-acting pumps are used to hydrostatically test equipment and systems. The direct-acting plunger pump is particularly well suited for this service because, with the drive fluid (air, steam, or gas) pressure regulated, the pump will stall at a predetermined liquid pressure, not running unless the liquid pressure falls.

3.14. CHEMICAL INJECTION

In a large number of industrial applications, it is necessary to meter a precise amount of liquid into a process stream. This service is usually performed by a "controlled-volume" or "metering" pump. This unit is a power pump especially constructed with an adjustable stroke length (or equivalent).

3.15. LUBRICATION INJECTION

The "mechanical lubricator," used to lubricate pistons, plungers, rods, and other sliding surfaces in reciprocating pumps and compressors, is a small single-acting, plunger, controlled-volume pump. The plunger seal is simply a close fit with the liquid cylinder, and leakage from this simple "stuffing box" drops back into the lubricator sump. The mechanical lubricator is discussed more in Chapter 10–Auxiliary Equipment.

REFERENCES

1. Hoyt, J. W. and J. J. Taylor, *Effect of Nozzle Shape and Polymer Additives on Water Jet Appearance,* New York: ASME Paper 77-FE-16, 1977.
2. A water jet that cuts like a knife, *Industrial Production,* 1972.

Chapter 4
STUFFING BOX DESIGNS

The function of the stuffing box assembly is to provide a seal around the piston rod or plunger so as to minimize leakage of pumpage to the atmosphere. Although the stuffing box on the drive end of a direct-acting pump also restricts leakage (of motive fluid), this chapter discusses only liquid end boxes. The terms "plunger" and "piston rod" are used interchangeably in this discussion. A variety of stuffing box designs are included, from the simplest nonlubricated arrangement to various lubrication and bleed-off schemes to extend packing and plunger life and minimize leakage. Also included are stuffing boxes designed especially for segmental ring packing and close-clearance throttle bushings.

4.1. THE SIMPLEST BOX

The simplest stuffing box is a groove in the cylinder opening containing a single ring of packing (Fig. 4.1). Analyzing the forces on this ring will enable us to better understand the forces in the packing sets discussed later.

This ring is braided from fibers of a material such as cotton, asbestos, or a plastic, and is fairly compliant. When installed in the groove, it is firmly seated in the bottom of the groove. The ring is also a snug fit on the plunger, so that inserting the plunger into the packing requires a strong push. (The end of the plunger should be chamfered or radiused so that the packing will not be cut or gouged by a sharp corner of the plunger.)

4.1.1. Pressures on the Packing Ring

Referring to Fig. 4.2, during operation, the left face of the ring is exposed to liquid pressure. This pressure pushes the ring against the right-hand face of the groove. This squeeze on the ring forces it tighter against the O.D. of the groove and against the plunger, helping to achieve a seal at these two critical surfaces. The pumpage provides the only lubrication and cooling available to this packing.

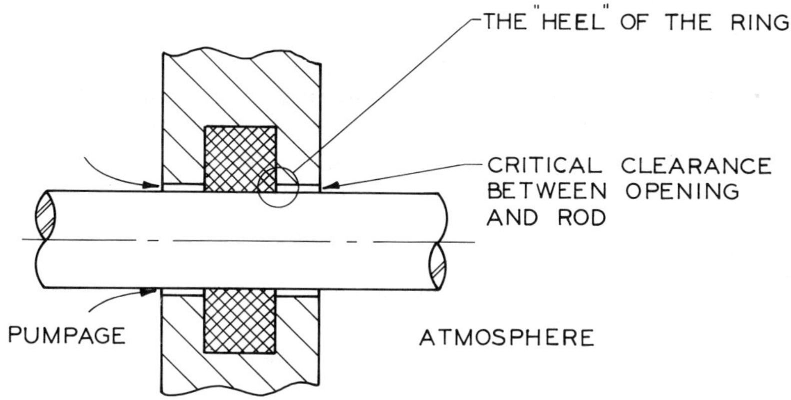

FIG. 4.1. The simplest stuffing box.

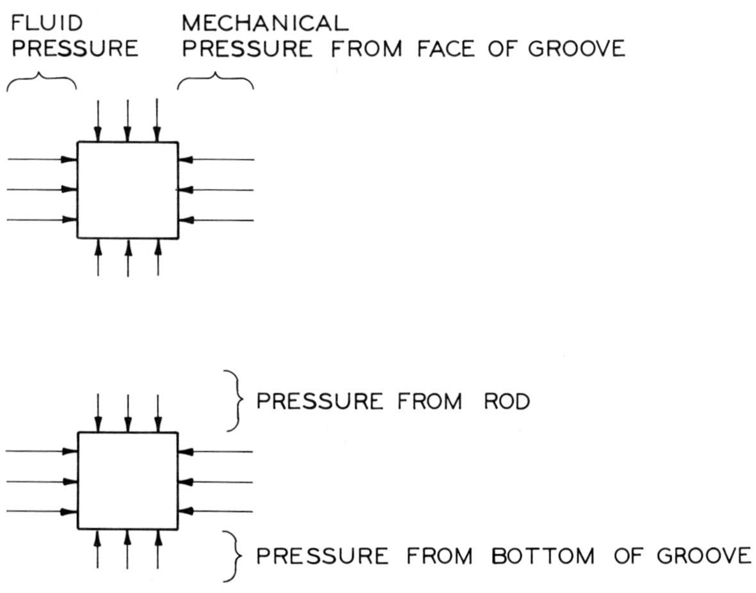

FIG. 4.2. Pressures acting on the ring in Fig. 4.1.

4.1.2. Extrusion of the Packing Ring

To produce an effective seal at the plunger and stuffing box, the packing must be soft and compliant enough to deform to these metal surfaces. Because it is pliant, pressure from the pumpage, and the movement of the plunger to the right, attempt to extrude the packing through the clearance gap between the plunger and the cylinder. If the "heel" of the packing extrudes into the gap, adjacent packing material will flow into the heel area, loosening the seal against the plunger and allowing increased leakage (Fig. 4.3). Preserving the integrity of the heel is critical in maintaining effective sealing, and the gap between the plunger and the packing support member is critical in preserving the heel.

4.2. A STANDARD STUFFING BOX

The simplest stuffing box actually found on most reciprocating pumps is similar to that shown in Fig. 4.4. The stuffing box is a cavity machined into the liquid cylinder—or a separate part bolted to the cylinder.

The bolt-on box has the advantage of being replaceable when worn, and providing for an easier change in plunger diameter, when required for higher capacity or discharge pressure. Some pumps are designed so that the entire box assembly, including the plunger, is removed from the pump

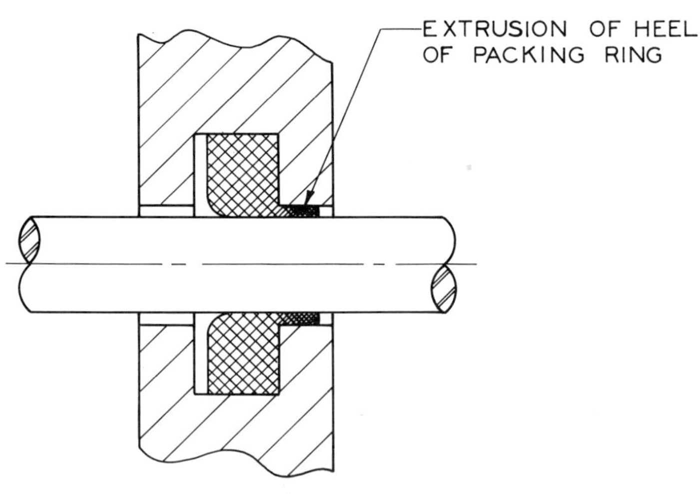

FIG. 4.3. Extrusion of a portion of the packing ring into the clearance gap on the atmospheric side.

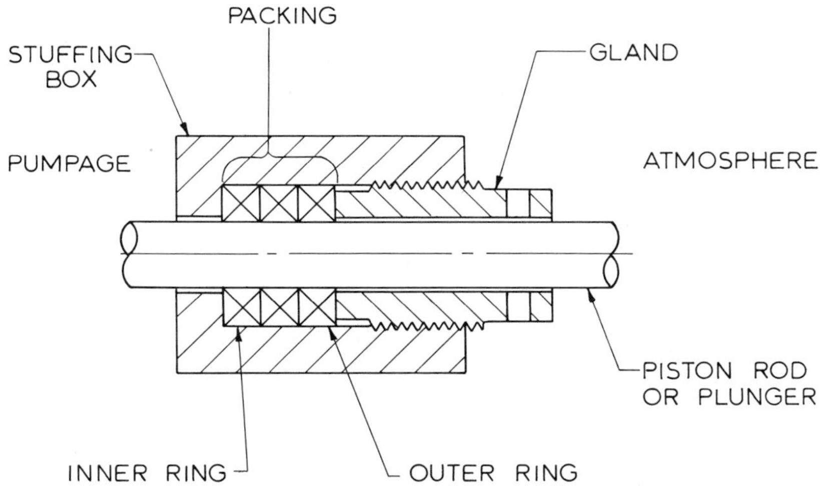

FIG. 4.4. A simple stuffing box assembly.

for packing replacement, enabling service personnel to carefully examine all components prior to repacking, and to more easily install new parts.

Disadvantages of the bolt-on box are that an extra sealing joint is created for each box, and alignment of the box with the power end is not always properly restored when the box is reinstalled on the pump.

The standard box assembly contains a gland which threads into or onto the box, or is held to the box with bolts or studs and nuts. The threaded gland has the advantage of remaining square with the stuffing box during adjustment, providing uniform loading on the packing, and providing uniform clearance over the plunger (it cannot cock). The bolted gland requires less torque to tighten. The gland provides for initial setting of the packing during installation, and for restoration of sealing effectiveness as the packing compresses or extrudes during operation. Experience has shown, though, that little of the force from the gland gets transmitted to the innermost ring of packing. Most of the force from the gland is absorbed by the outer ring, an unfortunate situation, as we will see later.

4.3. FORCES ON THE RINGS OF PACKING

Let us now examine the axial forces on the center ring of packing in Fig. 4.4. These forces are illustrated in Fig. 4.5. Both faces of the ring are exposed to fluid pressure, but because there is a decay in pressure from left to right, the left face feels a higher pressure. This hydraulic imbalance

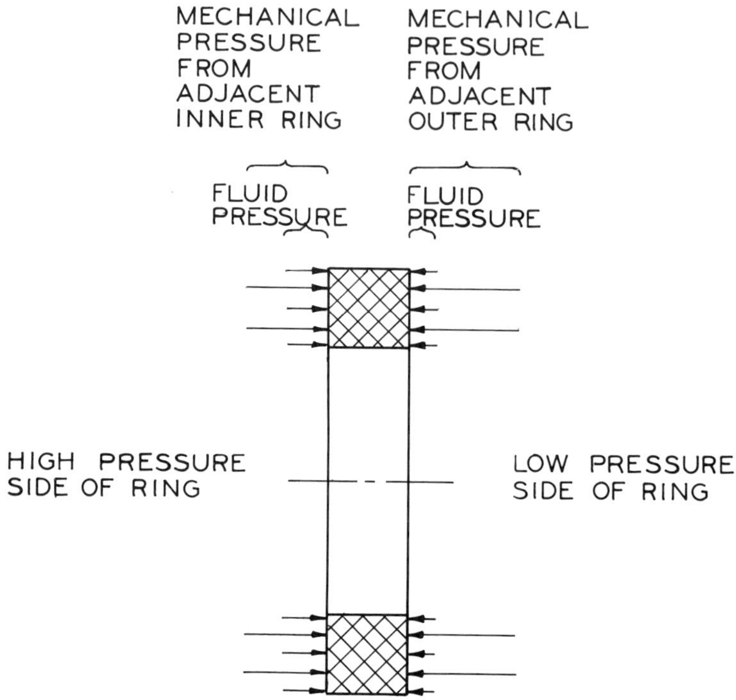

FIG. 4.5. A ring of square braided packing is exposed to both hydraulic and mechanical loads.

is offset by a higher force from the outer ring than from the inner ring. Because of this accumulation of axial forces in a stack of packing, the outermost ring feels the highest mechanical loading. Because of this loading, this ring seals tighter against the rod and stuffing box bore. Because of the tighter sealing, the outer ring experiences the greatest pressure drop. The resulting pressure gradient is approximately as shown in Fig. 4.6. The inner ring is exposed to discharge pressure (on the discharge stroke), but is the loosest of all rings, and therefore achieves very little sealing. Only a small pressure drop exists across this ring. The next ring, being exposed to slightly higher mechanical loading, seals a little tighter against the rod and box bore, and therefore achieves a pressure drop higher than the inner ring. Each successive ring seals tighter, and therefore experiences an increase in pressure drop. The outer ring feels the total accumulated loads of all other rings in the stack of packing, deforms the most, and feels the largest differential pressure. This ring therefore

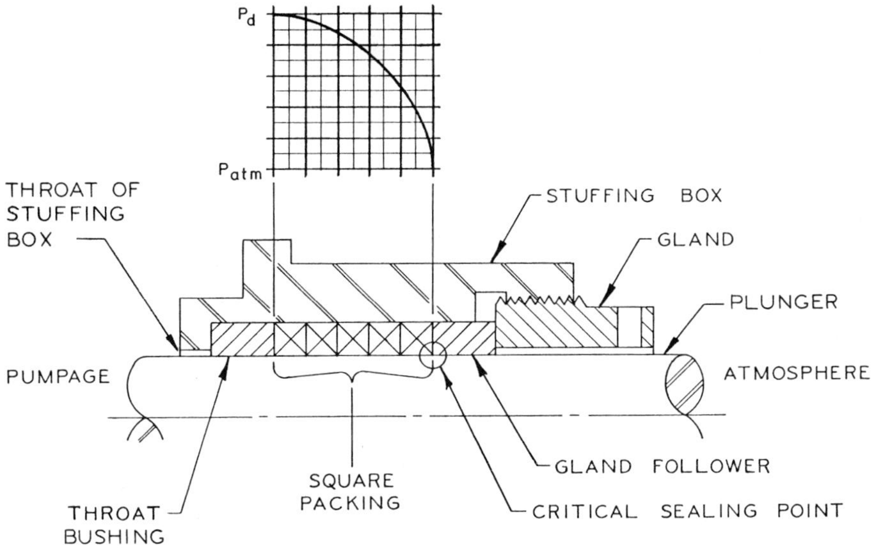

FIG. 4.6. Approximate pressure gradient across a set of packing.

seals tightest against the plunger, generates the most friction, and wears out first, often partially extruding through the gap between the plunger and the back-up bushing or gland. Because of its high loading, this outer ring is the one which *needs lubrication most*. If lubricated only by the pumpage, it is the ring which gets the *least* lubrication and cooling because it is farthest from the pumpage. It would be desirable to impose a higher load on the inner ring, to get it to share more of the task of sealing. It enjoys advantages not shared by the outer ring, such as greater cooling and lubrication from the pumpage, and a zero gap for extrusion (the adjacent ring has zero clearance over the plunger), but, alas, as we tighten the gland, the outer ring absorbs most of the load because of the higher differential pressure, and because the drag in the box bore inhibits its movement. (It stays put.)

Examination of a stack of packing which has operated a number of hours will confirm the above observations. The outer ring will be compressed most and, in some cases, will be partially extruded. The adjacent ring will show some deformation, but to a lesser degree. As we examine each successive ring, we'll find less and less deformation. The inner ring, in most cases, will be in like-new condition. It often serves primarily to deprive its neighbors of cooling and lubrication, and just adds a little friction to the box assembly—to add insult to injury.

Although we used square packing to illustrate these characteristics, chevron ("V") packing performs in a similar manner.

In Chapter 5, we'll see noncrushable packing which is designed to alleviate some of the problems inherent in conventional packing. Later in this chapter, we'll see more sophisticated stuffing box designs which partially alleviate these shortcomings.

4.4. THROAT AND GLAND BUSHINGS

Figure 4.7 shows the addition of a throat bushing and gland follower bushing to the box assembly. The function of these bushings is to retain the packing. Since the stuffing box and gland no longer must perform these functions, the box and gland can be made with generous clearances over the plunger. Since close clearance between the bushings and plunger is necessary for long packing life, (clearance of the follower is particularly important), these parts are often made of a nongalling material, such as bronze, to minimize damage to the plunger in case the plunger rubs against a bushing during operation. The bushings can be replaced more readily than the box or gland when clearances increase to an unsatisfactory value.

4.5. THE PARTS THAT WEAR

It is obvious to most service personnel that, when gland adjustment can no longer reduce leakage to an acceptable rate, the packing must be

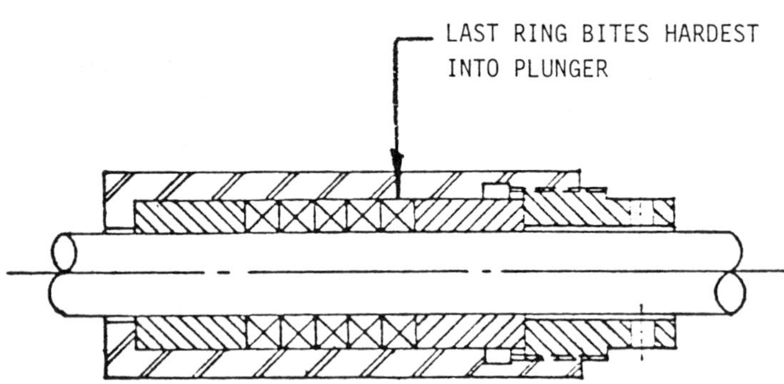

1. GOOD DESIGN.
2. SUITABLE FOR COOL WATER AND FLUIDS WITH COMPARABLE LUBRICITY.
3. TOTAL PACKING LENGTH MUST BE LESS THAN PLUNGER STROKE LENGTH TO PROPERLY WET LAST RING OF PACKING WITH PUMPAGE.

FIG. 4.7. Standard nonlubricated stuffing box.

replaced. It is not so obvious that the plunger, bushings, and even the stuffing box bore, wear out and need periodic replacement or reconditioning to restore acceptable packing life and leakage rates. The alignment of the plunger with the stuffing box bore is also a requirement that is often missed. These subjects will be covered in detail in Chapter 12.

4.6. THE STANDARD NONLUBRICATED STUFFING BOX (SUMMARY)

The standard nonlubricated stuffing box is typically one with three to five rings of square packing. It is a satisfactory design for many applications. It is suitable for cool clean water (up to about 6000 psig) and fluids with comparable lubricity. The total stack of packing relies on the pumpage for its lubrication and cooling. Because it bites hardest on the plunger, the outer ring requires lubrication and cooling more than any other ring in the box. Because it is farthest from the pumpage, the outer ring gets *less* lubrication and cooling. For this reason, fewer rings in a stuffing box will often last longer than a larger number of rings. Chesney[1] reported that, on numerous salt water injection pumps operating at pressures above 4000 psig, packing life was only two weeks when twelve rings of packing were installed in each stuffing box. With only three rings installed in each box, packing life increased to approximately six months.

A Gulf coast refinery complained of short packing life on a 5-inch-stroke triplex pumping cool lubricating oil. (What better conditions could packing ask for?) Upon investigation it was found that the lubricator and lantern rings (which shouldn't have been furnished in the first place) had been disposed of and each box was being packed with 15 rings of machined glass-filled PTFE chevron packing (excessive rings—excessive friction—excessive heat—short life). A unique close-clearance bronze bushing containing two small square rings of PTFE, installed in place of each stack of 15 rings, significantly outlasted the conventional packing.

A basic rule for nonlubricated boxes is that the total length (stack height) of the packing should not exceed the length of the plunger stroke. If this rule is followed, a portion of the plunger which is wetted by the pumpage on the discharge stroke will pass through the outer ring of packing on the suction stroke, hopefully dragging some of the pumpage to that critical sealing area.

4.7. LEAKAGE RATE

It has often been stated that packing must leak to be provided with cooling and lubrication. This statement is true, though, only for the nonlubricated (pumpage-lubricated) stuffing box, and not always then. How much should such a box be allowed to leak? That is a function of the plunger size, speed, and conditions of service, but a general rule is that a stream is not necessary. Ten to sixty drops/min/plunger seems to bracket most services.

If the pumpage has a vapor pressure above atmospheric, or is toxic, leakage to the atmosphere cannot be tolerated. If a product with a vapor pressure above atmospheric is allowed to leak to the atmosphere, it will vaporize and cool, causing the packing to cool. Many packings will shrink and harden when cooled. As the packing shrinks, leakage and flashing will increase, causing more cooling. Hard packing is difficult to adjust, but if, by brute force, the gland is adjusted enough to squeeze the packing and reduce the leakage rate, the packing will begin to warm up—and expand—with no place to go. The packing will then impose a high frictional load on the plunger, causing both to get hot. The packing will attempt to expand further, locking the gland so tightly that it will be almost impossible to loosen. If the pump is not stopped immediately, the packing will begin to disintegrate and the plunger will be scored. The nonlubricated stuffing box should therefore not be used for pumpage with a vapor pressure above atmospheric pressure, or for toxic fluids.

Lip-type packing, such as chevron and noncrushable, are not recommended for use in a nonlubricated stuffing box, even if the pumpage is a good lubricant (unless only one sealing ring is used), because the inner ring(s) wipe the plunger too dry, starving the outer ring of lubrication.

4.8. THE LUBRICATED STUFFING BOX

If the pumpage has insufficient lubricity for satisfactory packing life, it is necessary to supply lubrication to the packing from an external source. Sometimes grease is injected, with a hand-operated grease gun, into a fitting on the box; but it is more common to use a mechanical lubricator (actually a small plunger power pump) to inject oil into a lantern ring, as shown in Fig. 4.8. For most applications, steam cylinder oil has been found to be the best lubricant, providing the longest packing life and requiring the lowest injection rate. Because the pressure at the lantern ring is near pump discharge pressure (during the discharge stroke), the mechanical lubricator, plus tubing and fittings, must be suitable for pump discharge pressure.

Because a mechanical lubricator is a positive-displacement plunger pump, it can create a very high discharge pressure, limited only by the strength of the components exposed to the oil. The pumping rate of a lubricator is adjustable. If the rate is set higher than the lubricant can leak through the packing, lubricant pressure increases until something fails. It

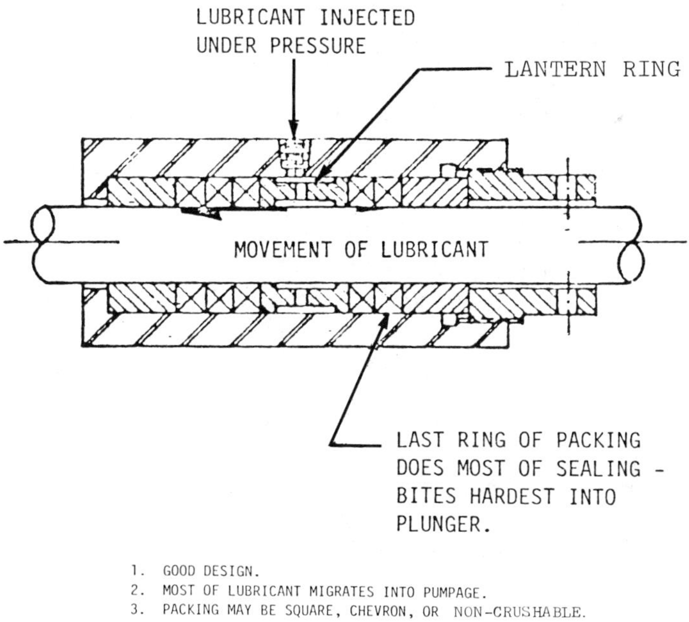

FIG. 4.8. Standard lubricated stuffing box.

is not uncommon to find packing that has been extruded and, in extreme cases, the stuffing box split by excessive pressure from the lubricator.

What is the proper lubrication rate? The proper rate is a function of the effectiveness of the lubricant, the type of packing, the characteristics of the pumpage, the size and stroke length of the plunger, and pump speed. Typically the rate will vary from 10 to 30 drops/min/plunger. The proper lubrication rate may be determined by monitoring the stuffing box temperature and ultimately, packing and plunger life.

The mechanical lubricator is discussed further in Chapters 10 and 12.

Continuing reference to Fig. 4.8, the lantern ring accepts the lubricant at the bore of the stuffing box and distributes it to the surface of the plunger. It should be located nearer to the outer packing ring, than the inner ring, to provide the lubricant where it's needed most (at the outer ring). In some earlier designs, the lantern ring was placed nearer the throat of the box, apparently with the thought that product leaking through the packing would carry the lubricant through all packing outboard of the lantern ring. But with new packing and a good plunger, a stuffing box often does not leak. This is especially true with lip-type packing (chevron and the noncrushable types). The lubricant is therefore wiped from the plunger sealing surface by packing rings inboard of the outer ring, leaving no lubricant for the outer ring. In this case almost all of the lubricant flows through the looser rings in the bottom of the box, through the box throat, and into the pumpage. By placing the lantern ring nearer the outer packing ring, less oil flows into the pumpage, and more to the outer ring, but even then, most of the oil usually flows into the pump.

It has been determined from field experience that oil dripped onto the plunger, just outboard of the outer ring, is usually as effective as the injection system (in the standard box), and less oil migrates into the product, usually reducing oil contamination to an unmeasurable level. This discovery led to the arrangement shown in Fig. 4.9. The lantern ring is used in place of the gland follower bushing. If desired, a mechanical lubricator can still be used, but it no longer needs to be suitable for pump discharge pressure. Often, an atmospheric drip-type lubricator is used, sometimes with a solenoid valve to stop oil flow when the motor driver is shut off. We also eliminate the possibility of over-pressuring the lubrication system and damaging the packing, stuffing box, or lubricator. This

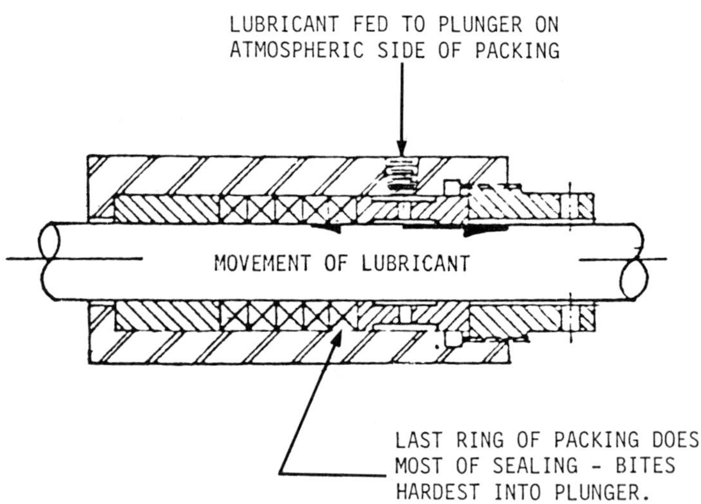

1. GOOD DESIGN - PERFORMS BETTER THAN STANDARD ARRANGEMENT IN MOST CASES.
2. PUTS LUBRICANT UNDER LAST RING - WHERE IT'S NEEDED MOST.
3. ALLOWS USE OF LOW-PRESSURE AND DRIP-TYPE LUBRICATORS.
4. VERY LITTLE LUBRICANT MIGRATES INTO PUMPAGE.
5. PACKING MAY BE SQUARE, CHEVRON, OR NON-CRUSHABLE

FIG. 4.9. An alternate method of lubricating packing.

lubrication arrangement can sometimes deceive us, though. The lubricant can cling to the box bore, flow around the lantern ring, and out through the gland without contacting the plunger. A special lantern ring or lubrication fitting can minimize the chance of this occurrence.

Other problems can occur with outboard atmospheric drip systems. If the lubrication tube is just suspended over the plunger surface outboard of the gland, it can be bumped sideways, with oil hitting only on the side of the plunger, or missing it entirely. Also, if the length of the gland and follower bushing exceed the stroke length of the pump, very little oil will be carried from the outboard side of the gland to the last ring of packing.

Advantages of lubricating the packing are increased life of the packing and plunger, and the reduction of product leakage. (The lubricant forms a seal between the packing and plunger to reduce leakage.) Another advantage, seldom considered, is the reduction in power consumption. Lubricating the packing will often increase pump efficiency as much as 2 points. With a 100 horsepower pump, this amounts to a power saving of about 2 horsepower, a significant savings over the life of the machine.

4.9. AN IMPROPER APPLICATION

We should not go on without mentioning how a standard lubricated stuffing box should *not* be used. Figure 4.10 illustrates an arrangement which is occasionally tried, usually with unhappy results. The reasoning seems sound, but the packing life is short. We'll see why. Here's the logic: Let's bleed off all leakage at the lantern ring, pipe it back to the suction line, and the two outer rings will be exposed only to suction pressure, leaking very little.

If the rings felt only liquid pressure, this arrangement would work fine—but they don't. Since the total differential pressure now occurs across the first three rings, the third ring becomes the "outer" ring—fully-loaded and tigntly sealing (and generating friction). This full force is also transmitted through the lantern ring into the fourth and fifth rings, which also rub hard against the plunger and generate heat. The friction in rings 4 and 5 is even greater than ring 3, because there is insufficient pressure at the lantern ring to force enough product between the plunger and packing to provide adequate lubrication. The result is often a hot box, charred packing, and a scored plunger.

4.10. A SATISFACTORY BLEED-OFF ARRANGEMENT

With the application of the knowledge gained above, we can now devise a bleed-off scheme without the above shortcomings. Assuming that the pumpage is an adequate lubricant for the packing, we need only to elimi-

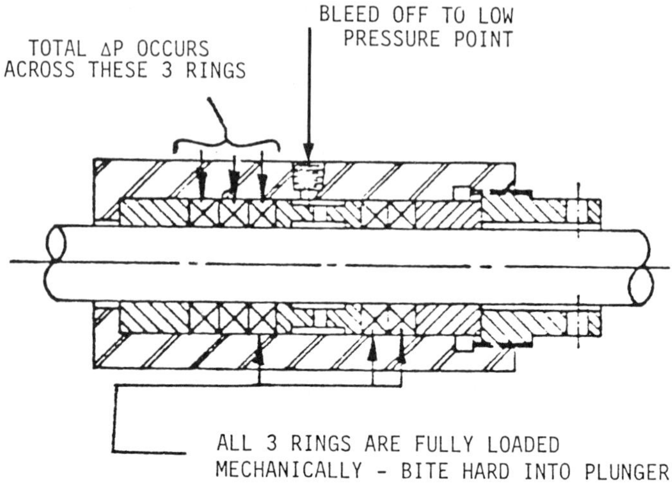

1. HIGH FRICTION CAUSES EXCESS HEAT.
2. SHORT LIFE OF PACKING AND PLUNGERS.
3. IMPROPER USE OF STANDARD BOX - POOR APPLICATION.

FIG. 4.10. A standard lubricated box used to bleed pumpage to a low pressure (bad application).

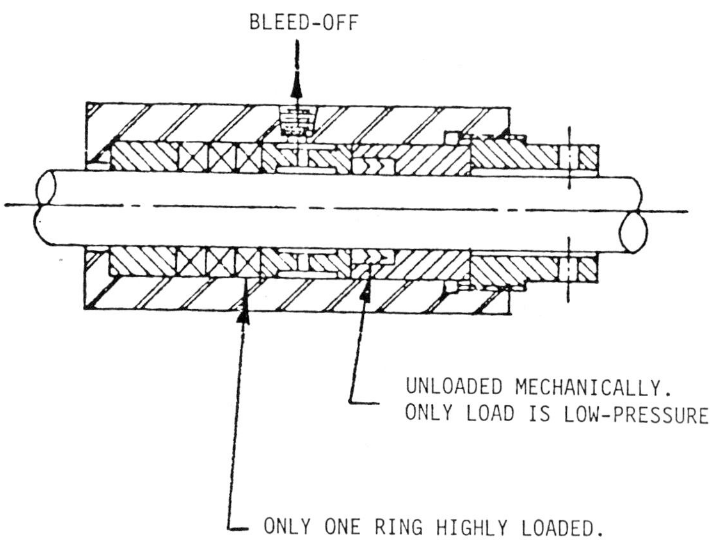

1. LESS FRICTION AND LOWER TEMPERATURE THAN FIGURE 4–10.
2. LONGER LIFE OF PACKING AND PLUNGERS.
3. BIG IMPROVEMENT OVER FIGURE 4–10, BUT SECONDARY PACKING NOT ADJUSTABLE TO COMPENSATE FOR WEAR.

FIG. 4.11. A modified gland follower allows bleed-off to a low pressure.

nate the mechanical load applied to rings 4 and 5. This can be done, as shown in Fig. 4.11, by machining a long gland follower with a recess to accept one ring of chevron packing (with its male and female adapter) or one or two rings of square packing. The axial force from the lantern ring now gets transmitted *around,* rather than through, the low-pressure ("secondary") packing. The secondary packing therefore rides lightly on the plunger, creating minimal friction and heat. Shortcomings of this design are that the secondary packing is small (therefore with minimal wear will lose its sealing ability) and is not adjustable to compensate for wear.

4.11. THE TWO-GLAND STUFFING BOX

If we make the secondary packing full size, and make it adjustable, we have a stuffing box as shown in Fig. 4.12. Note that the assembly includes two glands—one inside the other. The large gland adjusts the primary

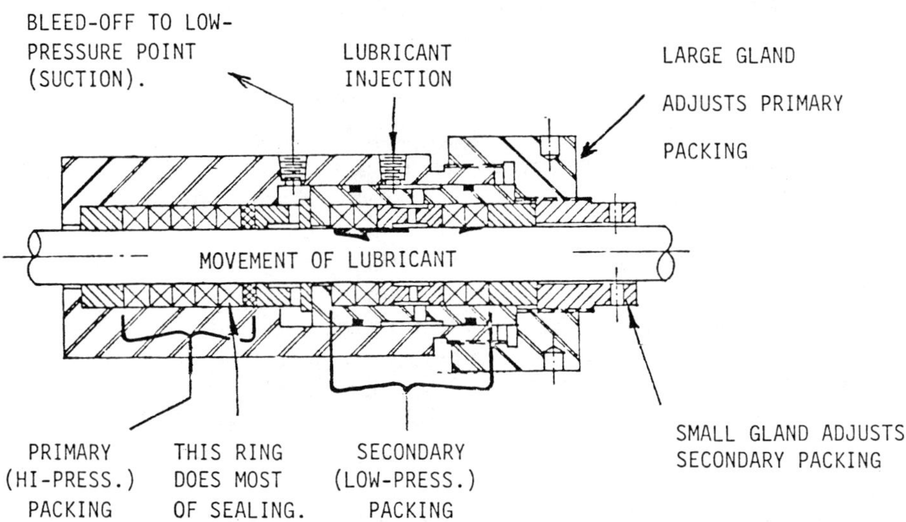

1. THE OLD STANDARD FOR HIGH-PRESSURE, CRITICAL SERVICES.
2. PROVIDES INDEPENDENT ADJUSTMENT OF PRIMARY AND SECONDARY PACKING. (PROPER ADJUSTMENT REQUIRES SKILLED MECHANIC.)
3. FULL-SIZE SECONDARY PACKING.
4. POSITIVE PACKING LUBRICATION.
5. NEGLIGIBLE LEAKAGE TO ATMOSPHERE.
6. LONG PACKING AND PLUNGER LIFE.
7. EXCELLENT FOR VOLATILE FLUIDS.

FIG. 4.12. A two-gland stuffing box, lubricated.

(high-pressure) packing. The small gland adjusts the secondary (low-pressure) packing. Examination of the drawing will reveal that we have two separate, practically-independent stuffing boxes arranged in tandem. The hydraulic load on the primary packing is transmitted around the secondary packing (through the body of the secondary stuffing box) and into the large gland. The hydraulic load on the secondary packing (suction pressure only) is transmitted into the small gland (which transmits it to the large gland).

The packing normally used in this box is braided PTFE impregnated with a PTFE suspensoid. The rings are quite dense and hard.

The outer ring of the primary packing is the key packing ring in this box assembly, doing most of the sealing. If adjustment of only the primary packing is required, only the large gland need be adjusted. As the large gland is tightened, the entire secondary packing assembly (secondary box, packing, rings, and small gland) slides forward, without altering the setting of the secondary packing. If only the secondary packing requires tightening, only the small gland is tightened.

The annular space, between the primary and secondary packing, collects leakage from the high-pressure packing and serves as a reservoir for the lubricant which migrates forward from the secondary packing. This space is vented to a low pressure, preferably suction pressure. The bleed-off line is normally piped to the suction line. If the pumpage has a specific gravity less than the lubricant, such as ammonia, the bleed-off should be from the top of the box. If the pumpage has a specific gravity greater than the lubricant, such as carbamate, the bleed-off should be from the bottom of the box. In this way the lubricant remains in the annular space, providing adequate lubrication to that hardworking outer ring of the primary packing.

The secondary packing seals only the low pressure of the annular space (normally at suction pressure), a relatively easy task since this pressure seldom exceeds 300 psig. The secondary packing also accepts the lubricant, which is distributed throughout the length of the secondary packing, with most flowing into the annular space to provide lubrication for the primary packing.

It is important, when pumping volatile liquids, that the pressure in the annular space not be reduced below vapor pressure (which is normally just below suction pressure). If the pressure drops below vapor pressure, product leaking through the packing will flash and cool the packing. If the packing gets cool, it shrinks and allows more leakage which causes more chilling of the packing. The lubricant also "curdles" and provides minimal lubrication and sealing. The problem is the same as that described under the nonlubricated stuffing box. If suction pressure (or higher) is maintained in the annular space, no chilling will take place as the product leaks through the primary packing. As a matter of fact, this area of the stuffing box, and the bleed line, should be at a temperature 30° to 40°F above pumping temperature. Monitoring these temperatures is the method used to determine if adjustment of the large gland is required.

It should be noted that leakage of a volatile product through the secondary packing to the atmosphere will chill the secondary packing and lubricant, but since the secondary packing seals only suction pressure (normally less than 300 psig) preventing leakage through this packing is a relatively easy task. The secondary packing also often contains one or more rings of soft packing, making it more sensitive to gland force, and therefore easier to adjust.

Each bleed-off line should contain a check valve to prevent loss of product from the suction line to the atmosphere in case of a failure of the secondary packing (although such an incident is not known to have occurred). A pressure drop across this valve (such as from a spring-loaded ball) of 100 psi or so will be of no detriment to the operation of the stuffing box, possibly even extending the life of the primary packing by slightly reducing the differential pressure imposed thereon.

The two-gland stuffing box has found application for decades in services handling volatile liquids such as hydrocarbons, ammonia, and ammonium carbamate, at pressures up to 6000 psig. It is an excellent design, and performs extremely well when properly adjusted. One fertilizer plant with a pair of 5-inch-stroke triplex pumps, equipped with such boxes, running at 300 rpm, pumping ammonia at 4000 psig, was operating without detectable leakage to the atmosphere after 2 years of *continuous* operation. Only a few minor adjustments had been made to the glands during the 2 year period. Only *one person* was allowed to make those adjustments.

4.12. SPRING-LOADED PACKING

The most significant progress in stuffing box arrangements in recent years has been the expanded application of spring-loaded packing. Although discussed in the literature for many years, and used by one manufacturer for at least 20 years, only recently has this arrangement received the attention that it deserves. Although applied almost exclusively to V-ring (chevron) packing, spring-loading has been found to work equally well with square packing rings.

Figure 4.13 illustrates a simple spring-loaded V-packing arrangement. A cut-away of such a stuffing box assembly is shown in Fig. 4.14, and a cut-away of the packing and adapters, in Fig. 4.15. Because this is lip-type

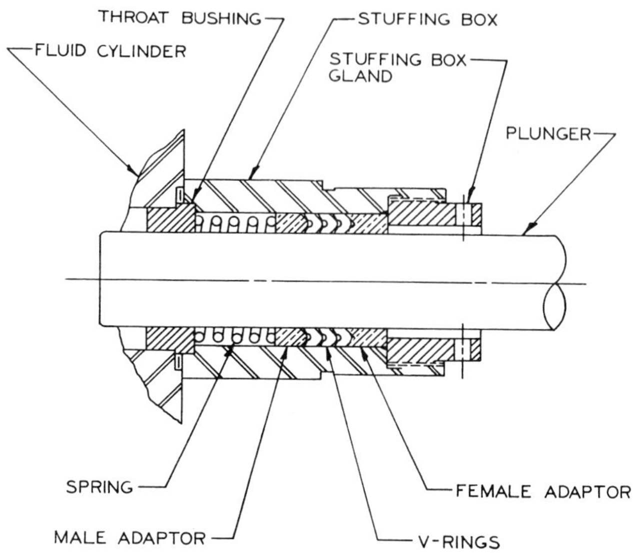

1. MINIMAL LEAKAGE.
2. BECAUSE NON-LUB, NORMALLY LIMITED TO INTERMITTENT DUTY.
3. SELF-ADJUSTING.

FIG. 4.13. Spring-loaded V-ring packing, nonlubricated.

FIG. 4.14. Cutaway of a stuffing box assembly equipped with spring-loaded packing, as illustrated in Fig. 4.13. (*Courtesy Union Pump Co.*)

FIG. 4.15. Cutaway of two V-rings with hard male and female adapters—Spring-loaded. (*Courtesy Utex Industries*)

packing and has no external lubrication, it is recommended only for intermittant operation. (Otherwise packing life is too short.)

The spring is always on the pressure side of the packing, located in the pumpage, or in a barrier fluid if one is provided. The spring may be a single large coil (normally guided by the bore of the stuffing box, as illustrated), or any of a variety of loading devices such as a multiple-coil assembly, wave-washer, belleville washer, or a thick rubber washer. The force of the spring on the packing is quite low (equivalent to one to two atmospheres of pressure) compared to the force imposed on the packing by the fluid. The major function of the spring is to provide a small preload to help set the packing, and to hold all bushings and packing rings in place during operation.

Note that the gland is tightened until it bottoms, and is locked in that position. No gland adjustment is ever required. Therefore one of the biggest variables affecting packing life (operator skill) is eliminated. If the packing at any time expands, due to higher friction during break-in, or an increase in pumping temperature, the spring allows for expansion with only a slight increase in axial loading.

As the packing wears, compresses, or contracts because of a temperature reduction, the spring maintains an axial load, automatically adjusting the packing from *inside* the box. The problem of transmitting the gland force through the outer packing ring, to adjust the inner packing rings, as discussed on gland-adjusted packing sets, is therefore eliminated.

When used on slurry applications, the spring cavity can be used for injection of a clean barrier fluid to help isolate the packing from the slurry.

To make this arrangement suitable for continuous operation, it is necessary that the "V" packing rings be lubricated from an external source, preferably with an oil, such as steam cylinder oil. Borrowing the idea discussed earlier, the lubricant can be supplied to the plunger surface on the atmospheric side of the packing. Figure 4.16 illustrates such an arrangement, with a lantern ring directing the lubricant onto the plunger (adjacent to the packing) plus providing that critical support for the female adapter.

Figure 4.17 illustrates a stuffing box equipped with spring-loaded *square* packing. The operation of this box is the same as the spring-loaded

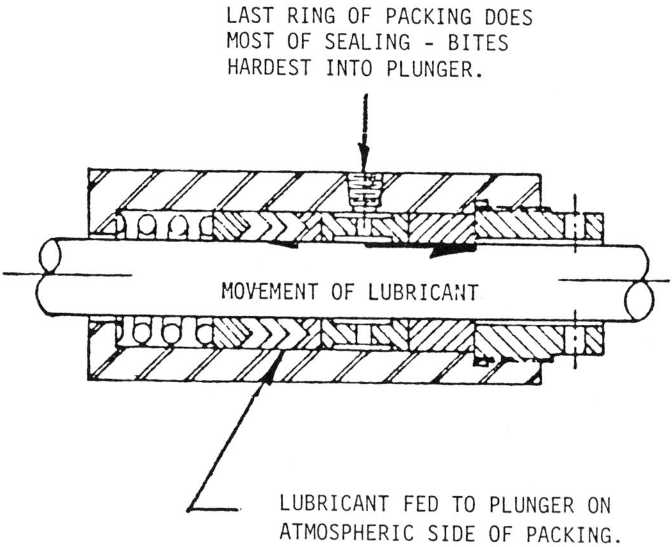

1. GOOD DESIGN - LONG LIFE - MINIMAL LEAKAGE.
2. PUTS LUBRICANT UNDER LAST RING - WHERE IT'S NEEDED MOST.
3. ALLOWS USE OF LOW-PRESSURE AND DRIP-TYPE LUBRICATORS.
4. NO ADJUSTMENT REQUIRED (SELF ADJUSTING).

FIG. 4.16. Spring-loaded V-ring packing, lubricated.

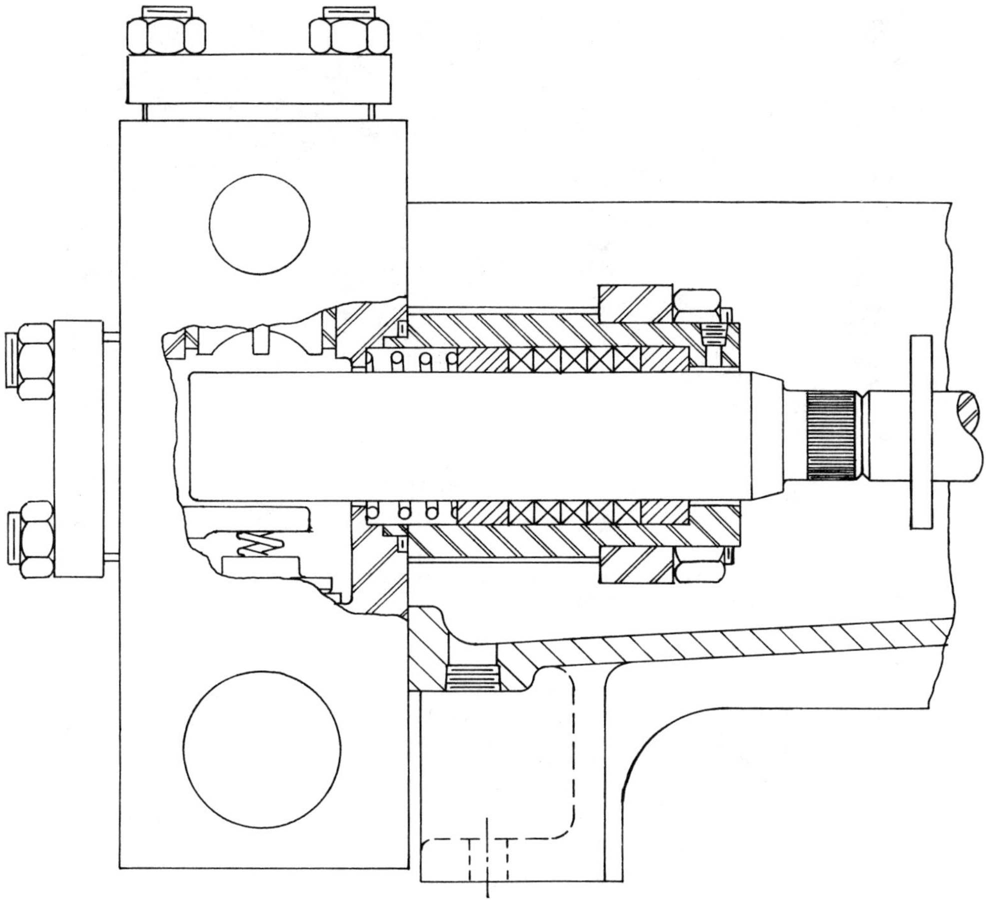

FIG. 4.17. A glandless stuffing box with spring-loaded square packing. (Lubrication is optional through connection at atmospheric end of box.) (*Courtesy Union Pump Co.*)

V-packing, except that the square rings don't wipe the pumpage from the plunger surface as completely as do lip-type rings. Therefore, if the pumpage is an adequate lubricant, this arrangement can be used for continuous duty without external lubrication.

It will be noted that this stuffing box has another unique feature: no gland. Because the spring automatically adjusts the packing, a gland is no longer required. To replace the packing, it is necessary to remove the stuffing box. The entire box assembly is removed by removing four bolts, as shown in Figure 4.18. The unit is then disassembled and reassembled

Fig. 4.18. A spring-loaded glandless stuffing box assembly being removed from a liquid end. (*Courtesy Union Pump Co.*)

on a suitable work surface. If the pumpage is a poor lubricant, or the packing is of the "V" style, an external lubricant is required. This stuffing box is lubricated by feeding oil through the small hole in the top of the box onto the plunger surface on the atmospheric side of the packing.

It should be noted that noncrushable packing, as discussed in Chapter 5, is designed to be clamped tightly in a stuffing box by the gland. There is no advantage to spring-loading such packing.

Spring-loaded packing is the reciprocating pump's equivalent to the mechanical seal, as used extensively in centrifugal and rotary pumps. Leakage is low, life is extended, and manual adjustment is eliminated. When leakage exceeds an allowable level, the packing (and any other worn parts) must be replaced. As with a mechanical seal, no adjustment is possible.

4.12.1. Disadvantages of Spring-Loaded Packing

The disadvantages of spring-loaded packing are associated partly with the cavity containing the spring. Since this cavity communicates directly with the pumping chamber, the additional clearance volume causes a reduction in volumetric efficiency if the pumpage is sufficiently compressible. This cavity also provides a spot for vapors to accumulate, and if the pump design does not provide for venting, a reduction in volumetric efficiency may result.

Another potential problem of spring-loaded packing became apparent when a user complained of intermittent shocks, sometimes at a frequency as high as pump rpm, being experienced in the discharge line of his pump. An examination of the liquid end internals revealed that the end of one plunger had been dragging heavily in the close-clearance bore of the long bronze male adapter bushing. Apparently the bushing would be pushed forward by the plunger (compressing the packing spring) during the early portion of the discharge stroke, then would slam back against the packing. When the bushing would slide forward, the liquid discharge velocity would increase significantly. When the pressure would slam the bushing back against the packing, the discharge velocity would drop significantly below normal. All of this would occur in just a few milliseconds, causing a shock wave in the discharge line. Replacing the plunger and bushing solved the problem.

4.12.2. Tandem Spring-Loaded Packing

Now, we can combine the general configuration of the two-gland stuffing box with spring-loaded packing to produce the tandem spring-loaded assembly, as shown in Fig. 4.19. This is equivalent to the tandem mechanical seal for a rotating shaft. We obtain all of the desirable characteristics of the two-gland stuffing box while eliminating the one big disadvantage: the skill required to properly set and adjust the packing. With tandem spring-loaded packing only one gland is required, and there is no adjustment. Figure 4.20 shows a power pump equipped with such a stuffing box.

Although the drawing shows V-packing, the more common arrangement, the tandem spring-loaded stuffing box can also utilize square packing.

Figure 4.21 illustrates a modified version of the tandem spring-loaded packing design which was developed for pressurized-water nuclear reactor charging pumps. Because an oil lubricant cannot be tolerated by the system, the lantern ring has been deleted. The annular space between the two packing sets is used to circulate demineralized water, the only coolant and lubricant available to the packing, at atmospheric pressure.

The water circulates through the stuffing box, driven by a simple thermal-siphon system, as shown in Fig. 4.22. No auxiliary pump is required. The water, heated by the friction of the plunger sliding through the packing, rises from the box and into the upper portion of the water tank. Cool water from the lower portion of the tank flows into the bottom of the box. There is sufficient heat lost from the tank and tubing to the surrounding air

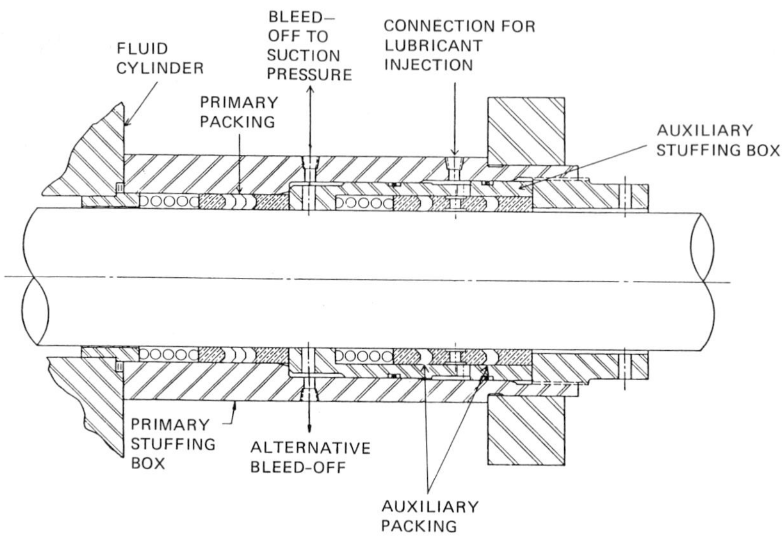

1. THE BEST DESIGN AVAILABLE FOR MOST HIGH-PRESSURE, CRITICAL SERVICES.
2. COMBINES BEST FEATURES OF TWO-GLAND BOX AND SPRING-LOADED PACKING - NEGLIGIBLE LEAKAGE, LONG LIFE, AND THE PACKING IS SELF-ADJUSTING.

FIG. 4.19. Tandem spring-loaded packing—with bleed-off and lubrication. (*Courtesy Union Pump Co.*)

to provide adequate cooling for the boxes. Leakage from the primary packing leaves the system through an overflow connection in the tank. Leakage from the secondary packing is negligible.

Because the coolant is driven only by the low energy available from the friction heat, a gas bubble at any high point in a coolant line will restrict coolant flow. Care must therefore be exercised in installing the tubing to avoid any high point in any tube. Upper tubes should slope upward to the tank.

One such system installed on a 5-inch-stroke quintuplex pump, pumping a 2% boric acid solution, was operated under carefully-controlled conditions for 2000 hours. Speed was 200 rpm, and discharge pressure was 2500 psig. Leakage through the primary packing, at the end of the 2000 hour test, was only 0.003% of the pump capacity. Leakage through

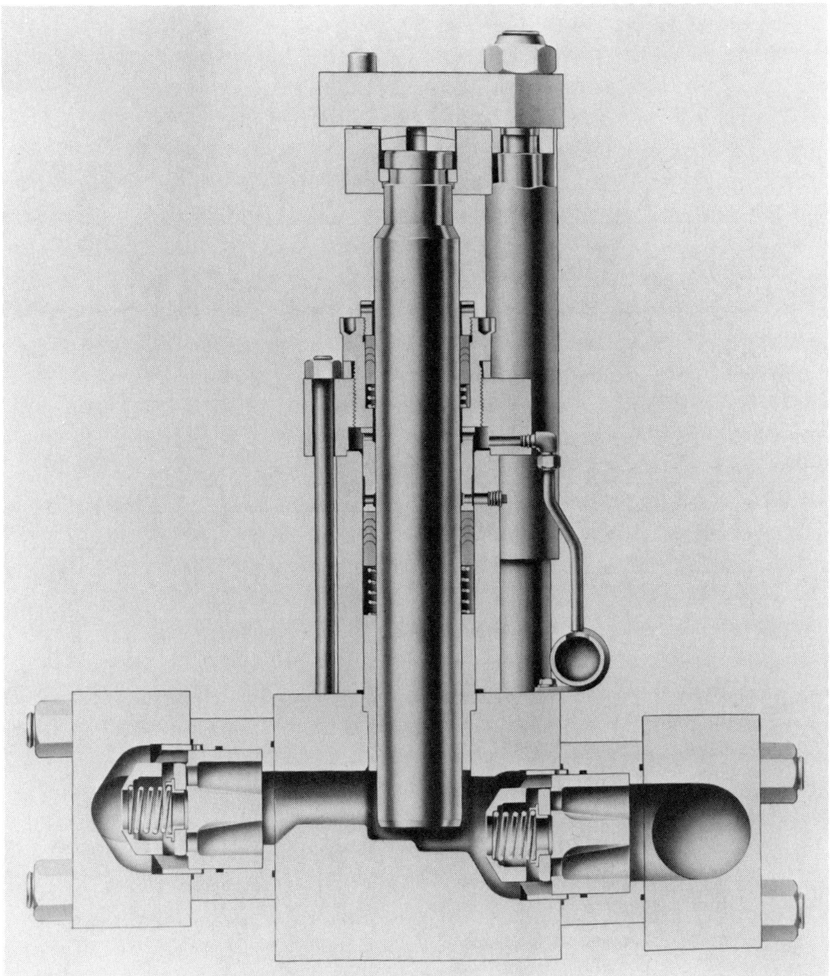

FIG. 4.20. A power pump equipped with tandem spring-loaded packing—with bleed-off and lubrication connections. (*Courtesy Worthington Pump Div., Dresser Industries, Inc.*)

the secondary packing was so low that it evaporated on the gland, leaving only a light coating of the white powder used to make the acid solution.

Figure 4.23 is a photograph of a charging pump equipped with such a system. The lower portion of the elevated tank is seen in the upper left hand corner of the photo.

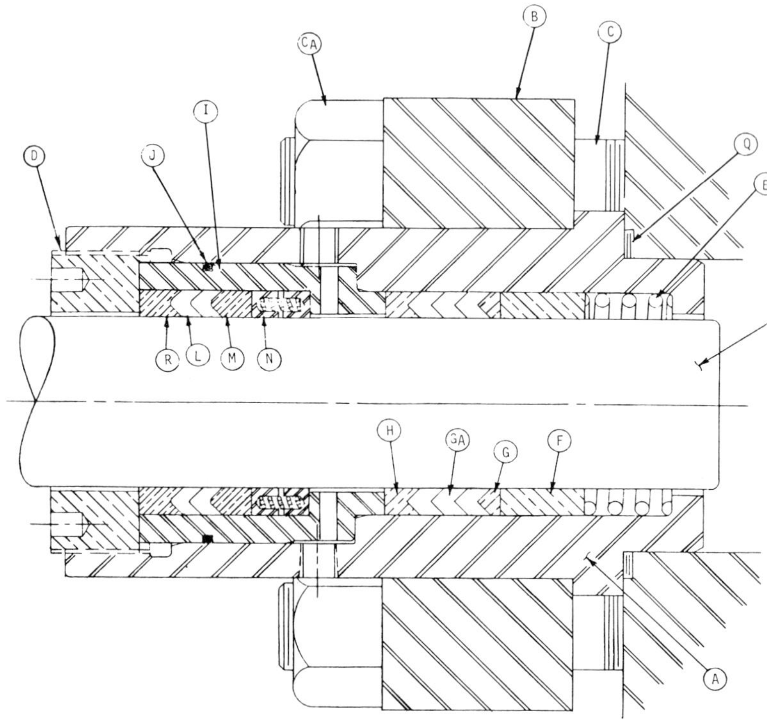

ITEM	PART NAME	ITEM	PART NAME
A	Stuffing box	I	Gland follower
B	Flange	J	"O" ring
C	Flange studs	L	Secondary male adapter
C_A	Flange nuts	M	Secondary packing
D	Gland	N	Secondary spring
E	Primary spring	Q	Gasket
F	Throat bushing	R	Secondary female adapter
G	Primary male adapter		
GA	Primary packing		
H	Primary female adapter		

FIG. 4.21. Tandem spring-loaded packing for nuclear charging pumps. (*Courtesy Union Pump Co.*)

Because of any of a number of various conditions, packing and plunger life between adjacent stuffing boxes is seldom uniform. One or more stuffing boxes may fail prematurely. It is recommended, therefore, that any sealing system be capable of monitoring the leakage rate from each individual stuffing box so that the recalcitrant box can be identified and *only* that packing changed. In the above-described coolant circulation system, it is beneficial to provide a separate tank for each stuffing box, or

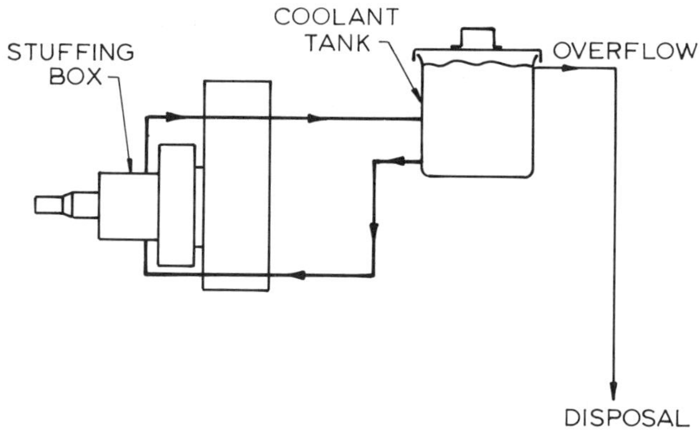

FIG. 4.22. Operation of the nuclear tandem stuffing box coolant circulation system. (*Courtesy Union Pump Co.*)

FIG. 4.23. A triplex pump equipped with tandem spring-loaded packing and coolant circulation system for nuclear charging service. (*Courtesy Union Pump Co.*)

a segmented tank that allows the measurement of leakage from each box. Otherwise the failure of only one set of primary packing will require the replacement of all sets in the pump.

4.12.3. Segmental Garter-Spring-Loaded Packing Rings

A type of stuffing box which has found limited use on reciprocating pumps since about 1960 is illustrated in Figure 4.24. It is the tangentially-spring-loaded segmental ring stuffing box, found more commonly on reciprocating *compressors*.

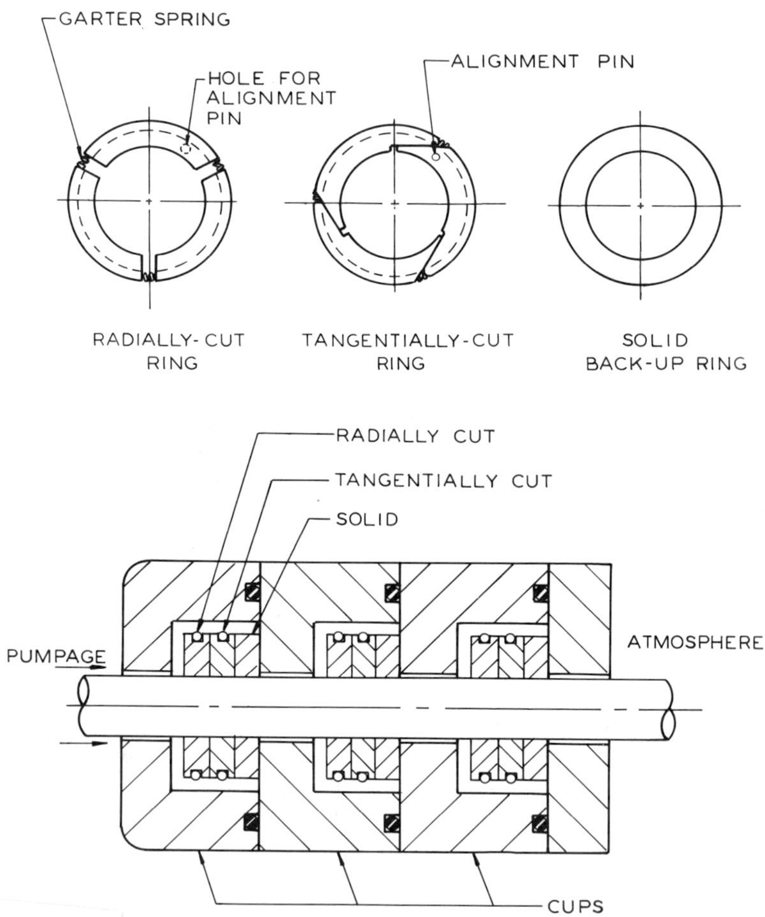

ntal rings, loaded by garter springs, and a typical segmental stuffing

This stuffing box typically consists of a stack of "cups," each containing two or three of the different types of rings shown in the top of the illustration. These rings are made of a hard material such as metal, carbon, or a filled PTFE. Since they are not very compliant, the first two rings are made in segments so as to maintain contact with the plunger as they wear.

The radially-cut ring is placed at the pressure end of each cup. Its function is to provide a seal across the tangential cuts of the second ring. The tangentially-cut ring performs the major sealing. The solid ring (often bronze) provides support for the second ring. The liquid pressure against the face of the first ring is also at the O.D. of all three rings, pressing the two segmental rings against the plunger. Circumferential ("garter") springs provide an additional small load and hold the rings against the plunger when liquid pressure is low. The garter spring for the first ring is sometimes mounted adjacent to the face of the ring, providing an axial load to keep all three rings pressed against the face of the adjacent cup. The faces of the adjacent cup and the solid ring are lapped to provide an adequate seal at that joint.

Because of the generous clearances between the cups and the plunger, and around the outside of the sealing rings, radial misalignment of the plunger and stuffing box can be more readily tolerated than with the more conventional stuffing boxes discussed earlier in this chapter.

This type of stuffing box has been used satisfactorily in light hydrocarbon services. Lubrication is required.

4.13. SOLID THROTTLE BUSHING

Some plunger pumps utilize a stuffing box sealing arrangement that consists of a throttle bushing, as shown in Figs. 4.25 and 4.26. The bushing is often a hardened stainless steel, and is lapped with the plunger, resulting in a clearance of less than 0.001 inch. Each plunger and bushing must be retained as a set. The leakage rate is high, compared to conventional packing, but life is longer, particularly at the higher pressures which tend to extrude soft packing. Leakage is sometimes pumped back into the suction line with a scavenging pump. Application is limited to products with vapor pressures below atmospheric pressure, and is more satisfactory with, but not limited to, high viscosity liquids.

4.14. THE POT SEAL

The "pot seal" is shown in Figs. 4.27, 4.28, and 4.29. The features of this system are similar to those of the two-gland and tandem spring-loaded

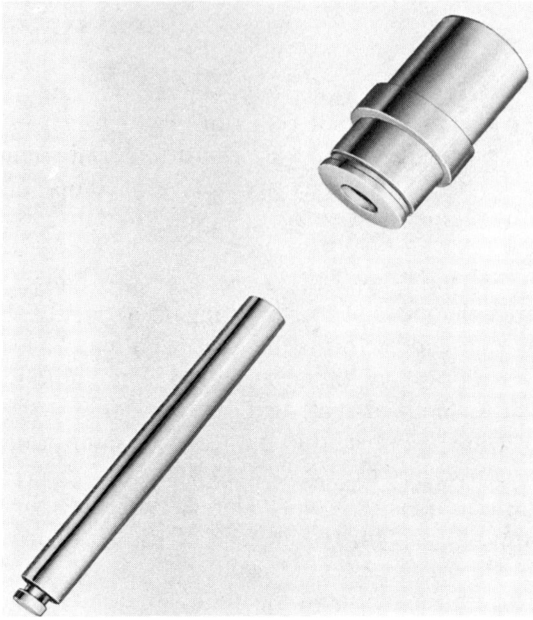

FIG. 4.25. A solid throttle bushing (liner) and plunger. (*Copyright Kobe, Inc. Used with permission.*)

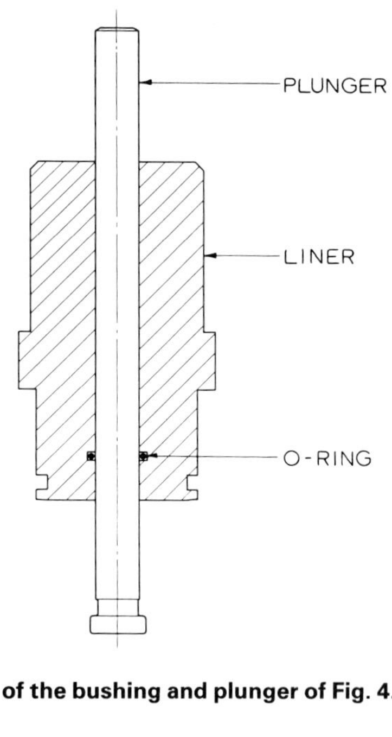

FIG. 4.26. A cross-section of the bushing and plunger of Fig. 4.25. (*Copyright Kobe, Inc. Used with permission.*)

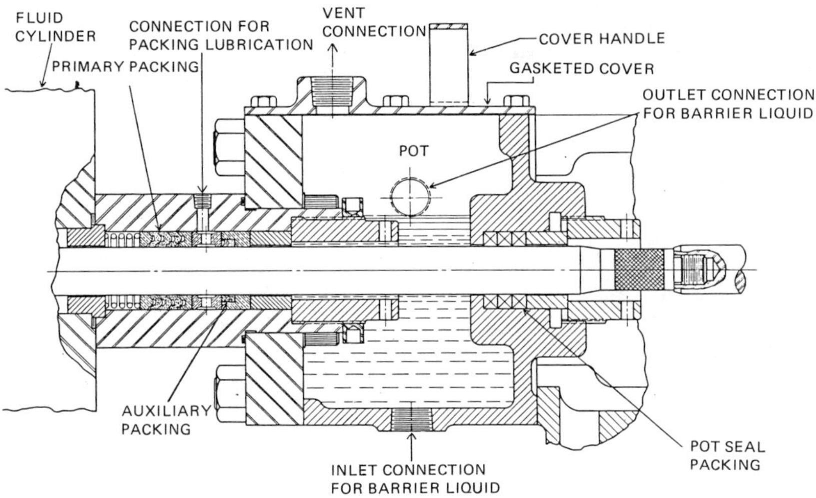

FIG. 4.27. A pot seal stuffing box arrangement for sealing toxic liquids. (*Courtesy Union Pump Co.*)

FIG. 4.28. A triplex pump for detergent service equipped with an open pot seal. (*Courtesy Union Pump Co.*)

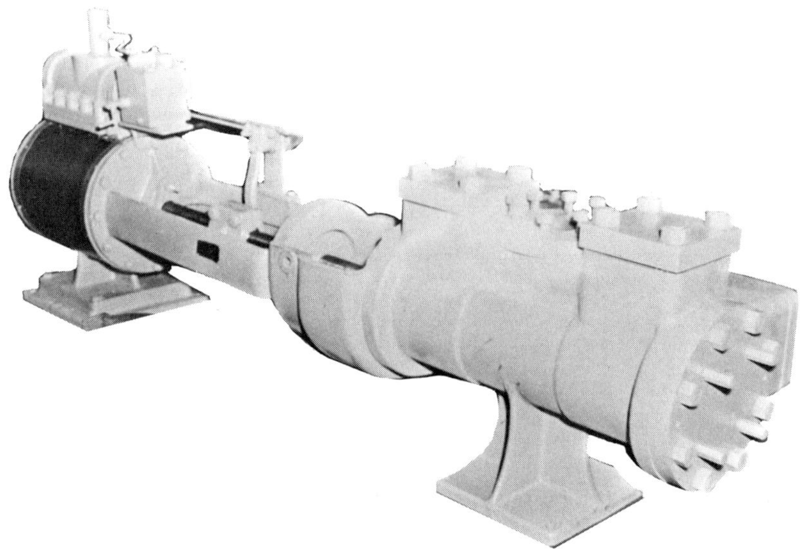

FIG. 4.29. An open pot seal on a direct-acting pump for detergent slurry service. (*Courtesy Union Pump Co.*)

designs. There are two independent stuffing boxes for each plunger. The gap between the primary and pot seal boxes normally exceeds the stroke length of the pump so that the surface of the plunger passing through the high-pressure packing does not enter the pot seal packing. Sometimes the diameter of the plunger is different at the two boxes. The space between the two boxes (the "pot") is filled with a low-vapor-pressure liquid (often oil or water). Leakage from the primary packing disperses into the seal liquid—or separates, depending on the compatibility of the two fluids. A small flow of sealing liquid through the pot prevents an accumulation of product.

The pressure in the pot seal is maintained at, or very near, atmospheric pressure, seldom exceeding 5 psig. Therefore the primary packing seals the total pressure, from discharge to atmosphere, and the pot seal packing seals only the small head of liquid in the pot. Many pots, such as in detergent slurry service, are completely open at the top. Others on pumps handling toxic fluids such as carbon disulfide, are (as shown in Fig. 4.27) equipped with a gasketed cover and a vent connection for piping noxious vapors to a disposal system.

4.15. PLUNGER WELL INCORRECTLY USED AS VAPOR TRAP

Occasionally, the plunger well on a standard power pump has been used to trap vapors emitting from stuffing boxes. This is accomplished with a tight-fitting cover over a plunger well. Such an arrangement traps vapors, as intended, but also forces some of these vapors through the crosshead extension seals and into the power end. If the vapors contaminate the oil, problems may result with bearings and other power end parts. If the vapors do not contaminate the oil, they will pass through the power end breather to the atmosphere, partially defeating the purpose of trapping them in the plunger well.

One small triplex experienced repeated failures of the power end. Inspection of the pump in operation revealed a large quantity of water mixed with the oil in the power end. The pumpage?—hot water! Lifting the solid plunger cover revealed all internal surfaces of the plunger well to be wet. In fact, water was dripping from the undersurface of the cover onto the crosshead extensions. The solution? Replace the solid sheet steel plunger cover with one made of expanded metal, allowing the air in the plunger well to stay much dryer.

A 3-inch-stroke quintuplex pump experienced failure of power end bearings. Although no leakage of product (lean absorption oil) was evident from the stuffing boxes, power end oil was found to be diluted with lean oil. Apparently vapors were leaking through the packing, being trapped by the plunger cover, migrating through the extension seals, then condensing in the oil. Replacement of the solid cover with expanded metal was undesirable because it was feared that blowing sand in the area would be deposited on plungers and extensions, damaging related parts. A compromise arrangement was to elevate the cover on spacers. This reduced, but did not eliminate the contamination problem. Maintenance personnel found it necessary to change the oil weekly.

In summary, the plunger well should not be used as a trap for vapors. It is quite satisfactory for catching and draining away liquid leakage, but not vapors. If the pumpage emits vapors, the plunger well should be ventilated. If the vapors must be trapped, they should be trapped in a special stuffing box or a pot seal as discussed earlier.

4.16. SLURRY STUFFING BOXES

Power and direct-acting pumps are used in a variety of slurry applications. If the slurry is low in abrasivity, such as detergent, no effort is made to keep the slurry from entering the primary packing, but with detergent slurry a water-filled pot seal is normally used to dissolve and flush the detergent from the plunger surface.

If abrasive particles wedge themselves between the packing and plunger, the packing will be cut and/or the plunger scored. For slurries which are abrasive, such as coal and metal ores, stuffing boxes are provided which attempt to minimize the migration of the particles into the packing.

The simplest arrangement places a packing ring with a soft rubber lip adjacent to the slurry. The lip acts as a soft scraper to remove most of the particles from the plunger surface as the plunger retracts from the pumping chamber on the suction stroke.

4.17. LONG INJECTION THROAT BUSHING

Another arrangement is as shown in Fig. 4.30. The length of the stuffing box is increased and the box equipped with a throat bushing longer than the stroke length of the pump. Therefore, that portion of the plunger wetted by the pumpage does not pull back into the packing. The outer end

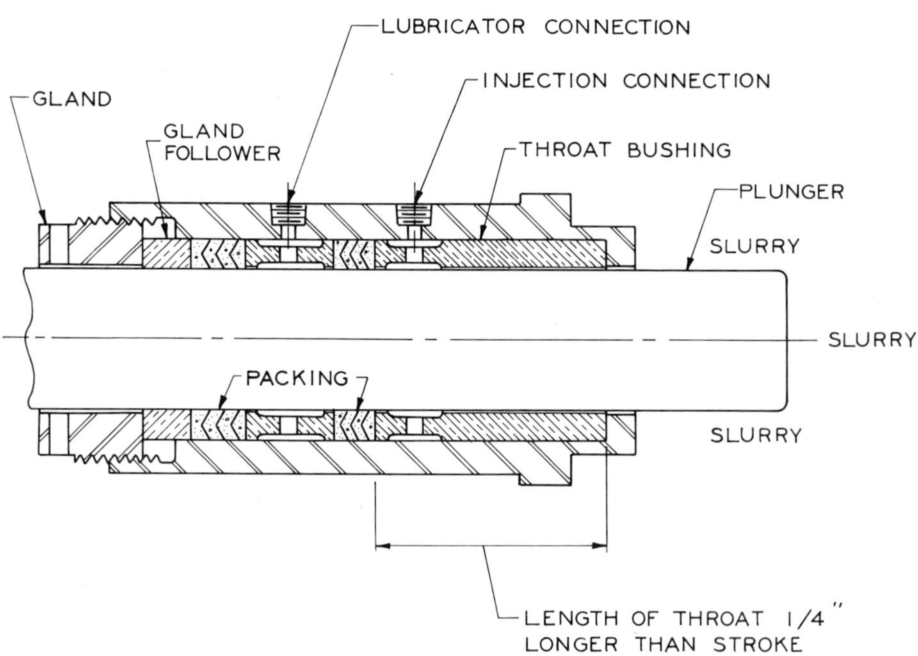

FIG. 4.30. A stuffing box with a long throat and injection connection for slurry service.

of the throat is also machined as a lantern ring, into which is metered a small quantity of clean liquid, the rate normally being less than $\frac{1}{2}\%$ of the capacity of the pump. Some installations utilize a small power pump to inject the clean liquid. Some installations drive the small injection pump from the shaft of the large pump, running both pumps at exactly the same speed. The injection pump is timed to inject the clean liquid as the large plunger is pulling back into the packing on its suction stroke. This requires the injection pump to have a separate discharge connection for each of its pumping chambers, and individual lines from each pumping chamber to each large stuffing box.

A simpler arrangement consists of a common injection header at a pressure above pump suction pressure. One (or two in series) check valve(s) is provided at the injection connection of each stuffing box. When the plunger starts on the discharge stroke, the check valve closes to prevent back flow from the pump into the injection system. As the plunger starts on the suction stroke, the check valve opens to allow clean liquid to flow into the throat bushing and along the plunger to wash particles from the plunger surface. Monitoring and controlling of the injection flow rate into each stuffing box is required to maintain the desired injection rate. An excessive rate can cause the check valve to slam each time the plunger starts on its discharge stroke.

If the packing is spring loaded, the spring cavity may be used as a point to inject the clean liquid.

4.18. SURGE-LEG CONSTRUCTION

To further reduce the possibility of solids migrating into the packing, the packing may be separated even further from the slurry by a liquid end design called a "surge leg," as shown in Fig. 4.31. The surge leg may be considered an elongated stuffing box or elongated pumping chamber. Figure 4.31 shows the fluid cylinder, containing the valves, located away from the stuffing box, and at a lower elevation. The cylinder is connected to the box with an "S"-shaped leg. (A separate leg is required for each plunger.) This particular construction is used for slurries which are denser than the injection liquid (the common case). The slurry remains in the bottom portion of the leg, the top of the leg being filled only with clean injection liquid. The injection liquid tends to keep the interface depressed in the leg, and provides a make-up for liquid lost as leakage through the packing. Any gases which migrate into the surge leg from the pumpage, or through the packing, must be vented from the top of the pumping chamber. Because the clearance volume of the liquid end is quite high, the compressibility of the pumpage and injection liquid is magnified signifi-

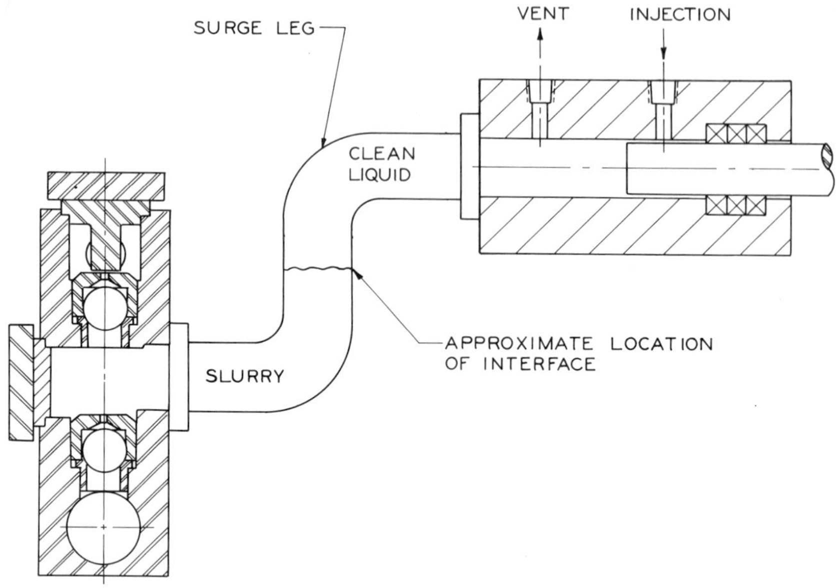

FIG. 4.31. A surge-leg liquid end for slurry service.

cantly, reducing volumetric efficiency and possibly contributing to pressure pulses in both suction and discharge piping. (See Chapters 6 and 9 for a discussion of volumetric efficiency.)

Also because of the increased inertia of the oscillating column of liquid (compared to a standard pump), the acceleration head of the column reduces the maximum allowable operating speed of the pump. (See Chapter 7 for a complete discussion of acceleration head.)

The surge leg configuration has been supplied for both power and direct-acting pumps, with both piston and plunger pumping elements.

4.19. LIFE EXPECTANCY OF PACKING

Of all expendable items on a reciprocating pump (such as bearings, valve assemblies, gaskets, seals, packing, and plungers), packing is the one requiring most maintenance. Packing life ranges from a few days to over two years (18,000 hours), with standard packing in a typical power pump installation normally lasting about 2500 hours (about 3 months of continuous operation).

4.20. CAUSES OF SHORT PACKING LIFE

In summary of the foregoing discussions in this chapter, short packing life can be caused by any one of, or a combination of, the following:

1. *Improper packing* for the application. The packing must be chemically compatible with the pumpage, mechanically compatible with the plunger (rod) surface, and suitable for the operating temperature of the stuffing box and the method of lubrication. A condensed packing selection guide is contained in Chapter 5.
2. *Insufficient lubrication*. If a lubricator is being used, it may be malfunctioning on one or more feeds, or may be set at too low a feed rate.
3. *Excessive lubrication*. If a mechanical lubricator is injecting oil into a lantern ring bracketed by packing, the injection rate may be set too high, causing the lantern ring and packing to be overpressured.
4. *Improper lubrication*. If relying on the pumpage for lubrication, it may be inadequate for the packing being used. If an external lubricant is being used, it may be the improper type, or may be contaminated. For many applications, steam cylinder oil has been found to outperform most others. It is compounded with tallow to give it a tenacity for metal surfaces, a characteristic most appropriate for lubricating a plunger surface sliding through a packing ring. If the operating or ambient temperature is too low to maintain sufficient fluidity (if the steam cylinder oil gets too viscous), the second choice is rock drill lubricant, similarly compounded to lubricate sliding surfaces.
5. *Misalignment* of the plunger (or rod) with stuffing box. If the plunger is pushed off-center in the box, pumpage will leak through the "loose" side, and excessive friction may occur on the "tight" side, possibly leading to premature failure of the packing. If the plunger rubs against the stuffing box bushings, the plunger and bushings may become scored. (See Chapter 12 for a more detailed discussion).
6. *Worn plunger* (or rod). If the plunger is worn, the packing has difficulty conforming to the irregular surface. The greater distance between the box bore and plunger surface makes it more difficult for the packing to expand radially enough to seal against both surfaces. The increased gaps between the plunger and box bushings allow the packing to more readily extrude.
7. *Rough plunger surface*. If the plunger has been recoated or otherwise refinished, the surface must be ground to a mirrorlike 8 microinch RMS finish. Circumferential finish marks are more damaging to packing than axial marks. Circumferential grooves on a reciprocating rod act like file teeth on the packing.

8. *Worn stuffing box bushings*. If the bores of the bushings increase, due to wear or corrosion, the packing will extrude more readily. This is more true for the follower bushing. A gap between the bushing O.D. and box bore will also allow packing to extrude, but not as readily as at the plunger surface.

9. *Worn stuffing box bore*. This is an area often overlooked. The wear rate is much slower than at the plunger surface, but because the packing flexes on each discharge stroke, there is a small amount of rubbing between the packing and stuffing box. This rubbing will wear the box bore. If the pumpage is corrosive to the stuffing box, the wear rate is higher. The lip-type packings, such as chevron and noncrushable, will wear a box faster because the loading is more concentrated.

10. *Improper packing installation or start-up*. Packing must be carefully installed, and in some cases "set" and broken-in. There are myriad ways of doing it wrong. Correct packing procedures are partially covered in Chapter 12.

11. *Too much packing in the box*. Unless profusely lubricated, a large number of rings in a stuffing box create excessive friction and quickly fail from high temperature. More than six rings can be excessive in a standard box, and one to three is often adequate.

12. *Packing gland too tight*. With gland-adjustable packing, an excessively-tight gland will cause excessive friction between the packing and plunger, resulting in high temperature, deterioration of the packing, and scoring of the plunger.

13. *Packing gland too loose*. If the gland is too loose, the packing may "chuck" back and forth in the box, wearing the packing on the O.D. and wearing the stuffing box bore.

14. *Excessive pressure*. Excess pressure in the discharge line, or in the pumping chamber caused by a faulty or partially plugged discharge valve, will shorten packing life.

15. *Solids*. Abrasive particles, such as rust or grit, which find their way to the interface between the plunger and packing will cut the packing and score the plunger. These particles may come from the pumpage, may be carried into the packing by the lubricant, may be borne by the air and deposited on the plunger outboard of the gland, or may simply be dropped onto the plunger from overhead work.

16. *Pumping chamber gas-bound*. For a stuffing box which is pumpage lubricated, if a pumping chamber is not primed during start-up, or if the chamber ingests gas and quits pumping, the packing will fail due to a lack of lubrication and cooling.

17. *Shock conditions*. Packing will fail prematurely if subjected to shock conditions. Shocks (pounding) can be caused by gas entrained in

the pumpage, gas that flashes from the pumpage (dissolved gas—or flashing of the pumpage itself), a broken or weak valve spring, or excessive acceleration head in the suction or discharge piping.

18. *Low temperature*. Low temperature of the stuffing box can result from either a low temperature of the pumpage, or flashing of a volatile pumpage as it leaks to the atmosphere. Low temperature may cause a packing lubricant to "curdle." If the box or plunger temperature is below 32 °F, ice crystals may form on the plunger surface or in the packing, resulting in cut packing and/or a scored plunger.

As can be seen from the above list, short packing life may result from improper stuffing box design, improper packing selection, improper system design, improper maintenance procedures, or improper operation. Long packing life indicates that the pump and its environment are designed, applied, maintained, and operated properly.

REFERENCE

1. Chesney, A. J., Jr., *Water Injection—Pump Development,* New York: ASME Paper 68-PET-11, 1968.

Chapter 5
PACKING AND PLUNGERS

5.1. INTRODUCTION

Packing is used to provide a seal around a rod, plunger, or piston. For a piston, the packing is contained in a groove in the piston, and moves with the piston. For a rod or plunger, the packing is stationary and is contained by a stuffing box.

5.2. PACKING FOR STUFFING BOXES

Packing for stuffing boxes can be divided into the following categories:

1. Square braided
2. "Plastic" (bulk)
3. Lip-type
4. Segmental ring
5. Special

5.2.1. Square Braided Packing

Braided packing falls into the general category of "compression" packing, a square ring which deforms when subjected to an axial load from the gland or pumpage. Three major types of braids are supplied: interlocking (Fig. 5.1), braid-over-braid (Fig. 5.2), and plaited (Fig. 5.3). Each of these types is made by a different type of braiding machine. The interlocking braid is braided in a square shape, whereas the braid-over-braid is braided round and calendered to the square shape.

Braided packing is supplied in a variety of materials including cotton, flax, asbestos, PTFE filament, carbon filament, and aramid filament. Lubricants such as paraffin, oil, graphite, and PTFE suspensoid are applied to the packing material before, during, and/or after the braiding process.

A popular packing for a variety of services is an interlocking braided asbestos packing impregnated with PTFE suspensoid, but due to the increasing concern for the potential health hazard of asbestos, use of this packing is diminishing. The PTFE provides some lubrication, these two

FIG. 5.1. Interlocking braided packing.

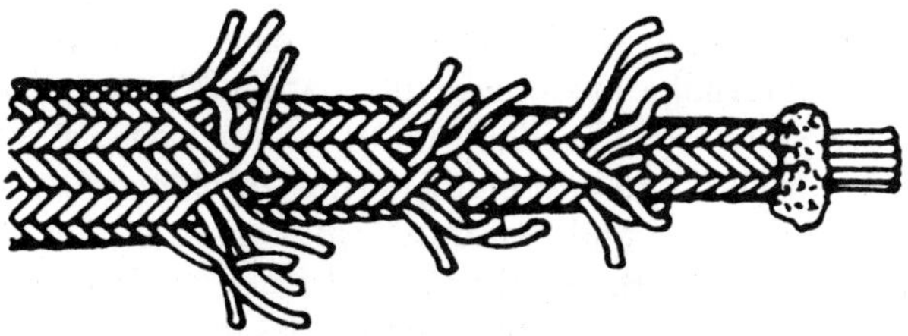

FIG. 5.2. Braid-over-braid packing.

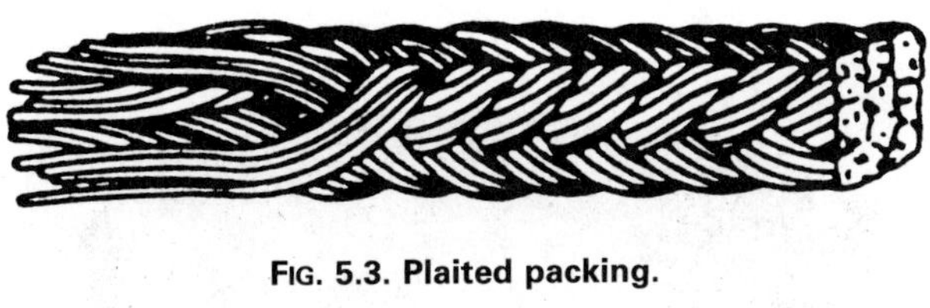

FIG. 5.3. Plaited packing.

materials are inert to most liquids, they are suitable for a wide range of temperature, and the low thermal coefficient of expansion provides for an easy break-in. Cost is low. This packing is often confused with the all-PTFE ring described below.

Although similar in construction and appearance, the all-PTFE ring is braided of PTFE filaments and impregnated with a PTFE suspensoid. This ring is tougher than the asbestos ring, and will last longer if carefully broken-in and adjusted. Having a higher thermal coefficient of expansion, though, it is more difficult to break-in and adjust. The cost is also considerably higher.

A ring combining fibers of PTFE and aramid has proven itself in such tough services as amine and hot water—even without cooling or external lubrication. The PTFE fibers provide the low friction and primary sealing, while the aramid fibers, located at the corners of the ring, provide an anti-extrusion barrier to retain the PTFE which alone has a proclivity to "cold flow" into the bushing clearances.

5.2.2. Combination Packing Sets

Some sets of square packing consist of two or more types of packing rings. For example, a mixture of hard and soft rings may be used, the soft rings providing more internal lubricant and performing most of the sealing, with the hard rings acting as support for the soft rings. One example is the combination set shown in Fig. 5.4. This particular packing is used for cool water at pressures to 6000 psig without external lubrication, and to 10,000 psig with external atmospheric lubrication. The set consists of one or two (two are shown in the illustration) center rings of soft lead foil, twisted and die-formed, two intermediate rings of plaited flax impregnated with paraffin (die-formed) and two outer rings of hard phenolic plastic.

The heart of this set is the lead ring in the center. It readily deforms, and seals tightly against the plunger and box. Its low coefficient of friction and high conductivity contribute to a low operating temperature. Because of its high degree of compliancy, it must be retained by a zero-clearance ring. That is the function of the flax ring, which also sacrifices some of its

FIG. 5.4. A combination set of packing consisting of two center rings of lead foil, two intermediate rings of flax, and end rings of hard phenolic. (*Courtesy Utex Industries*)

wax as a lubricant. The end rings provide no sealing whatever, their sole function being to retain the flax rings. End ring clearance over the plunger is about 0.008 inch diametrally.

Other combination sets contain washers of virgin PTFE between rings of braided asbestos, soft rings of braided PTFE interspersed among hard rings of braided PTFE, and "plastic" (bulk) rings bracketed by braided asbestos rings.

5.2.3. Cutting of Packing Rings

Since braided packing is made in a long "rope," it must be cut into lengths for installation in a stuffing box. The rope is wrapped around a mandrel the same diameter as the plunger, then cut so that no gap exists between the ends of the ring when placed in the stuffing box. Some packing suppliers and users prefer an angle or "skive" cut, which is shown in Fig. 5.5. Others prefer a square or "butt" cut, illustrated in Fig. 5.6. In both cases, there is no gap when installed, so there is no leakage path. The author prefers the butt cut because it is easier to install and the joint seems to be stronger than the skive cut.

5.2.4. Die-Forming of Packing Rings

For a better fit in the box and on the plunger, the ring should be formed in a die of the same dimensions as the plunger and box. Die-forming reduces installation time, and minimizes the time required to set and break-in a set

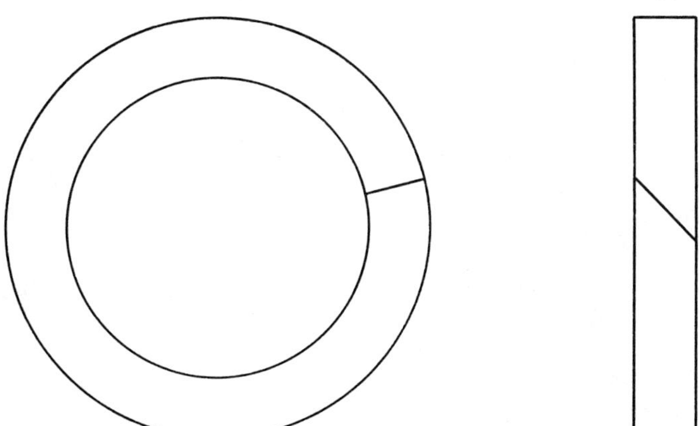

FIG. 5.5. A ring of packing cut at an angle ("skive" cut).

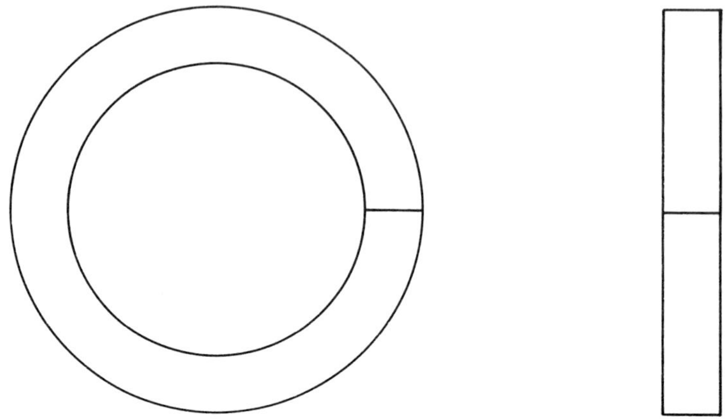

FIG. 5.6. A ring of packing cut square ("butt" cut).

of new packing. Some users state that for optimum performance, the pressure used to die-form each ring should at least equal the discharge pressure of the pump. For example, with a 2 inch diameter plunger in a $2\frac{7}{8}$ inch diameter stuffing box operating at 4000 psig discharge pressure, the minimum force used to form the ring would be calculated as follows:

$$F = \frac{\pi}{4}(2.875^2 - 2^2)\ (4000) = \underline{13{,}400}\ \text{lbs}$$

A force of about 7 tons would, therefore, be used to form each ring in the die.

5.2.5. 'Plastic' Packing

The packing industry has unfortunately adopted the term "plastic" to designate soft bulk packing which can be readily molded, regardless of the material from which it is made. Even though various types of packing are made of plastic material, in the packing industry vernacular, the term is used as an adjective to designate the soft, bulk, moldable packings.

Plastic packing is made of chips or particles of various materials including asbestos, wood, metal (such as babbit or lead), and PTFE, which are usually lubricated with graphite, mica, oil, or grease. This product is supplied in various forms, such as bulk for hand insertion and tamping into the stuffing box, sticks for injection into the box through a special fitting with a special injection gun, enclosed in a thin open jacket (to facilitate handling only), in a tightly-woven jacket to help retain the packing in the stuffing box, and in die-formed rings for easier installation and setting.

5.2.6. Lip-Type Packing

A lip-type packing ring is one which includes a lip which faces toward the pressure and seals against the plunger or rod. The ring usually also includes a lip at the O.D., which also faces toward the pressure and seals against the bore of the stuffing box.

V-Ring Packing. The most common lip-type packing is the V-ring, often called a "chevron" ring. This packing obtains its name from the shape of its cross-section. Figure 5.7 shows a typical V-ring with its male and female adapter. Figure 5.8 is a photograph of a V-ring set. These illustrations show V-rings and adapters made of woven cloth, impregnated with a material such as rubber, or PTFE, and laid in layers prior to forming in a hot die. V-rings are also made of materials such as leather, homogeneous rubber, virgin PTFE, filled PTFE (sometimes molded, sometimes machined), polyphenylene sulfide resin, and various combinations of these and other materials. The throat and gland-follower bushings in a stuffing box are normally flat-faced adjacent to the packing. A V-ring will not

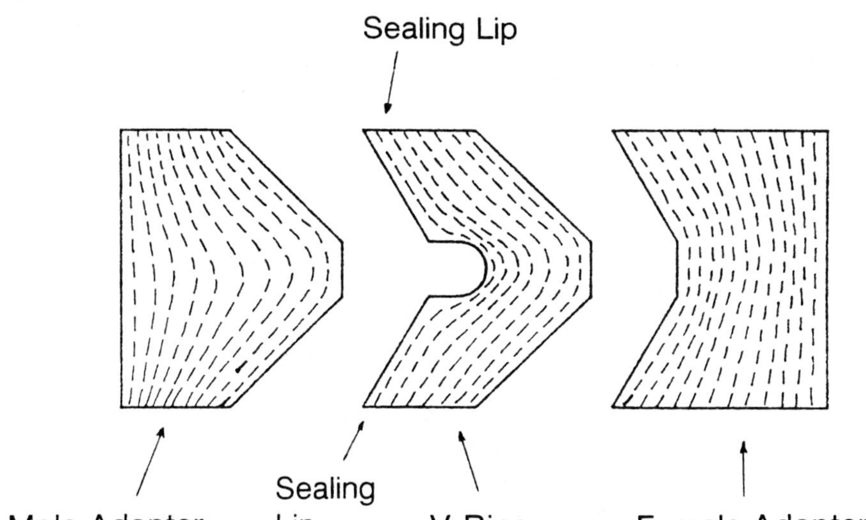

FIG. 5.7. The cross-section of a "V" (chevron) packing ring with its male and female adapters. (*Courtesy Utex Industries*)

FIG. 5.8. A cutaway set of three V-rings with a male and female adapter. (*Courtesy Utex Industries*)

perform satisfactorily if mated against a flat surface. It is therefore necessary to provide a "female" adapter adjacent to the follower bushing which provides support for the outer packing ring. It is also necessary to provide a "male" adapter adjacent to the throat. If the box includes a flat-sided lantern ring located between packing rings, an additional male and female adapter must be provided on each side of the lantern ring. For lower pressures, these adapters are often made of the same material as the V-rings. For higher pressures, the adapters are made of harder, stronger materials, such as metal or plastic. Because it must bridge the gaps between the plunger and stuffing box bore to prevent extrusion of the outer ring, the female adapter must have minimal clearances with both the plunger and stuffing box. For pressures in the range of 5000 psig, it is not unusual to find diametral clearances at the inside diameter (I.D.) of 0.004—0.006 inch, and at the O.D. of 0.002—0.004 inch. (The O.D. clearance is usually smaller so as to center the ring in the box.) Because of the close running clearance with the plunger, metal adapters are usually made of a nongalling material such as bronze.

Split Rings or Endless Rings? V-rings are available either split (cut) or endless (not cut). Cut rings have the advantage of ease of installation, especially around a piston rod, and of being able to cut off material from a ring made for a larger diameter plunger for use on a smaller plunger which utilizes the same packing cross-sectional size. The disadvantages of the split (cut) ring are that the cut provides a leakage path and weakens the ring. Endless rings outperform split rings, and are to be preferred.

What Mating Angles? There are two distinctly different approaches taken by packing manufacturers in designing the mating angle between adjacent V-rings. Figure 5.9 shows two V-rings that contact each other at the I.D. and O.D. of the rings, i.e., at the tips of the lips. These rings are molded or machined with parallel I.D.'s and O.D.'s. The O.D. is slightly smaller than the box bore and the I.D. is slightly larger than the plunger diameter. To achieve a seal, it is necessary to compress the rings so that

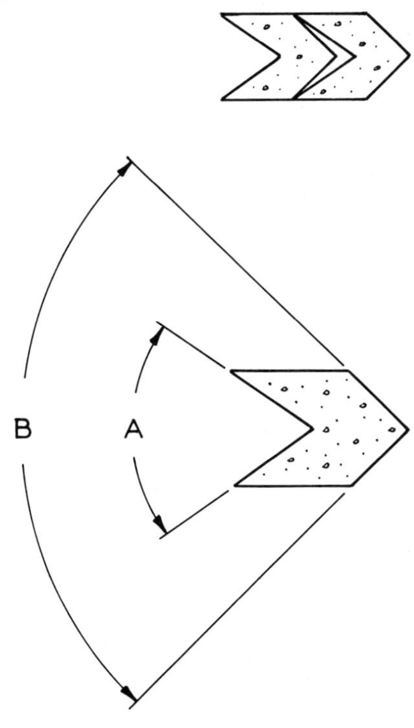

FIG. 5.9. V-rings that mate at the tips of the sealing lips (B > A).

the lips are forced against the sealing surfaces. Judicious gland adjustment is required to achieve the proper lip deformation and sealing force: too loose and the packing leaks, too tight and it overheats.

The second approach to V-ring mating angle is shown in Fig. 5.10. The female angle, A, is made larger than the male angle, B. Therefore, two adjacent rings contact each other at, or near, the apex of the V's. The I.D. and O.D. surfaces also are not parallel to each other. The lips are made to have a slight interference with both the box bore and plunger, although the heels of the ring each have a small clearance. This type of ring seals on both the I.D. and O.D. without any load from the gland or fluid. Also, gland and/or hydraulic loading of the ring set does not increase the lip sealing force as much with this type ring as with the other type, resulting in less friction and a lower operating temperature. The latter ring also tends to be better suited for spring-loading, since it does not rely on the spring force to deflect the lips to achieve a seal.

For purposes of communication, the author proposes defining the mating angle as the difference between the female angle and the male angle, or A-B. The packing in Fig. 5.9, therefore, has a negative mating angle, and the packing in Fig. 5.10 has a positive mating angle.

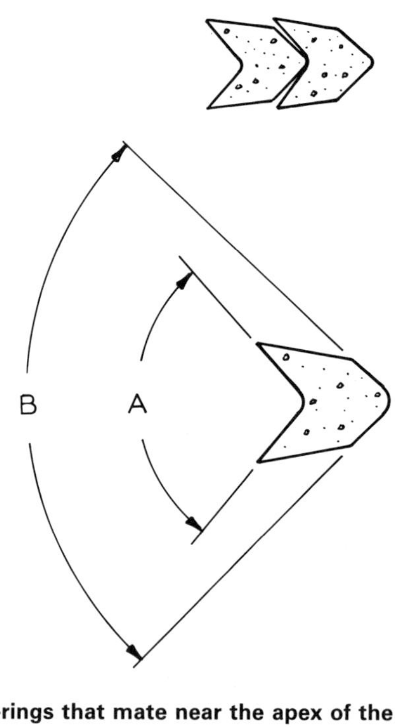

FIG. 5.10. V-rings that mate near the apex of the V's (A > B).

Noncrushable (Nonadjustable) Packing. Another popular type of lip-type packing, referred to as "noncrushable" or "nonadjustable," is illustrated in Fig. 5.11. This design further extends the characteristic of the positive-mating-angle V-ring. An axial force imposed on the sealing ring has minimal effect on the two sealing lips. The sealing ring can be clamped tightly between the male and female adapters without effecting the sealing force between the inner lip and the plunger. With this type of packing, the stuffing box gland is kept tight at all times. No adjustment is required (or possible). If the gland is not kept tight, the rings may "chuck" back-and-forth in the box, generating friction at the box bore, and wearing both the outer sealing lips and box bore.

Figure 5.12 shows a cut-away of a set of three noncrushable sealing rings. Note that each sealing ring is bracketed by its own male and female adapter.

Another type of nonadjustable packing, as seen in Fig. 5.13, is a rubber lip seal as used in drilling mud pumps.

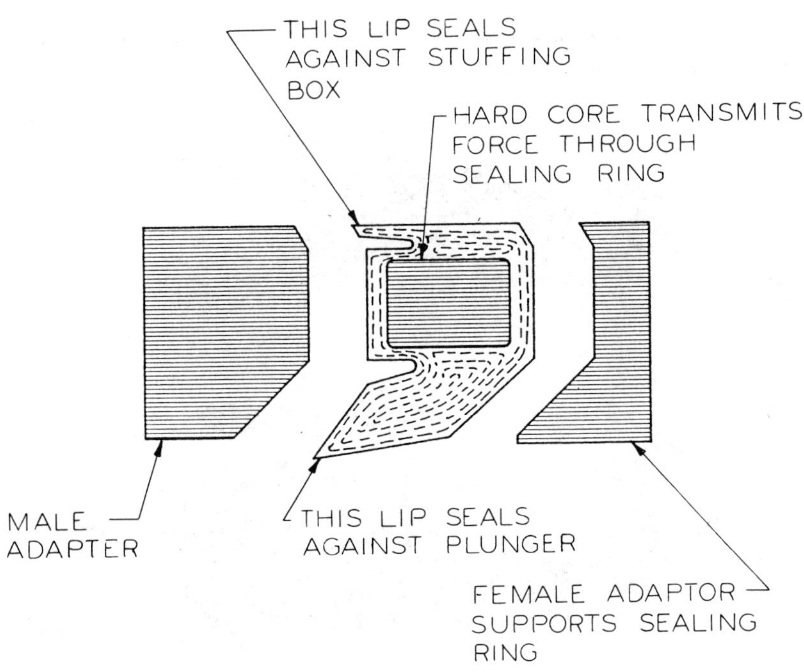

FIG. 5.11. A noncrushable (nonadjustable) packing ring with male and female adapters. (*Courtesy Utex Industries*)

FIG. 5.12. A set of 3 noncrushable sealing rings, each with its own male and female adapter. (*Courtesy Utex Industries*)

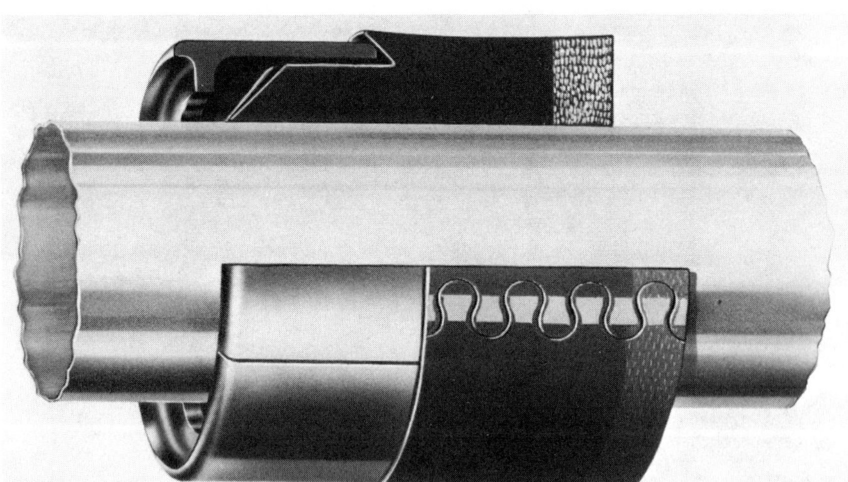

FIG. 5.13. Rod packing for a drilling mud piston pump. (*Courtesy TRW/Mission Drilling Products Div.*)

5.2.7. Segmental Ring Packing

Because of the special type of stuffing box required for this type of ring, both the packing and box are described in Chapter 4.

5.2.8. Special Types of Packing

This category covers reciprocating pump liquid end sealing devices which do not readily fall into the preceding four categories.

Solid Throttle Bushing. This sealing device is discussed in Chapter 4.

Trapezoidal Packing Rings. The unique trapezoidal cross-section packing ring, as shown in Fig. 5.14, has been used in some reciprocating pump applications.

5.2.9. Selection Guide

Table 5.1 is a condensed guide to packing selection which will cover many reciprocating pump services, particularly in petroleum-processing and related plants. It is only a general guide, and is not intended to cover all services, or any special conditions of any one service.

5.3. SEALING DEVICES FOR PISTONS

Sealing devices for pistons include square and lip-type packing, as used on plungers; special combination sets of hard rings; individual rings of

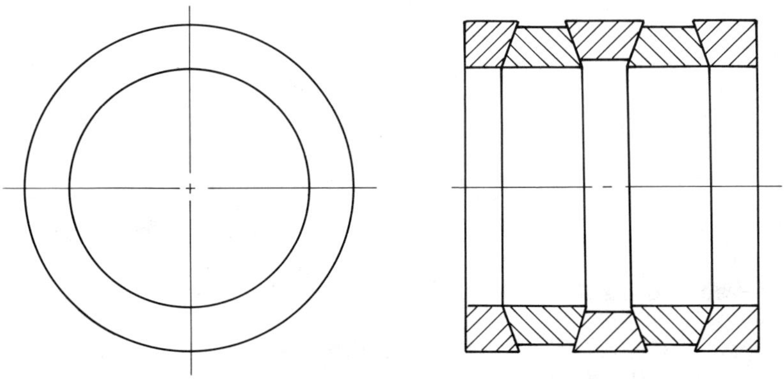

FIG. 5.14. Trapezoidal cross-section packing rings.

Table 5.1. Condensed Packing Selection Guide.

Fluid	Maximum Pressure (psig)	Maximum Temperature (°F)	Type Packing	Lubrication Required?
Water	3000	200	H	yes*
	6000	180	C	no
	6000	350	F	yes*
	14000	500	E	no
Amine or Glycol	2500	360	A	no
	3000	380	D	no
	14000	500	E	no
Gasoline, Kerosene, or Lean Oil	2500	180	H	yes*
	2500	200	A + B	no
Light Hydrocarbons	3000	150	H	yes*
	5000	200	G**	yes*

* Lubricant: Steam cylinder oil. If ambient or pumping temperature is too low for steam cylinder oil, rock drill oil may be substituted. Inject with a mechanical lubricator or drip onto plunger on atmospheric side of packing.
** Spring-loaded.
Square Packing:
A—Braided asbestos, impregnated with PTFE suspensoid.
B—"Plastic"—mesh jacket over asbestos fibers, soft metal particles, and graphite.
C—Three different rings—crimped lead foil, braided flax, and hard phenolic.
D—Braided PTFE—alternate hard and soft rings.
E—Braided fibers of PTFE and aramid.
V-Ring Packing:
F—Cotton duck and nylon impregnated with EPDM.
G—Cotton duck and nylon impregnated with nitrile and PTFE.
Noncrushable Packing:
H—Nylon impregnated with nitrile (with hard phenolic core).
J—Nylon impregnated with nitrile and PTFE (with hard phenolic core).
Note: This chart is intended only as a *general guide* to packing selection. Maximum pressures and temperatures are approximate. Factors beyond the scope of this guide may necessitate a different style of packing.

metal, hard plastic, or rubber; and special types of cups and discs which require special pistons.

For very viscous liquids, pistons are sometimes left smooth, or just grooved, without any rings. The close clearance between the piston and cylinder is sufficient to satisfactorily limit leakage with such liquids.

5.3.1. Packing for Pistons

Since pistons are usually limited to differential pressures (across the piston) of about 1500 psi, these devices are not subjected to the differential pressures imposed on many plunger seals. Some of the packing rings used on plungers are also used on pistons, and many of the comments relative to their use on plungers also applies to their use on pistons. A major difference in their application is that of adjustability. Piston packing is confined in a fixed-length groove or cavity, and relies solely on fluid pressure (or any built-in interference) to maintain contact with the cylinder bore (or liner bore). Figure 5.15 shows conventional square packing installed on a body-and-follower type of piston. The seal between the packing and cylinder is obtained by the deformation of the ring created by the liquid pressure differential across the packing set. Figure 5.16 shows conventional V-ring packing installed on the same type piston. This packing is molded with the lip O.D. larger than the liner bore, thereby achieving a seal without axial loading.

The piston is made with the follower as a separate piece so that one-piece rings (such as the V-rings) can be installed, and so that packing can be replaced without removing the piston from the rod or cylinder.

It should be noted that the dynamic seal occurs at the O.D. of packing on pistons, as opposed to the I.D. with plungers. Most of the wear and extrusion therefore occurs at the O.D., and the critical clearances to prevent packing extrusion are between the piston and liner and between the follower and liner.

Hard plastic or rubber composition rings are sometimes placed side-by-side in the larger open groove.

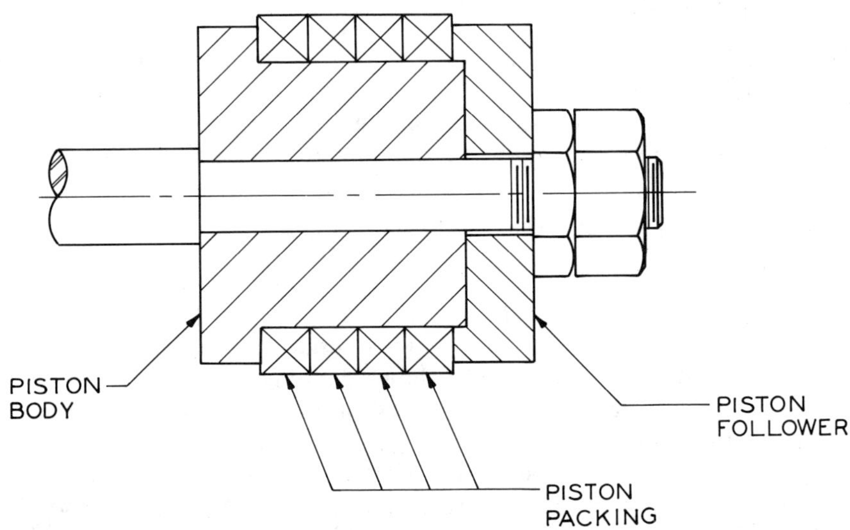

FIG. 5.15. A body-and-follower type piston fitted with square rings of packing.

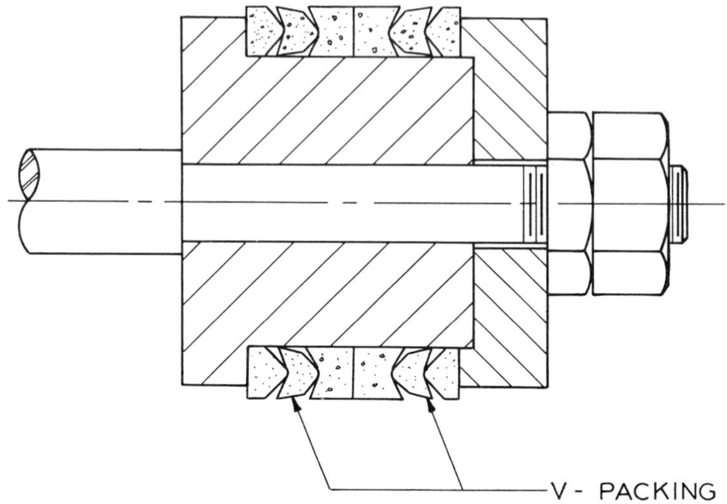

FIG. 5.16. A body-and-follower type piston fitted with V-packing and adapters for double-acting service.

The above sealing devices are normally limited to differential pressures of about 1000 psi.

5.3.2. Bull-Ring Packing

To obtain longer packing life, the so-called "bull-ring" construction is used with hard plastic or metal sealing rings (if the liner material can tolerate the harder rings). As shown in Fig. 5.17, the bull-ring is made to fit the width of the packing groove and support the sealing rings.

5.3.3. Individual-Groove Piston Rings

Figure 5.18 shows two different forms of individual-groove pistons. Figure 5.18(a) shows a piston of the type furnished in both the drive ends and liquid ends of direct-acting pumps which operate at differential pressures up to about 350 psi. Figure 5.18(b) shows the long piston used to reduce the clearance volume on the liquid end of a minimum-clearance pump.

Hard piston rings are made in various forms. Figure 5.19 shows three different type of cuts, or openings, supplied in one-piece snap rings. For installation into an individual groove, each ring must be expanded so that the I.D. slides over the piston O.D. The ring "snaps" back to its original size in the groove, hence the expression "snap ring." The ring O.D. is

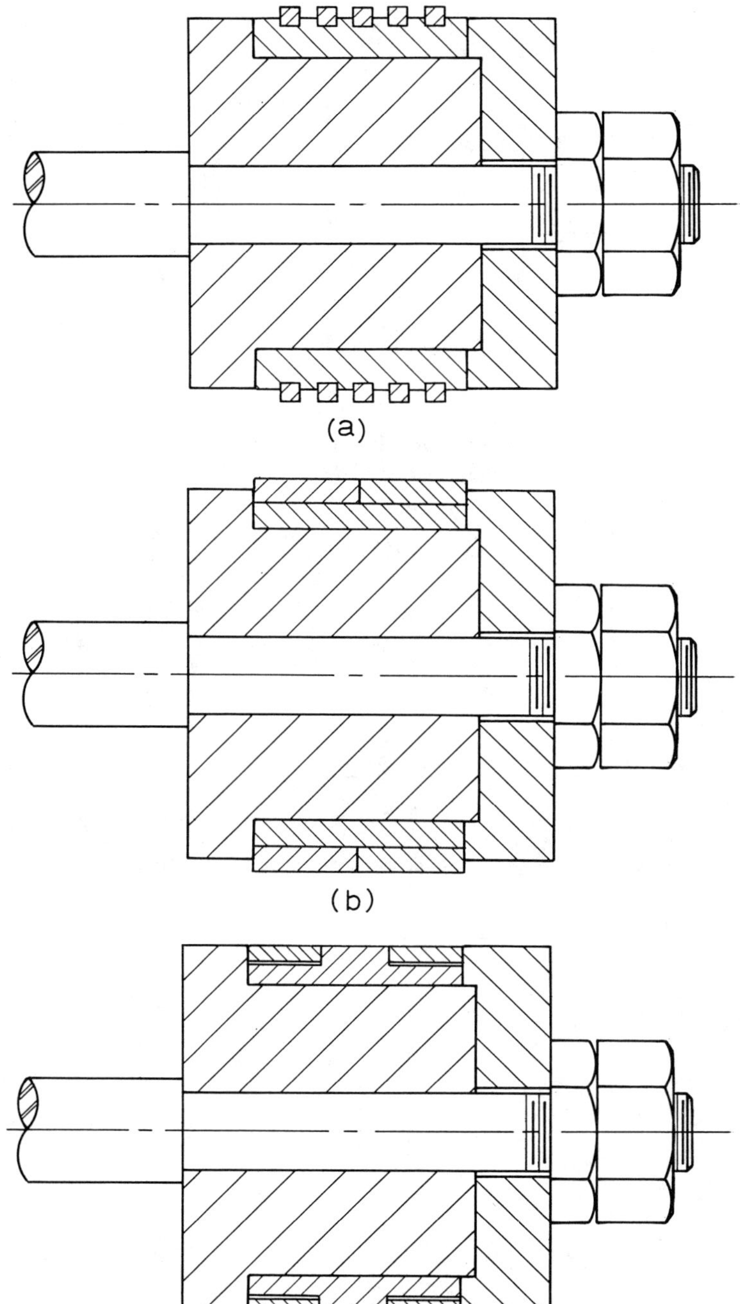

(a)

(b)

(c)

FIG. 5.17. Various arrangements of bull-ring packing.

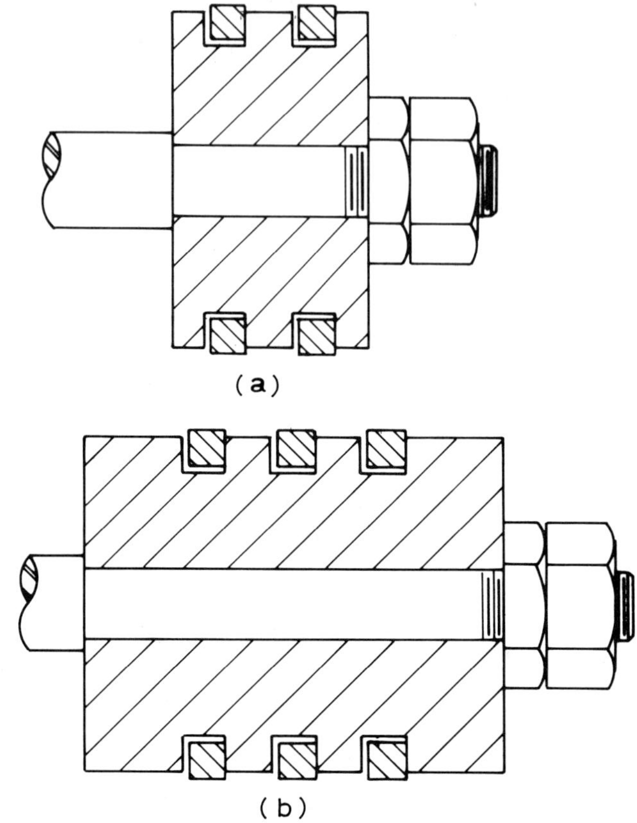

Fig. 5.18. Two forms of individual-groove pistons.

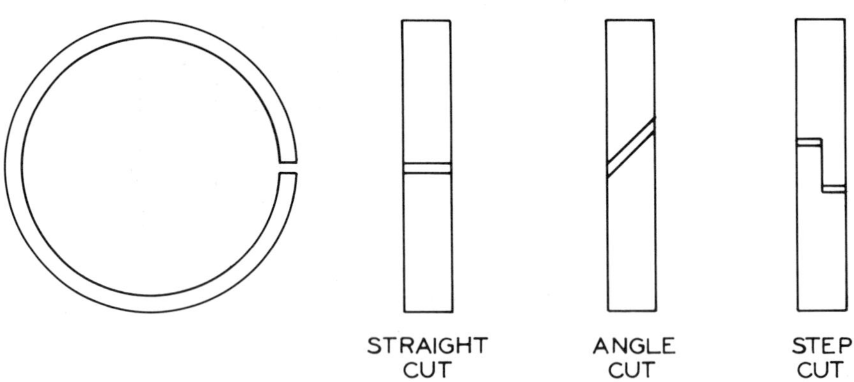

Fig. 5.19. Three forms of one-piece snap rings.

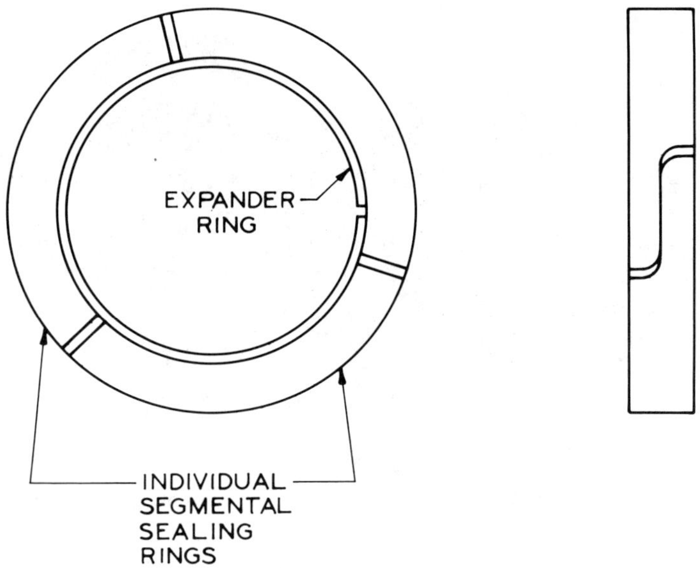

FIG. 5.20. Segmental ring packing with expander.

larger than the cylinder (or liner) bore so that it must be squeezed smaller as the piston is installed in the cylinder. This built-in compression provides a tight seal with the cylinder (or liner).

If ring configuration or material prevents its expansion for installation without plastic deformation, it can be supplied in segments as illustrated in Fig. 5.20. The expander ring, which provides the force to hold the ring against the cylinder wall, can be supplied in a material such as stainless steel, and is thin enough so that it can readily be expanded for installation without plastic deformation. The segmental ring can be supplied of phenolic or other type plastics, or of metal such as bronze or nickel-cast iron. The nickel-iron segmental ring is used for services in the 600–700°F temperature range.

5.3.4. Piston Cups

Although not limited to this service, cups and special thick discs are used extensively for slurry applications. This type of sealing service requires a special piston. The more conventional cups shown in Fig. 5.21 are used in piston assemblies as shown in Fig. 5.22. These units are used for utility services, for both clean and dirty liquids, and at pressures up to about 500

FIG. 5.21. Piston cups made of fabric impregnated with an elastomer. (*Courtesy Utex Industries*)

psig. A common material combination is synthetic rubber and cotton duck.

Figure 5.23 illustrates a piston used to pump slurries such as drilling mud and powdered coal. Figure 5.24 is a piston "rubber" used for such service. Note the layers of fabric located at the heel of the disc to limit extrusion at this critical sealing juncture. A rubber with a reverse sealing angle is shown in Fig. 5.25. These units are used at pressures up to about 5000 psig, but at these pressures, in slurry service, they experience a life of only 50 to 500 hours. The manufacturer reports that the rubber in Fig. 5.24, with the conventional "plowshare" sealing lip, lasted only about 55 hours, whereas the rubber in Fig. 5.25, with the reverse-angle sealing lip, lasting almost 500 hours and actually provided better sealing.

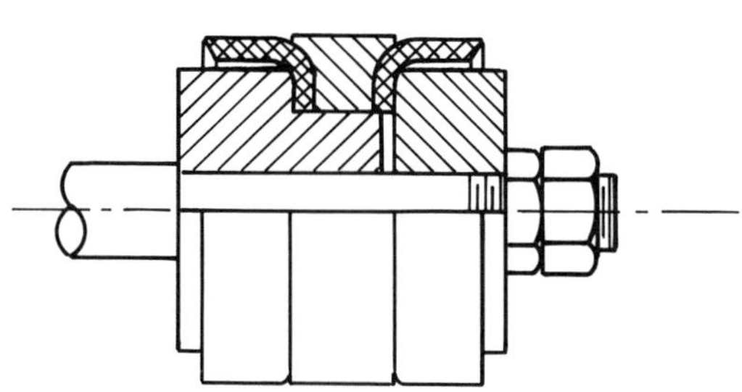

FIG. 5.22. A cup-type piston with cups installed.

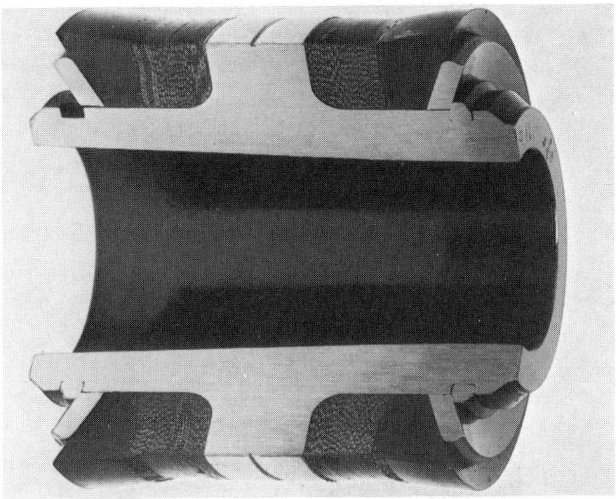

FIG. 5.23. A heavy-duty slurry piston with rubbers installed. (*Courtesy TRW/Mission Drilling Products Div.*)

FIG. 5.24. A piston rubber reinforced with woven cloth at the heel—conventional lip angle. (*Courtesy Utex Industries*)

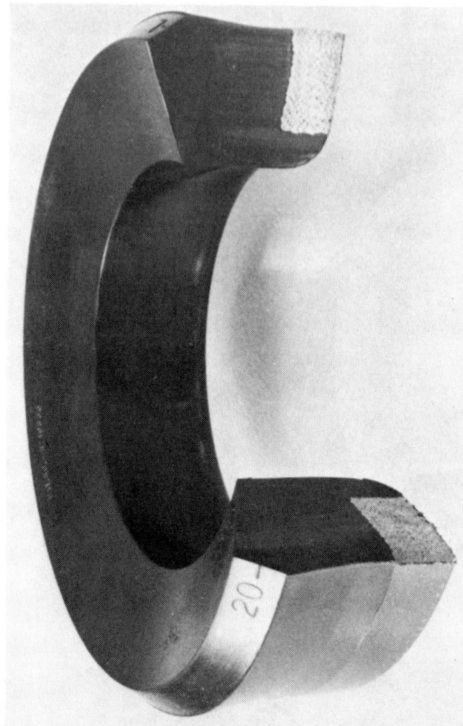

FIG. 5.25. A piston rubber with reverse lip angle. (*Courtesy Utex Industries*)

5.4. PLUNGERS

Next to packing, the plunger is the component of a power pump which normally requires most frequent replacement. The high speed and frictional load of the packing wear the plunger surface.

5.4.1. Coatings

For longer life, plungers are sometimes hardened, but more often are coated with a wear-resistant material. Some of the coatings used are chrome, various types of ceramic, nickel-based alloys, and cobalt-based alloys. Desired features of such coatings include hardness, smoothness, high bond strength, resistance to corrosion, and low cost. No one coating optimizes all these features. The ceramic coatings are harder than the metals, but are sometimes lower in bond strength, brittle, and porous. The porosity contributes to shorter packing life. Mixing of hard particles, such as tungsten carbide, into the less-hard nickel or cobolt alloys, has resulted in longer plunger life, but at the expense of shorter packing life (due to the abrading effect of the hard particles when the softer matrix wears down). The benefit is questionable.

Flame-spray coating should be applied with a radial thickness of about 0.020 inch. The application of such a coating actually requires the base metal and the coating to be fused together. This fusion process causes the outer portion of the base material and the inner portion of the coating to mix. Therefore, the inner approximately 0.010 inch of coating is diluted with base metal, and therefore softer. The outer 0.010 inch will remain undiluted and hard. If only 0.010 inch of coating thickness (radially) is applied, the outer coating will be an alloy containing some of the base metal, and the wear surface will be softer and have a shorter life.

These comments do not apply to coatings such as chrome and coatings applied by the detonation process, since they do not fuse with the base metal.

5.4.2. Base Metal

The base metal of plungers is usually carbon steel or stainless steel. In water and other mildly corrosive applications, the carbon steel will rust, but normally not an unacceptable amount during the normal plunger life (about 6 months in continuous duty). For more corrosive services, a stainless steel base material is used.

An 11–13% chrome stainless (such as 410 and 416) is sufficiently corrosion-resistant for many applications, but is often unsatisfactory for flame-sprayed coatings. Such coatings require the base material to be heated to about 1900°F during application. At 1900°F, 11–13% Cr-steel becomes austenite. At room temperature, it is martensite. Therefore, upon cooling to room temperature, the steel transforms from austenite to martensite. Such a transformation is accompanied by a decrease in density, a swelling, which occurs at about 700°F. The coating has already solidified at this temperature, and this expansion of the base material usually causes a multitude of small hair-line cracks in the coating. These martensitic stainless steels are therefore seldom used as base materials for plungers coated with a flame-sprayed material.

Instead, when a stainless steel base material is required for plungers, the austenitic grades, such as 304 and 316, are used. They are austenite at both 1900°F and room temperature. Because this welding process takes these materials through the sensitizing range of about 800°F, the low-carbon grades (such as 304L and 316L) are often used to prevent sensitization and preserve the material's resistance to corrosion.

5.4.3. Solid Ceramic Plungers

A popular plunger for salt water services is solid ceramic (usually alumina). The ceramic must be fitted with a metal adapter for connecting to the crosshead extension. The ceramic and adapter are joined with an epoxy-type cement. The plunger surface has a relatively long life and is corrosion-resistant. Problems include failure of the cement joint and breakage of the ceramic during shipment and handling.

5.4.4. Surface Finish and Alignment

Both proper finish on the plunger surface and proper alignment with the stuffing box are essential for extended packing life. A mirrorlike 8 micro-inch rms finish is recommended for the plunger surface. Close-tolerance machining and careful maintenance are required to obtain and maintain proper alignment between the plunger and stuffing box bore. Proper positioning of the *plunger* relies on proper machining and assembling of the plunger, crosshead extension, crosshead, and power frame. Proper positioning of the *stuffing box bore* relies on proper machining and assembling of the stuffing box, liquid cylinder, and support members for the cylinder.

Chapter 6
OPERATING CHARACTERISTICS AND RATINGS

6.1. LIQUID END

6.1.1. The Pumping Cycle

The pumping cycle of the reciprocating pump is as follows:

As the pumping element (plunger, piston, or diaphram) withdraws from the pumping chamber, the liquid within the chamber expands, and the pressure falls. Since most liquids are relatively incompressible, very little movement of the element is required to achieve a pressure drop. When the chamber pressure drops sufficiently below suction pressure, this differential pressure (suction pressure minus chamber pressure) begins to push the suction valve open. This occurs when the element is moving slowly so that the valve opens gradually and smoothly as the velocity of the element increases. Liquid flows through the valve assembly and follows the element on its suction stroke. As the element decelerates near the end of its suction stroke, the suction valve gradually returns to its seat. As the element comes to a stop, the suction valve closes.

The pumping element then reverses and starts on its discharge stroke. The liquid trapped in the pumping chamber is compressed until chamber pressure exceeds discharge pressure by an amount sufficient to begin to push the discharge valve away from its seat. As with the suction valve, the discharge valve is pushed higher as the pumping element accelerates. The valve is farthest from its seat when the pumping element is at its highest velocity. As the element decelerates, the valve moves back toward its seat. As the element stops, the valve closes.

6.1.2. Performance Curves

Those familiar with pumps know that each centrifugal pump model has a unique performance characteristic curve, as can be seen in Fig. 6.1. For this typical pump, head is a maximum at zero capacity, dropping as capacity is increased. The power requirement and efficiency both rise with increasing capacity, until peaks are reached, then start to drop.

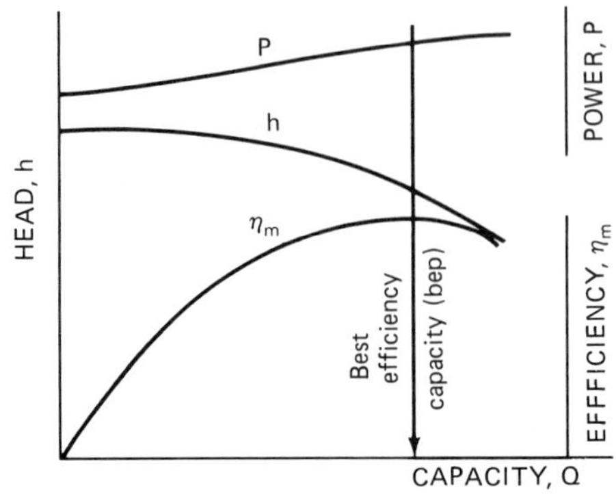

FIG. 6.1. A typical performance curve for a centrifugal pump.

For a constant-speed power pump, an attempt to use the same axes, as shown in Fig. 6.2, is unsatisfactory because capacity varies so little. Plotting characteristics against pressure, illustrated in Fig. 6.3, is more informative, but because power pumps are almost always purchased to operate at a fixed discharge pressure, performance characteristics are usually not shown as they are in Fig. 6.3, but as in Chapter 9 under "Selection Graph." As a matter of fact, power pumps are not normally

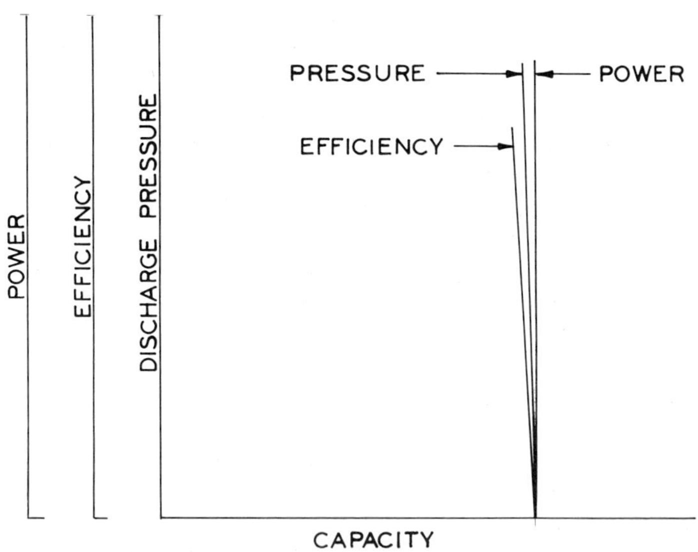

FIG. 6.2. Performance of a constant-speed power pump plotted on the same axes as a centrifugal pump.

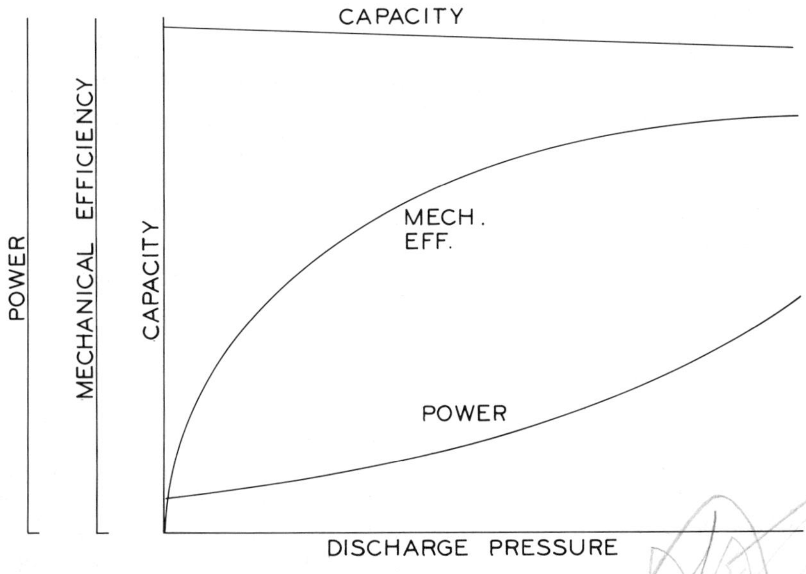

FIG. 6.3. Performance of a constant-speed power pump plotted against pressure.

tested in such a way as to provide the information to construct curves such as those in Fig. 6.3.

Important conclusions can be drawn from a comparison of the performance characteristics of the centrifugal and reciprocating pumps, see Figs. 6.1 and 6.3. The characteristics of a constant-speed centrifugal pump are a function of capacity. The characteristics of a constant-speed power pump are a function of the discharge pressure. For the centrifugal pump, the *system* establishes the capacity at which the pump operates, and the pump establishes the differential pressure. For the power pump—just the opposite. The *pump* establishes the capacity, and the system establishes the differential pressure.

With a centrifugal pump, if the differential pressure is less than that for which the pump was selected, the capacity will increase, and power will usually increase. With a power pump, a lower discharge pressure results in *reduced* power consumption, but negligible change in capacity.

6.2. DRIVE END

6.2.1. Power Pumps

The main driving element of the conventional industrial power pump is [illegible] crankshaft (or an eccentric). The crank drives one or more conne[illegible] rods which, in turn, drives one or more crossheads. In a horizonta[illegible] the direction of rotation of the crank is normally such that the force on the crosshead is down when the crosshead is being pushed toward the liquid end. When the connecting rod is pulling the crosshead away from the liquid end, the force on the crosshead is also down. The crosshead therefore rides on a film of oil in the bottom of its guide at all times.

The motion of the crosshead is purely reciprocal, the crankshaft is purely rotary, and the connecting rod is a mixture, transforming the pure rotary motion of the crank to the pure reciprocating motion of the crosshead.

6.2.2. Direct-Acting Pumps

In a direct-acting pump, the motive fluid (gas or liquid) acts directly upon the drive piston(s), pushing it first one way, then the other, as the main valve(s) shifts. The drive piston imparts this motion to the piston rod, first pushing it toward the liquid end, then pulling it back toward the drive end, providing the reciprocating motion to the pumping element in the liquid end.

Because the direct-acting pump contains no energy storage device, such as a flywheel, the pressure of the motive fluid must be maintained on the drive piston throughout the entire stroke. The main valve does not cut off the flow of fluid part way through the stroke, such as is done in a steam engine. Therefore, there is no expansion of the motive fluid during the power stroke, a characteristic which would be desirable for gas-driven units in order to achieve higher thermal efficiency. The motive fluid does not expand to a lower pressure until the end of the stroke, when the valve shifts, and the fluid expands rapidly to exhaust pressure. Although this characteristic results in a low thermal efficiency, it does keep the temperature of the exhaust gas higher, a desirable characteristic when exhaust steam is being used in a deaerator or other heating system.

The direct-acting pump is constructed primarily in two basic types—simplex and duplex. The manner of operation of the two different drive ends is quite different. A duplex pump contains two drive pistons and two drive (steam) valves, and as illustrated schematically in Fig. 6.4, each side
controls th ... ve on the opposite side. As side "A" nears the end
... system of levers it shifts the valve on side "B." As
... d of its stroke, it shifts the valve on side "A." This
... ends of each stroke, with one side accelerating as
... es, resulting in a smooth overlapping of strokes (if
... adjusted). After side "A" stops, it must wait for
... " valve before "A" runs again. Side "B" moves

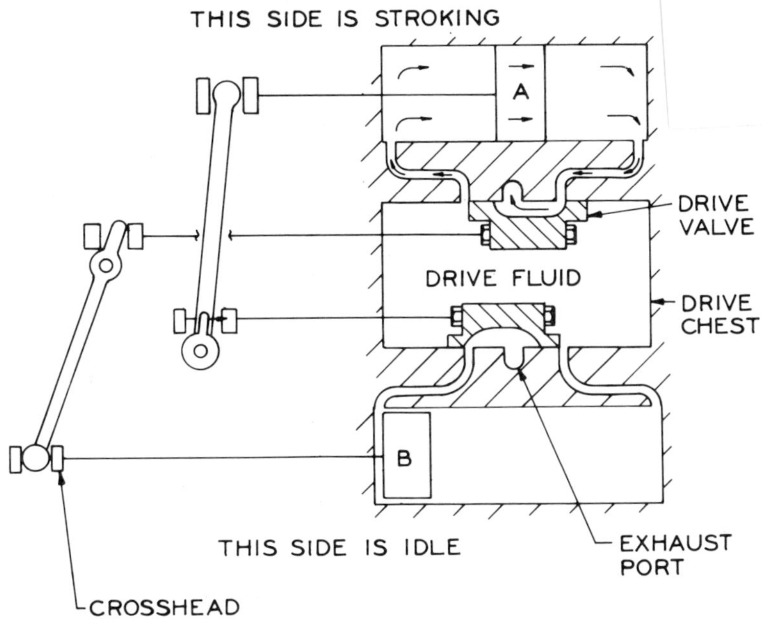

FIG. 6.4. A schematic of the control mechanism of a duplex direct-acting pump. Each valve is controlled by the crosshead on the opposite drive rod.

through most of its stroke before it shifts the "A" valve. Therefore side "A" is idle most of the time that "B" is working, and vice versa. Both sides are, therefore, idle almost half of the time during pump operation. This overlapping of strokes provides a smooth fluid flow in both suction and discharge pipes, and allows the unit to operate at a very low speed without stalling.

A simplex pump, as represented schematically in Fig. 6.5, has only one drive piston and main drive valve. It has no partner to shift its valve, so it must shift its own. If the crosshead (fastened to the piston rod) were to attempt to shift the main valve directly, the valve would move slowly through its stroke, and, while at the center of its stroke, the pump would stall. This is particularly true at low speed when the inertia of the reciprocating parts is minimal. It is therefore necessary that the simplex direct-acting pump be provided with a mechanism which causes the main valve to move quickly from one end of its stroke to the other even if the pump is running slowly. This can be accomplished mechanically, such as with an over-center spring, or "pneumatically" (with the drive fluid). The latter is the current common practice. The main valve is driven from one end of its stroke to the other with an auxiliary piston. The auxiliary piston is driven

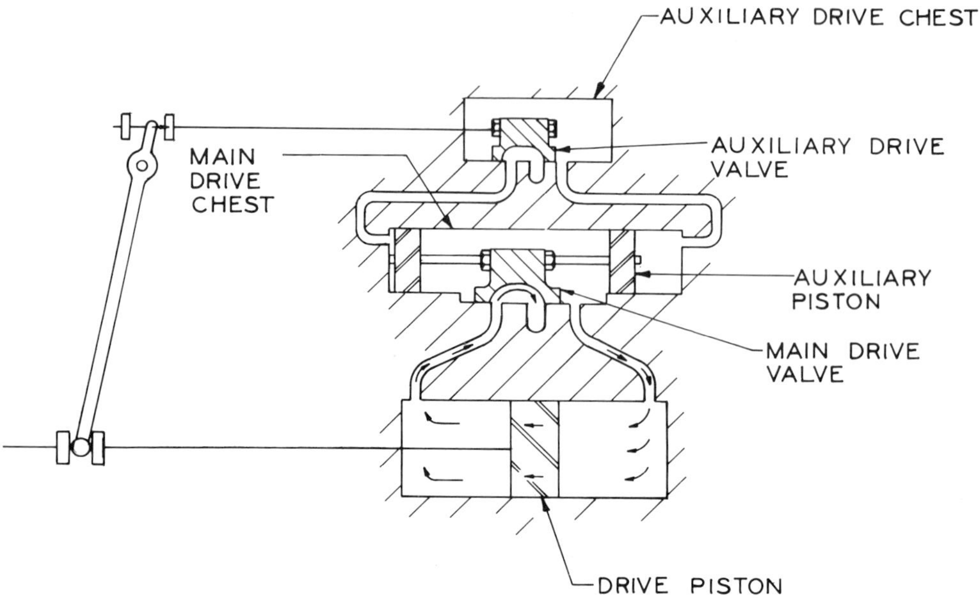

FIG. 6.5. A schematic of the control mechanism of a simplex direct-acting pump. An auxiliary valve and piston are required to provide snap action to the main valve to prevent stalling.

by gas from a small auxiliary valve, which in turn is driven through a lever by the crosshead. This mechanism assures that the main valve shifts positively from one extreme position to the other. To assure that sufficient pressure is always available to shift the main valve, these units are often provided with a separate gas connection to the auxiliary valve which is connected *upstream* of the throttling speed-control valve.

On a simplex direct-acting pump, the stroke length can be changed by adjusting the mechanism which controls the auxiliary valve. If the auxiliary valve shifts sooner, the stroke length is reduced. (If the stroke is reversed too soon, the reversal can be so quick that a hydraulic shock results.) On a duplex pump, adjusting the mechanism which controls each valve does *not* change the length of stroke. Since each side controls the drive valve for the other side, reducing the lost motion in the valve mechanism causes the other side to start sooner, but does not affect the stroke length of either side. On a duplex, the stroke length is established by the amount of "cushion" provided at each end of each cylinder. In large pumps, "cushion valves" are sometimes supplied to regulate stroke length. Otherwise, the stroke length of a duplex pump is not adjustable.

6.3. FLOW PATTERNS

6.3.1. General

Each pumping chamber of a reciprocating pump accepts pumpage at suction pressure and discharges it at discharge pressure. It does not function like a multistage reciprocating compressor or a multistage centrifugal compressor or pump, all of which pass fluid from one pumping element to the next, with each element adding a portion of the pressure difference. A reciprocating pump produces discharge pressure in each pumping chamber on each discharge stroke. Each pumping chamber of a reciprocating pump therefore acts in parallel with, and completely independently of, its neighbor. Table 6.1 lists the number of pumping chambers for various types of recips. For a recip pump to operate properly, all pumping chambers must be fully primed. If, for instance, a triplex, single-acting pump has only one chamber primed, it will pump only one-third of the rated capacity, and will run as rough as a simplex, single-acting unit. System design and the start-up procedure required to fully prime a recip pump are discussed in Chapters 11 and 13. The following discussion assumes that all chambers are fully primed.

Each pumping chamber accepts a fixed volume of pumpage from the suction manifold and discharges it into the discharge manifold. Ignoring the effects of compressibility and slip, both of which are normally small, the velocity in the suction pipe at any instant can be determined by adding the instantaneous velocities of all pumping elements on the suction stroke and multiplying by the ratio of the area of one pumping element to the

Table 6.1. Number of Pumping Chambers in Different Types of Reciprocating Pumps.

Type of Pump	Number of Pumping Chambers
Simplex, Single-Acting	1
Simplex, Double-Acting	2
Duplex, Single-Acting	2
Duplex, Double-Acting	4
Triplex, Single-Acting	3
Triplex, Double-Acting	6
Quintuplex, Single-Acting	5
Quintuplex, Double-Acting	10
Septuplex, Single-Acting	7
Nonuplex, Single-Acting	9

suction pipe area. The discharge pipe velocity can be likewise determined from all elements on their discharge strokes.

6.3.2. Direct-Acting Pumps

If such velocity determinations are plotted against time for direct-acting pumps, the results are as shown in Figs. 6.6 and 6.7. Figure 6.6 is for a *simplex* double-acting pump. Note the rapid acceleration at the beginning of each stroke, the constant velocity during most of the stroke, and the rapid deceleration at the end of the stroke. The liquid, in both suction and discharge pipes, comes to a dead stop twice per cycle of the pump. Figure 6.7 is for a *duplex* double-acting pump. Because of the overlap between the two sides (if the valves are set properly and the pump is in good condition), the resultant flow, in both suction and discharge pipes, is quite smooth.

6.3.3. Power Pumps

Pipeline flow velocities for a power pump can be determined by the same procedure. The velocity of a crosshead (or a plunger), plotted as a function of time (or crank angle), produces a sine curve, although it is a distorted sine curve because of the slider-crank mechanism of the power end. The degree of distortion is established by the ratio of the connecting rod length to the throw of the crankshaft. The smaller this ratio, the greater the distortion. Liquid velocities are proportional to the crosshead velocities.

Figure 6.8 shows a pure sine curve. It represents what the suction velocity or discharge velocity would be for a simplex single-acting power

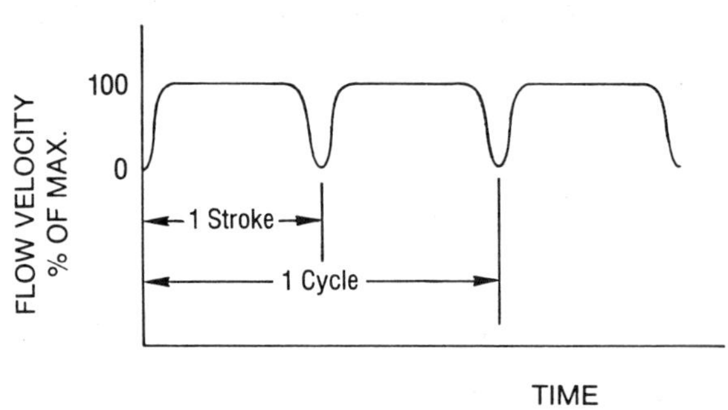

FIG. 6.6. Flow velocity of a simplex, double-acting, direct-acting pump. Flow rate drops to zero twice per cycle.

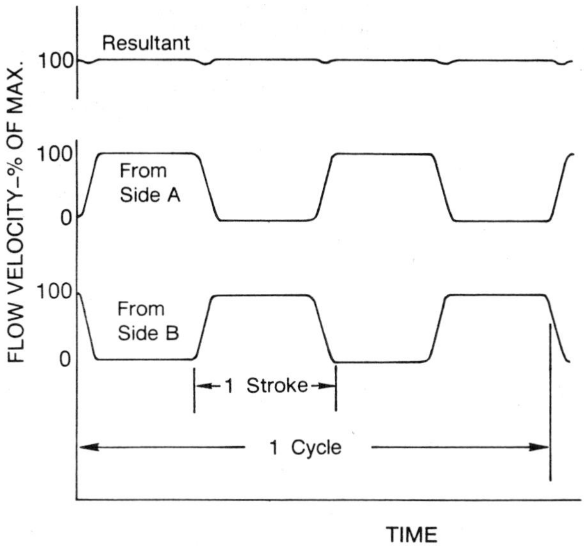

FIG. 6.7. Flow velocities of a duplex, double-acting, direct-acting pump. Flow rate is smooth if valves are set for proper overlap of the two sides.

pump if the rod length to throw ratio were infinity. Note that the maximum velocity is 3.14 (π) times the average. This means that such a unit, rated for 10 gpm (average), would discharge at a peak rate of 31.4 gpm. During half the time, there is no flow in the discharge pipe because the plunger is on the suction stroke (and vice versa).

Figure 6.9 shows the suction velocity of a *simplex single-acting* power pump with a typical rod-to-throw ratio of 5. Flow varies from 0 to 320% of the average.

Figure 6.10 shows the suction velocity of a *duplex single-acting* power pump. The throws on the crankshaft are offset 180°, so that while one plunger is on the suction stroke, the other is on the discharge stroke. The resultant velocity varies from 0% to 160% of the average.

As shown in Figure 6.11, the addition of another plunger, resulting in a *triplex* power pump, produces a considerably smoother flow. The throws on the crankshaft are offset 120°, resulting in a 60° overlap between plungers, on both the suction and discharge strokes. During each overlap, the flow from both plungers adds together, resulting in a secondary pulse or "hump" between the major pulses. The resulting flow therefore includes *six* pulses per revolution of the crankshaft. The velocity varies between about 82% and 107% of the average velocity, a variation of 25%. A triplex

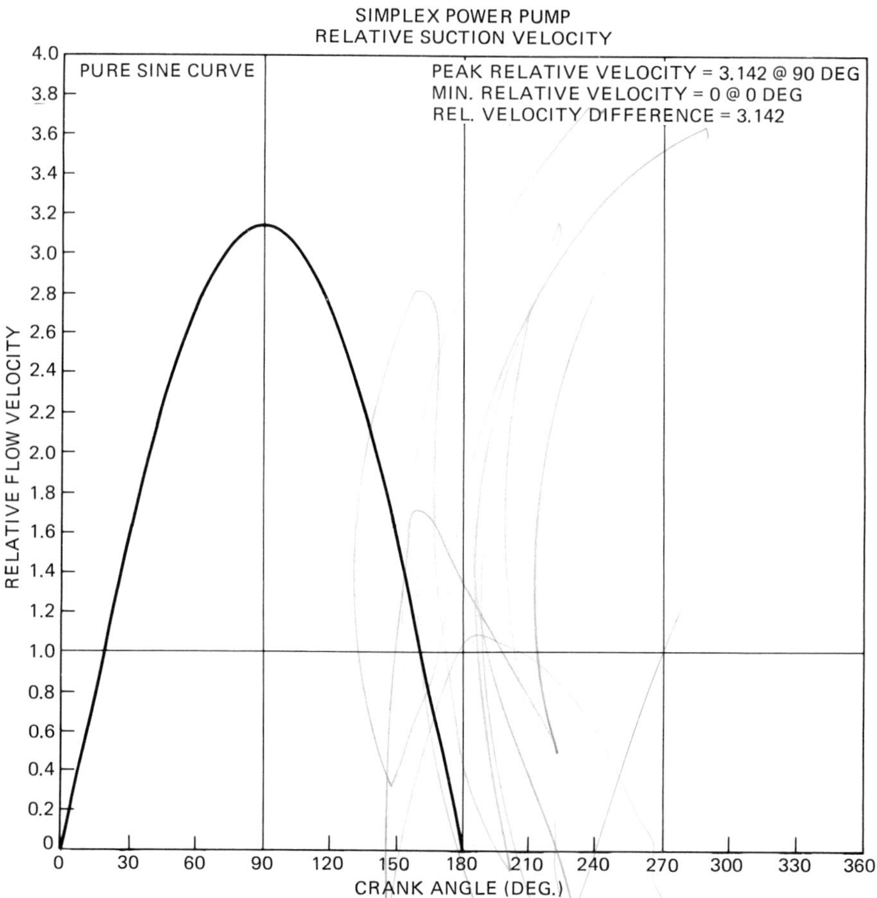

FIG. 6.8. A pure sine curve represents the suction or discharge velocity of a simplex, single-acting power pump with an infinitely-long connecting rod.

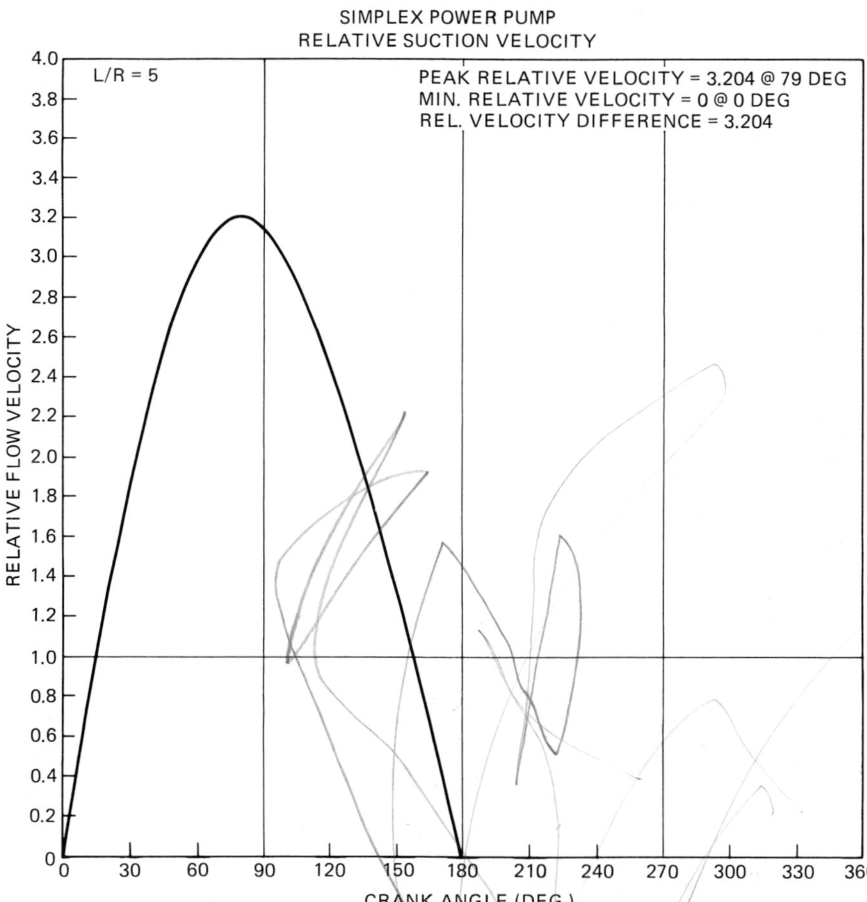

FIG. 6.9. Actual suction velocity of a single-acting simplex power pump. Because of the finite length of the connecting rod, the fluid velocity is a *distorted* sine curve.

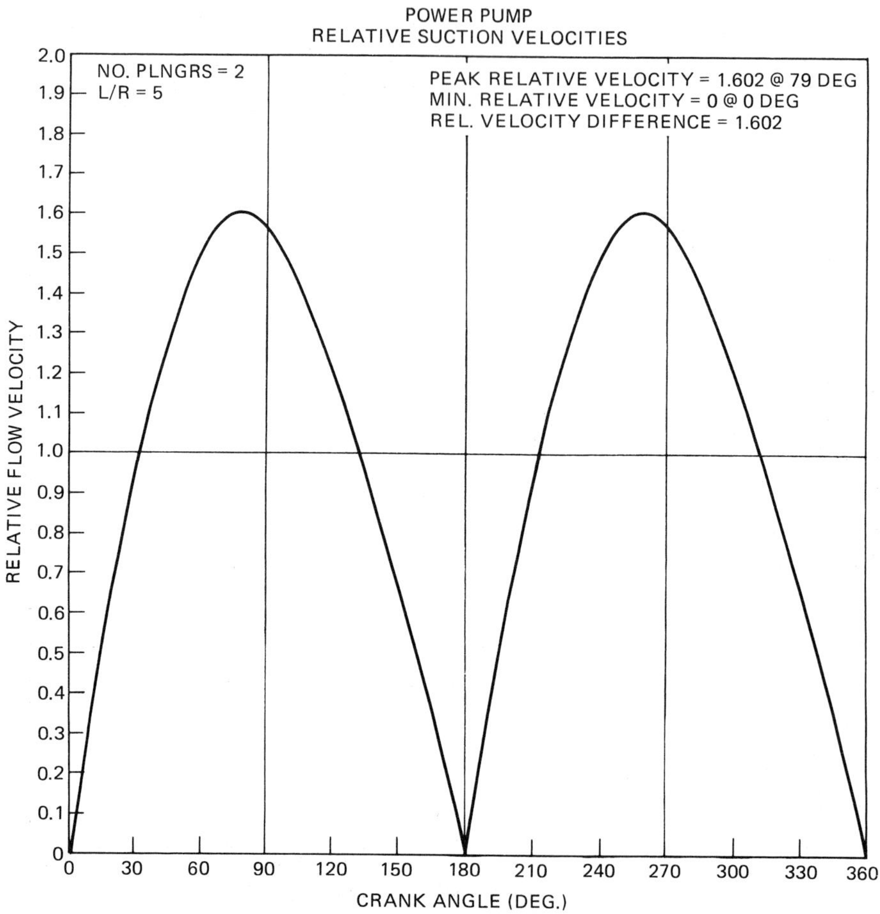

FIG. 6.10. Suction velocities of a single-acting duplex power pump.

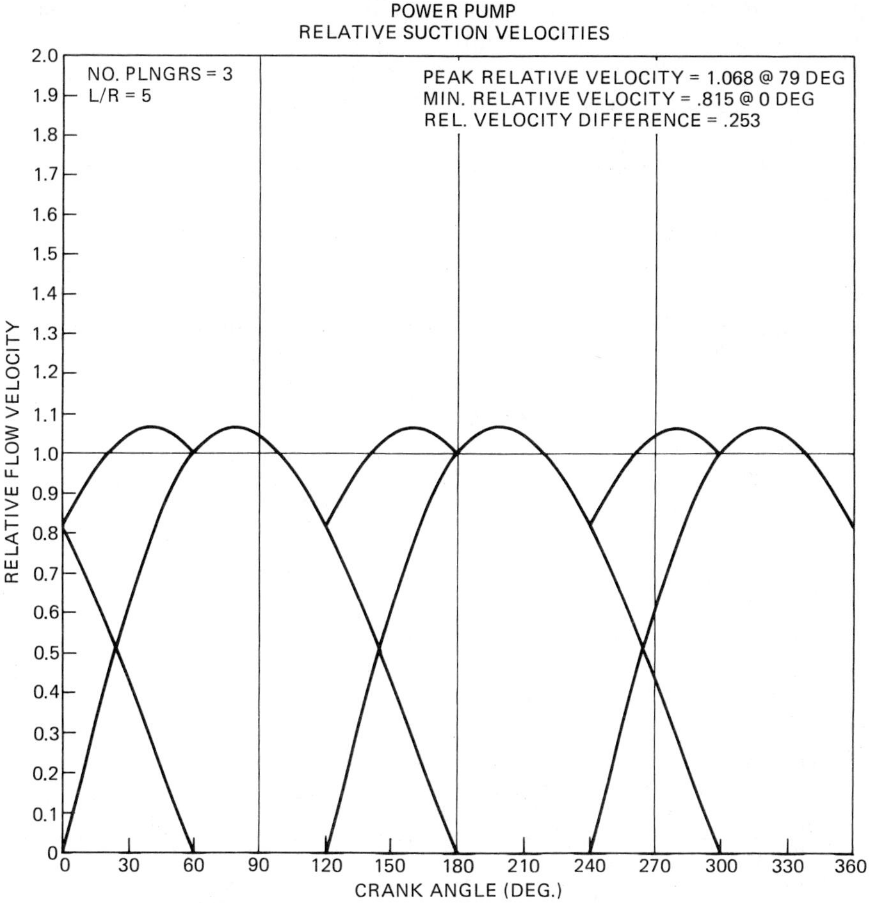

FIG. 6.11. Suction velocities of a single-acting triplex power pump.

pump, pumping an average 10 gpm, would have a minimum rate of about 8.2 gpm and a maximum of about 10.7 gpm.

A *quadruplex* (four-plunger) single-acting power pump produces the velocity variations shown in Fig. 6.12. Crankshaft throws are offset 90°, and the unit contains one more pumping element than the triplex, but flow varies from 79% to 111% of average, a variation of 32%—larger than the triplex. For this reason, quadruplex pumps are seldom produced.

The next single-acting power pump is the *quintuplex,* a five-plunger unit with suction velocities illustrated in Fig. 6.13. With the crank throws offset 72°, the resulting overlaps produce a velocity profile that varies from about 94% to 102% of the average, a variation of only 8%. As with

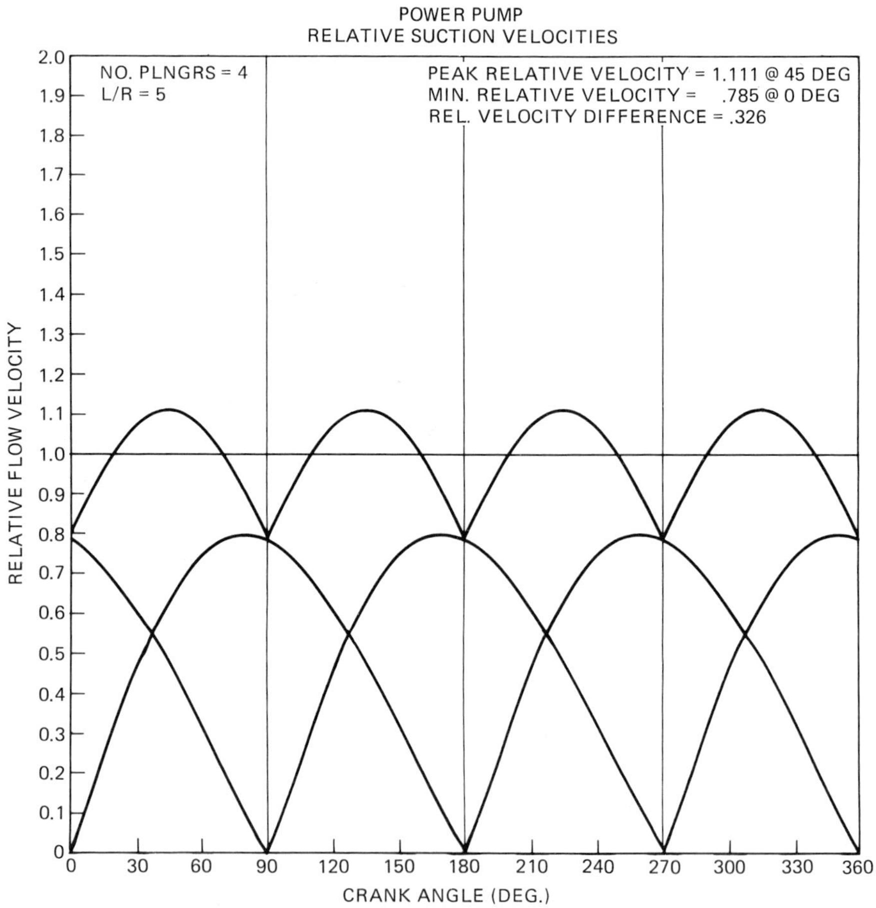

FIG. 6.12. Suction velocities of a single-acting quadruplex power pump.

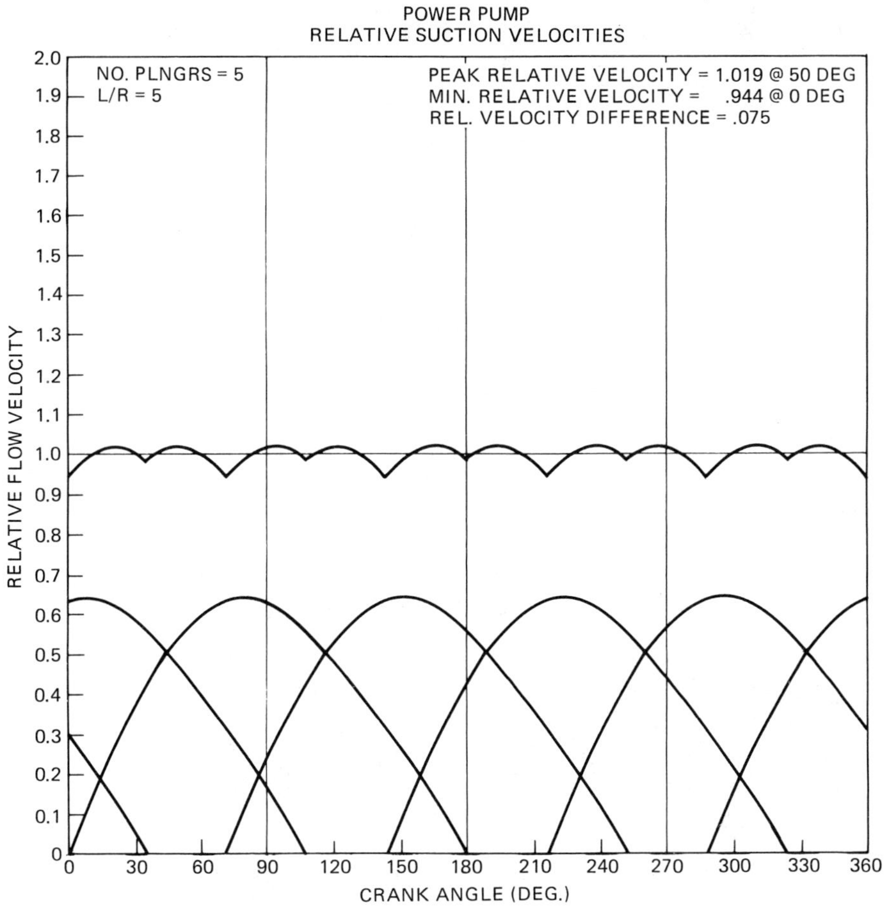

FIG. 6.13. Suction velocities of a single-acting quintuplex power pump.

the triplex, the quintuplex produces twice the number of pulses/revolution as it has pumping elements. It produces 10 pulses/revolution.

For the same reason that a quadruplex is seldom produced, pumps containing six or eight plungers are seldom built. They produce velocity variations greater than pumps with one less plunger. Pumps with seven or nine plungers are common, though, and their resultant velocity variations are about 4% and 2% respectively, as shown in Figs. 6.14 and 6.15.

Table 6.2 summarizes the foregoing. Comparing these numbers might lead one to conclude that the triplex pump would produce pressure variations about 1/6 that of the duplex, and the quintuplex about 1/3 that of the

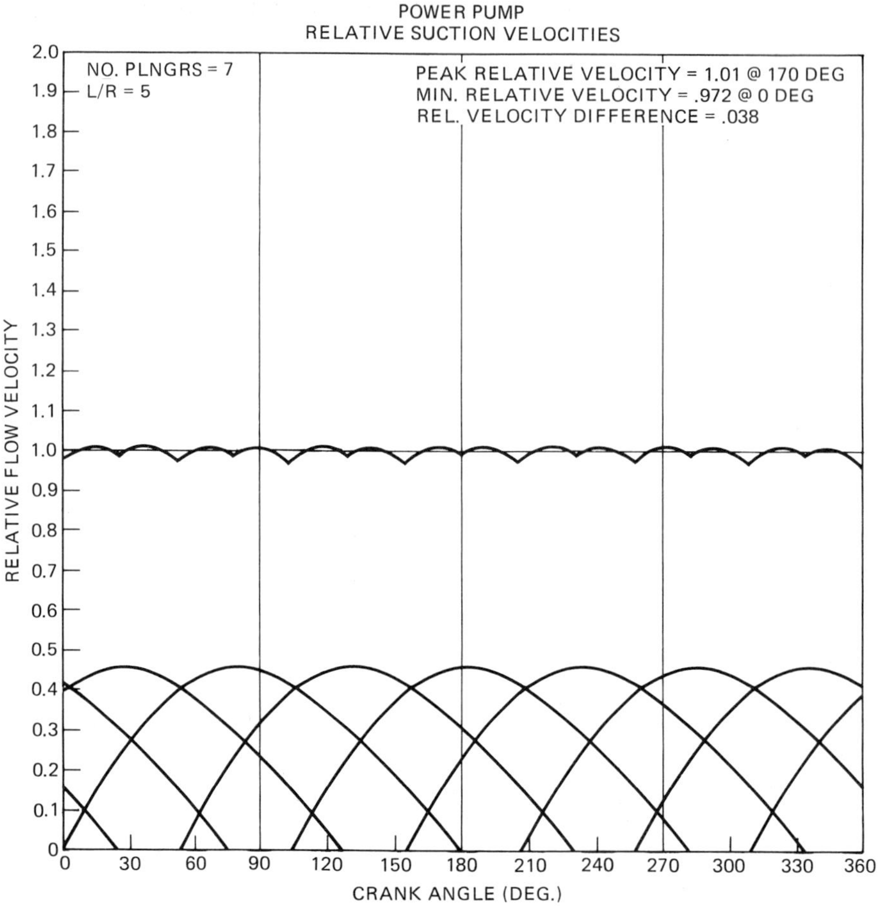

Fig. 6.14. Suction velocities of a single-acting septuplex power pump.

Table 6.2. Pump Types and Approximate Flow (Liquid Velocity) Variations as a Percent of the Average Flow.

Number of Plungers (or Equivalent)	Type of Pump	Flow Variations—Percent of Average
1	Simplex, Single-Acting	320
2	Duplex, Single-Acting	160
3	Triplex, Single-Acting	25
4	Duplex, Double-Acting	33
5	Quintuplex, Single-Acting	8
7	Septuplex, Single-Acting	4
9	Nonuplex, Single-Acting	2

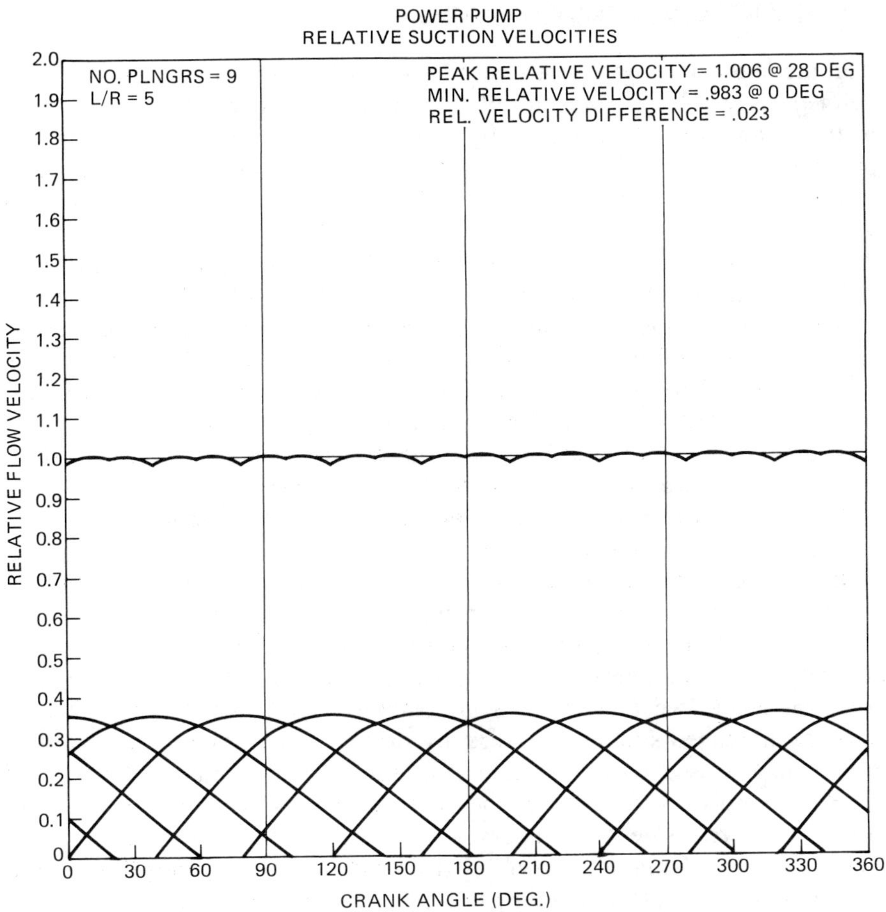

FIG. 6.15. Suction velocities of a single-acting nonuplex power pump.

triplex. If the flow were laminar, the liquid completely incompressible, and friction the only factor determining pressure, the pressure variation ratios would be proportional to these fractions. Except for low speed operation, though, such is not the case. At moderate and high speeds, flow is often turbulent, and another factor called "acceleration head" affects pressure variation more than friction. Acceleration head is covered in Chapter 7.

6.4. FRAME LOAD–POWER PUMPS

"Frame load" is synonymous with "rod load" and "plunger load." It is defined as the force transmitted to the power end by one plunger (or piston). Only the force created by the pressure of the fluid acting on the plunger is considered in this calculation. Inertial forces of the reciprocating parts and fluid are considered negligible, an acceptable assumption in most cases.

When a plunger is on the discharge stroke, it is exposed to discharge pressure plus the pressure drop across the discharge valve and friction loss in the discharge manifold. These two pressure drops are also considered to be negligible during plunger load calculations. "Discharge frame load" is therefore calculated by multiplying the pump discharge pressure by the cross-sectional area of one plunger. As an example, a pump with $2\frac{1}{2}$ inch diameter plungers and a discharge pressure of 2000 psig would be operating with a discharge frame load of 9817 pounds (see Fig. 6.16).

"Suction frame load" is the load on the plunger during the suction stroke, again ignoring the pressure drop from the suction nozzle to the face of the plunger. If the same pump is operating with a suction pressure of 500 psig, the suction frame load would be 2454 pounds.

The power end of a power pump is rated by the maximum frame load the power end is capable of absorbing. For single-acting units, this full-load condition is established for atmospheric suction pressure. (The effects of increased suction pressure are discussed in Chapter 9.) A triplex pump may have, for instance, a continuous-duty frame load rating of 10,000 pounds. Table 6.3 lists the maximum discharge pressures allowable for different sizes of plungers which may be offered in the pump. The pressures are calculated by dividing the 10,000 pounds by the plunger area in square inches. These maximum pressures must be reduced if the suction pressure is significantly above atmospheric, or if the working pressure of the liquid end is less than the calculated pressure.

Some power pumps are rated for continuous, around-the-clock operation, some for intermittent operation, and some are rated both ways.

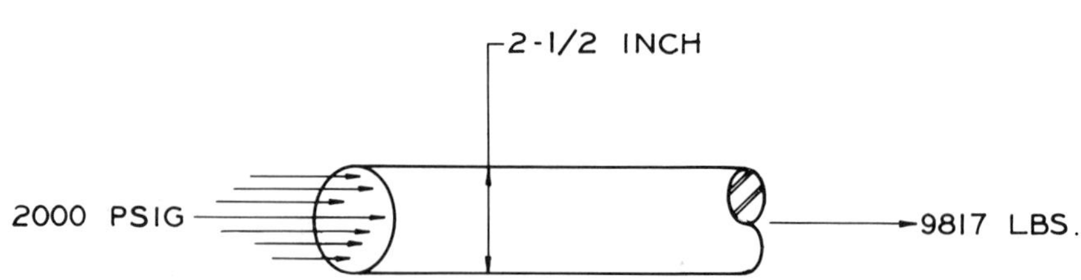

FIG. 6.16. Frame load is calculated by multiplying pressure times plunger area.

Table 6.3. Maximum Discharge Pressures for a Pump with a Maximum Frame Load of 10,000 Pounds.

Plunger Diameter (in.)	Maximum Discharge Pressure (psig)
1	12,730
$1\frac{1}{2}$	5,660
2	3,180
$2\frac{1}{2}$	2,040
3	1,410
$3\frac{1}{2}$	1,040
4	796
$4\frac{1}{2}$	629
5	509

6.5. SHAFT TORQUE–POWER PUMPS

6.5.1. Torque Fluctuations

A power pump, with *constant* discharge and suction pressures, and running at a *constant* speed, will require an input torque that pulses with the same magnitude as its flow velocity profile. A triplex, for instance, will impose 6 pulses/crank revolution on the drive system. The torque will vary from 82% to 107% of the average. A simplex, single-acting pump will require a torque input that varies from 0% to 320% of the average.

Simplex and duplex single-acting units, therefore, require more considerations when selecting drivers and drive systems. Even triplex pumps, intended to operate at speeds of about 40 rpm, may require special considerations when selecting drive components.

It should also be emphasized that, unless pulsation dampening equipment is very effective, suction and discharge pressures will *not* be constant. Particularly with high-friction systems, pressures vary as the flow pulses, magnifying the peak load on the pump and the peak torque on drive components.

6.5.2. Torque as a Function of Speed

If suction and discharge pressures are held constant, the *average* shaft torque required by a power pump is independent of speed (except for increases at very low and very high speeds). Therefore, a power pump will require the same average torque at, for instance, full, one-half, and one-fourth of its rated speed, and will therefore require, respectively, 100%, 50%, and 25% of the rated power.

Figure 6.17 illustrates the variation of average torque with speed for a typical triplex power pump. (This figure does not show the torque required to *accelerate* a pump to rated speed—only the torque required to run at any *constant speed* between zero and rated speed.) If the pump is started against full discharge pressure, break-away torque is about 150% of the rated torque. This will vary with the type of packing and plungers in the liquid end, the type of bearings in the power end, and the length of time that the pump has been idle. As speed is increased, and proper lubrication is established in the packing and power end, torque drops to the rated value, and is thereafter constant up to rated speed. Such a start-up imposes unnecessarily high loads on the pump and drive system. This and other undesirable characteristics of a start-up against full discharge pressure are discussed in Chapter 13.

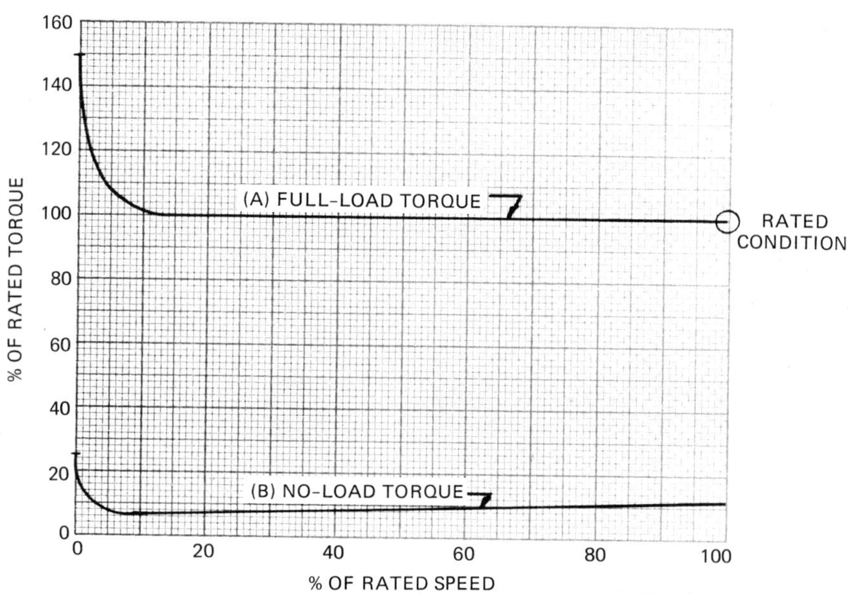

(A) FULL-LOAD TORQUE - FOR PUMP START-UP AGAINST RATED DISCHARGE PRESSURE

(B) NO LOAD TORQUE - FOR PUMP START-UP ON A BY-PASS OPEN TO PUMP SUCTION

FIG. 6.17. Torque vs speed for a typical triplex power pump.

For a start-up procedure that is easier on equipment, the pump discharge is piped back to the suction vessel, making discharge pressure near suction pressure. Curve "B" in Fig. 6.17 illustrates the resulting torque requirements imposed by the pump on the drive train. Break-away torque is approximately 25% of full-load torque. (This will vary for the same reasons stated above for the loaded condition.) As speed is increased, torque will drop to about 10% of full-load torque. The rise in torque indicated by curve "B" at higher speeds is due to the increase in friction head in the bypass system as flow rate increases.

6.6. POWER OUTPUT

The power output of any pump is the usable power it imparts to the fluid being pumped. A common term for this quantity is "hydraulic power," or in U.S. terminology "hydraulic horsepower." The equation for power output can be derived by considering the flow in the pipes leading to and from a pump. Using Fig. 6.18 for illustration, we can calculate the hydraulic power as follows. First we'll calculate the force required to move the fluid along the pipe.

$$F = pA \tag{6.1}$$

where

F = Force across the cross-section of fluid inside the pipe
p = Pressure of the fluid inside the pipe
A = Cross-sectional area of the fluid inside the pipe (pipe I.D.)

Power is force × distance/unit time. Distance per unit time is velocity. Therefore:

$$P = Fv = pAv \tag{6.2}$$

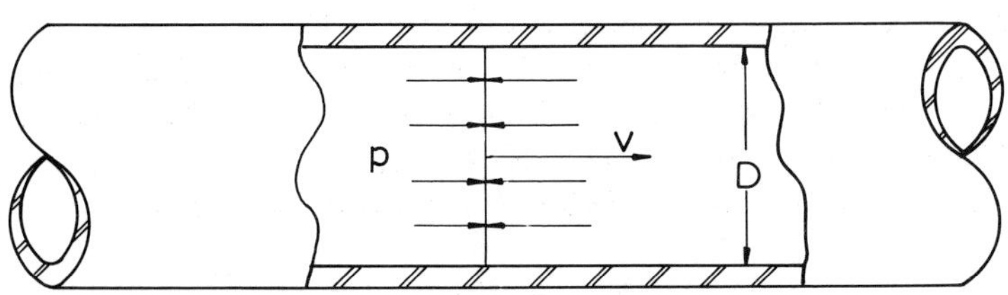

FIG. 6.18. Fluid flowing in a pipe represents hydraulic power.

where

P = Power
v = Velocity of fluid

Velocity can be expressed in terms of capacity, Q, as follows:

$$v = \frac{Q}{A} \tag{6.3}$$

Substituting into (6.2):

$$P = pAv = pA\frac{Q}{A} = pQ \tag{6.4}$$

The fluid power in both the discharge line and suction line can be calculated with Eq. 6.4 by using discharge and suction pressures. The power added to the fluid by the pump is the difference between discharge power and suction power. Therefore:

$$\begin{aligned} P_o &= P_d - P_s = p_d Q - p_s Q \\ &= Q(p_d - p_s) = Q\,\Delta p \end{aligned} \tag{6.5}$$

where

P_o = output power
P_d = discharge power
P_s = suction power
p_d = discharge pressure
p_s = suction pressure
Δp = total differential pressure = discharge pressure − suction pressure

It is necessary that we consider the units of the quantities in this equation. We will work in U.S. customary units. In Eq. 6.5, if Q is in ft³/min and p in lbs/ft², then the power is in ft–lbs/min. All of these are good engineering units, but not commonly used in industry. Let's get to U.S. gal/min (gpm), lbs/in² (psi), and U.S. horsepower (hp). Since there are 7.48 gal/ft³, 144 in²/ft², and 33,000 ft-lbs/min/hp, our equation becomes:

$$P_o = \frac{(Q)}{(7.48)}\frac{(\Delta p)(144)}{(33{,}000)} = \frac{Q\,\Delta p}{1714} \tag{6.6}$$

where

P_o = pump output power, U.S. hp
Q = pump capacity, U.S. gpm
Δp = pump differential pressure, psi*

Equation 6.6 can be used for *any* pump, be it reciprocating, rotary, or centrifugal, but since the differential pressure of a centrifugal pump is normally expressed as a dynamic head, and in U.S. units is normally expressed in ft–lb/lb, or just "feet," Eq. 6.6 is altered for centrifugal pumps by substituting the following equation:

$$\Delta p = \frac{(\Delta h)(SG)}{2.31} \tag{6.7}$$

where

Δh = pump total differential head, feet
SG = specific gravity of liquid, normally taken at suction conditions

Plugging into Eq. (6.6):

$$P_o = \frac{Q\ \Delta p}{1714} = \frac{Q\ \Delta h SG}{(1714)(2.31)} = \frac{Q\ \Delta h \mathrm{SG}}{3960} \tag{6.8}$$

Both Eqs. 6.6 and 6.8 will yield identical results for any type of pump.

It should be noted that a pressure gage attached to a pipe does not measure the total energy in a flowing stream. It measures only "static" pressure, and, unless attached to a pitot tube, does not correct for any velocity head (kinetic energy) of the fluid. Therefore, unless the suction and discharge pipes are the same diameter, the velocity head differs from the suction to discharge. With lower-differential-pressure pumps, and moderate-to-high fluid velocities, this difference in velocity head may need to be accounted for in pump testing. With high-differential-pressure pumps, the velocity head is a negligible portion of the fluid energy, and is usually ignored.

---

* Since this is a *difference* between discharge and suction pressures, it is neither "psi*g*" nor "psi*a*"—just "psi."

6.7. POWER INPUT–POWER PUMP

The power input of a power pump is the mechanical power transmitted from the shaft of the driver into the pump. Losses of internal gears or bolt-on gear units are included in power input. Losses of belt drives are often included. Losses of external gear units and variable-speed couplings are normally not included. Power input can be determined with such devices as torque meters, calibrated motors, and dynamometers.

6.8. MECHANICAL EFFICIENCY

6.8.1. Power Pumps

The "mechanical efficiency," also called " pump efficiency," of a power pump is defined as the ratio of power output to power input. In equation form:

$$\eta_m = \frac{\text{power out}}{\text{power in}} = \frac{P_0}{P_i} = \frac{(Q)(\Delta p)}{(1714)(P_i)} \tag{6.9}$$

where

η_m = pump mechanical efficiency (overall)
Q = capacity, U.S. gpm
Δp = pump total differential pressure, psi
P_i = input power from driver, hp

As shown in Fig. 6.19, mechanical efficiency of these units often exceeds 90% (sometimes exceeding the efficiency of the driving motor), and often varies with the plunger diameter. The larger plungers result in reduced efficiency because of increased valve and packing losses and reduced differential pressure. As speed is reduced, a reduction in efficiency sometimes occurs, apparently due to a slight increase in the friction coefficient in the bearings and packing. Other pumps have exhibited the opposite trend, though, apparently due to a reduction of valve losses. For each power pump there is probably a speed at which optimum efficiency occurs. The more significant fact, though, is that the mechanical efficiency of a power pump changes very little as speed is varied within the normal operating range.

As seen in Fig. 6.19, as the plunger load is reduced, mechanical efficiency drops. This is caused by the output power (hydraulic power) falling faster than friction losses, and therefore becoming a smaller part of the

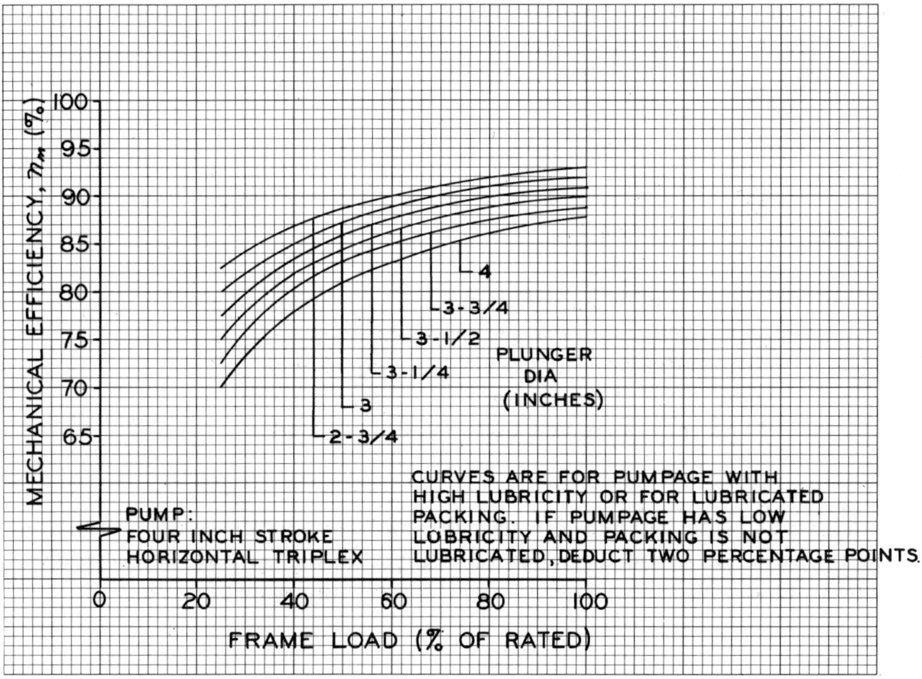

FIG. 6.19. Mechanical efficiency of a power pump.

power input. If these curves were extended to zero plunger load, the efficiency would be zero because the power output would be zero and some input power would be required to overcome friction losses.

Some pump applications people use a blanket 85% mechanical efficiency regardless of operating pressure. As can be seen from the curves, this is safe for most applications because a power pump is usually applied at a discharge pressure more than half its rating, and suction pressure is usually low. There are some services, though, where the differential pressures are low, and the use of 85% η_m has resulted in undersized drivers.

6.8.2 Direct-Acting Pumps

For a direct-acting pump, "mechanical efficiency," also called "pump efficiency," is the ratio of the *force* transmitted to the liquid by the piston or plunger, to the force transmitted to the drive piston by the motive fluid. For double-acting pumps (most direct-acting pumps are double-acting), the differential pressures are used at both ends of the pump; and since fluid friction losses in valves and ports must be charged to the pump, the pressures are measured at inlet and outlet ports (while the pump is moving through a stroke). In equation form:

$$\eta_m = \frac{(A_L)(\Delta p_L)}{(A_{dr})(\Delta p_{dr})} \tag{6.10}$$

where

η_m = pump mechanical efficiency
A_L = area of the liquid piston or plunger
Δp_L = differential pressure across the liquid end
A_{dr} = area of the drive piston
Δp_{dr} = differential pressure across the drive end

The area of the piston rod is usually small relative to piston area, and is therefore usually considered to be negligible; but it must be considered when the rod area becomes a significant portion of piston area.

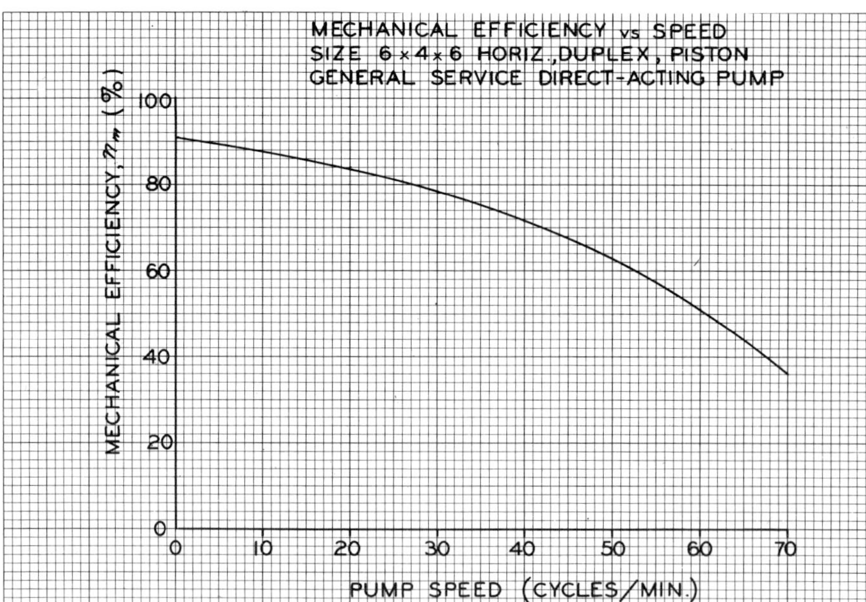

FIG. 6.20. Mechanical efficiency of a direct-acting pump.

Figure 6.20 is a plot of mechanical efficiency versus speed for a typical direct-acting pump. Note that η_m rises as speed is reduced. It is this characteristic that allows the unit to be controlled by throttling the motive fluid. Throttling the motive fluid reduces the differential driving pressure available to the pump and forces the pump to operate more efficiently, i.e., at a lower speed. It is the magnitude of the fluid friction losses in both the drive end and liquid end which determine the slope of the η_m curve. The drive end losses are usually considerably larger than liquid end losses. Figure 6.21 shows the points of intersection of gas pressures which establish the running speeds of a direct-acting pump.

If the pump is initially selected to operate at a lower speed, the inherently higher efficiency will allow selection of a smaller diameter drive end, resulting in lower cost and lower gas consumption. A smaller drive end will also assist in maintaining low operating speeds (the drop in efficiency, as speed increases, limits the "no-load" speed), thereby extending pump life.

Figure 6.22 shows the approximate mechanical efficiencies of direct-acting pumps at maximum rated speeds. Plunger pumps have lower efficiencies than piston pumps because of the additional friction of the plungers in the packing, and the side rods in their support bearings.

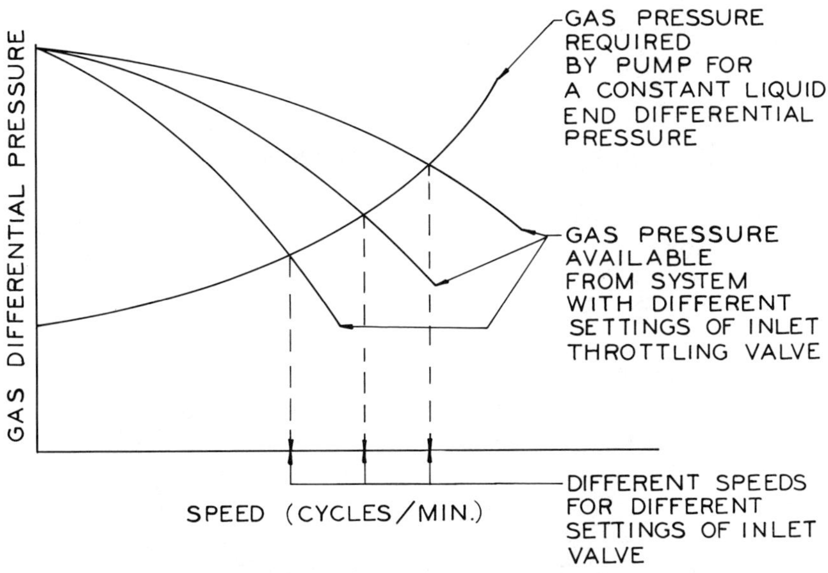

FIG. 6.21. Direct-acting pump speed is determined by the intersection of the lines for Δp gas available and Δp gas required.

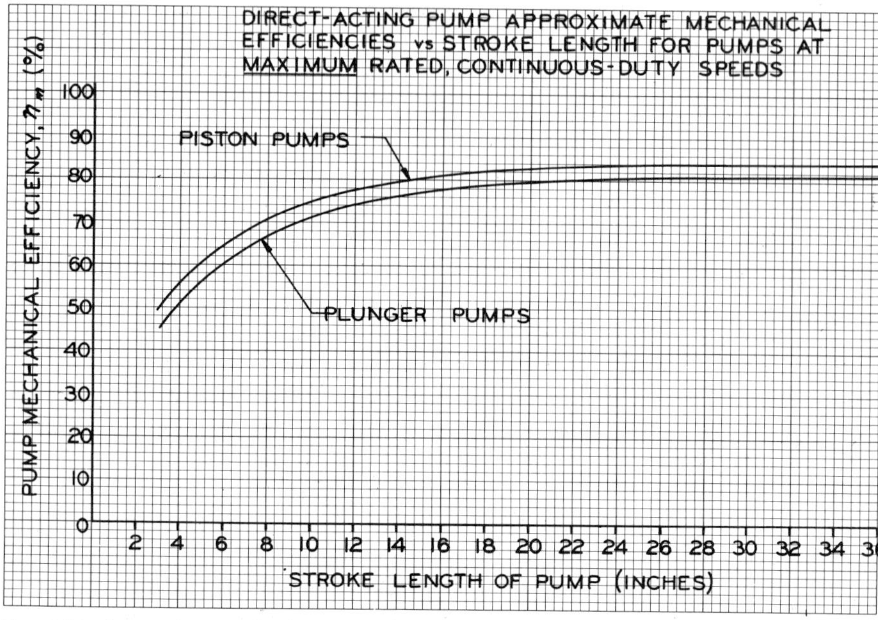

FIG. 6.22. Approximate mechanical efficiencies of direct-acting pumps, at maximum rated speeds, as a function of pump stroke length.

6.9. THERMAL EFFICIENCY—DIRECT-ACTING PUMPS

As mentioned in Chapter 1, the standard "simple" (a "single-stage," non-expansion unit) gas-driven end has a low thermal efficiency. The reason for the low efficiency is that the drive gas does not expand during the stroke. The gas remains at a constant pressure throughout the stroke. Therefore, no work is extracted from the gas. (If the drive fluid is liquid, efficiency is high because a liquid, being nearly incompressible, has little expansion work available.) If the pump were to be equipped with a flywheel or other energy-storage device, the gas valve could be set to cut off the inlet gas during the stroke, allowing gas expansion during the power stroke, extracting work from the gas, and thereby resulting in higher thermal efficiency. Even with its low thermal efficiency, a steam pump is sometimes more efficient than a steam-turbine-driven centrifugal pump, because the steam pump has such a high mechanical efficiency.

A way to extract expansion work from a gas, without resorting to an energy-storage device, is the "compound" drive end. A compound gas end consists of two different size drive pistons, mounted coaxially on a common piston rod. The small piston is driven by the inlet gas. The large piston is driven by the exhaust from the small piston. Since the displacement of the larger piston exceeds that of the smaller piston, expansion takes place during the stroke, extracting work from the drive gas, resulting in higher thermal efficiency. Gas consumption of a compound drive end is approximately 30% less than that of a standard "simple" drive end.

6.10. VOLUMETRIC EFFICIENCY

6.10.1. General

For a reciprocating pump, volumetric efficiency is defined as the ratio of capacity to displacement. In equation form:

$$\eta_v = \frac{\text{capacity}}{\text{displacement}} = \frac{C}{D} \tag{6.11}$$

The term "delivered capacity" is avoided here because of the connotation that the volume flow rate is to be measured at the *discharge* of the pump. Since many liquids compress a measurable amount as they move from suction to discharge conditions through a high-pressure reciprocating pump, the *volume* flow rate (gpm) is less at discharge conditions than at suction conditions, even though the mass flow rate is the same. In most every case, the process engineer will state the capacity requirements and liquid characteristics at pump suction conditions because the characteristics at the discharge are not known. Also, if the system contains a flow meter, it will usually be installed on the suction side to avoid the requirement for a high working pressure which would be required if mounted on the discharge side. For convenience and uniformity, the term "capacity" will therefore be used to indicate the volume flow rate in the suction pipe.

The other half of the η_v equation is pump displacement. The term "theoretical displacement" should be avoided, for there is nothing hypothetical about displacement. It exists, and is easy to measure. Displacement is simply the volume swept by the pumping elements per unit time. For a single-acting triplex plunger pump, the equation is:

$$D = (3)\,\frac{\pi}{4}\,d^2\,L\,N \tag{6.12}$$

where

D = displacement, in.^{3}/min
d = diameter of plunger, in.
L = stroke length, in.
N = crankshaft speed, rpm

EXAMPLE 6.1
For a 2 × 3 triplex power pump running 330 rpm,

$$D = (3)\left(\frac{\pi}{4}\right)(2)^2(3)(330) = \underline{9330} \text{ in.}^3/\text{min}$$

To convert cubic inches to gallons, we divide by 231.

$$D = \frac{9330}{231} = \underline{40.4} \text{ gpm}$$

On test, the capacity of this pump is measured at 39.2 gpm. The volumetric efficiency is therefore calculated as follows:

$$\eta_v = \frac{Q}{D} = \frac{39.2}{40.4} = \underline{0.97}$$

6.10.2. Slip

One reason that volumetric efficiency is not 100% is that leakage and backflow occur past the suction and discharge valves. Leakage occurs when a valve does not seal perfectly against its seat. (Valves seldom seal perfectly.) Backflow occurs when a valve remains open for a few milliseconds after the plunger reverses direction. In high speed power pumps, some valve lag is inevitable. This leakage and backflow is referred to as "slip." Slip also includes leakage from the high pressure side of a double-acting piston to the low pressure side. In a plunger power pump, slip is typically 1% to 3% of displacement.

6.10.3. Test Results

Figure 6.23 plots the results of a series of tests on a horizontal triplex power pump in six different configurations. The pump was tested with two different plunger diameters and three different valve springs. The $2\frac{3}{4}$-

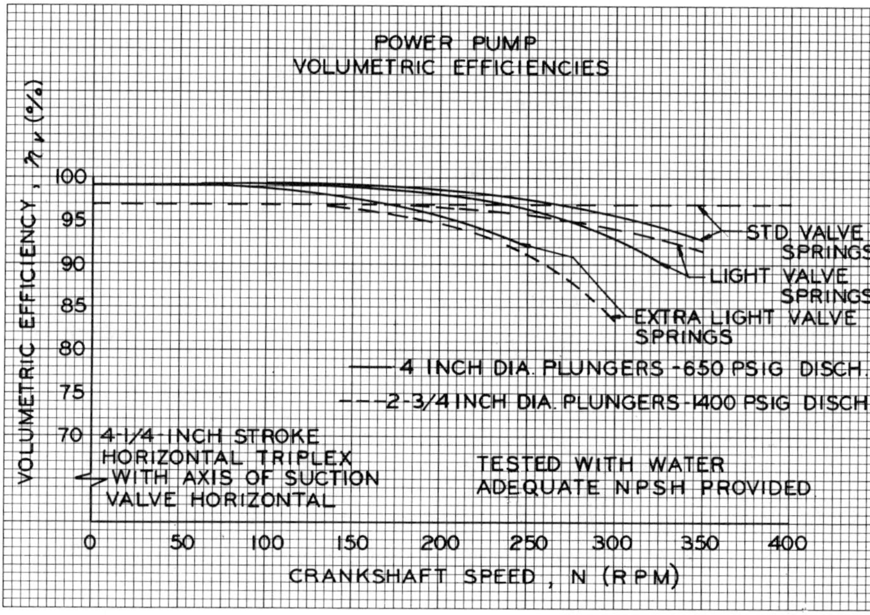

FIG. 6.23. Effects of speed, plunger diameter, and valve spring stiffness on the volumetric efficiency of a power pump.

in. plunger is the smallest recommended for this particular liquid end. The 4-in. plunger is the largest. The 4-in. plunger displaces more than twice that of the $2\frac{3}{4}$-in. plunger. The solid lines represent the performance obtained with the 4-in. plungers. The dashed lines represent the same series of tests, but with $2\frac{3}{4}$-in. diameter plungers.

A number of characteristics can be noted from these curves. It can be seen that, at low speed, valve spring stiffness has no effect on volumetric efficiency. The $2\frac{3}{4}$-in. plungers obtained about 97% η_v with all three types of valve springs. The 4-in. plungers, 99%. The higher η_v of the larger plunger probably occurred because of the lower pressure differential across the valves (less leakage and backflow) and/or the reduced effect of liquid compression (the discharge pressure was less than half, plunger displacement was more than twice, and clearance volume was less than that for the smaller plungers). (The effect of fluid compressibility and clearance volume on volumetric efficiency is discussed in Chapter 9.)

It can also be noted that η_v dropped as speed increased in five of the six tests. This was caused by backflow through the valves. Each valve was still partially open when the plunger reversed, resulting in reverse flow through the valve before the valve sealed off against its seat. Only the small plunger, in combination with the stiffest springs, exhibited no reduction in η_v as speed increased. This indicates that the standard spring is adequate to get the valve back to its seat quickly enough to prevent any measurable effect from backflow, when used with the small plunger. It also indicates that a stiffer spring would probably increase the η_v of the large plunger at higher speeds.

Yet another point to ponder is the fact that the large plunger had a higher η_v than the small plungers at speeds below about 275 rpm, but at higher speeds, had a lower η_v. This was caused by the higher valve velocities, associated with the larger plungers, lifting the valves further from their respective seats, and allowing some backflow at the beginning of each stroke.

Although volumetric efficiency has an affect on mechanical efficiency, the two do not necessarily move in unison. It is possible to have a high η_m with a low η_v (with a compressible fluid) or a low η_m with a high η_v (such as would result in a high-suction-pressure, low-differential-pressure, application).

6.11. PLUNGER SPEED

Plunger speed, or piston speed, is a rough indication of the service life which can be expected from pump parts such as packing and plungers. It is more commonly used as a yardstick for direct-acting pumps than power pumps. For direct-acting pumps, it is also used in the gas consumption equations, as will be shown later. Plunger speed is also used in one method of calculating valve velocity, as will be shown below. Plunger speed is calculated as the average velocity over the complete pumping cycle. In equation form:

$$S_p = \frac{2LN}{12} = \frac{LN}{6} \tag{6.13}$$

where

S_p = plunger (or piston) speed, ft/min
L = stroke length, in.
N = rpm of the crankshaft (power pump) or number of complete cpm (cycles per minute) (direct-acting pump)

EXAMPLE 6.2

Using the triplex pump from Ex. 6.1, $L = 3$ in. and $N = 330$ rpm.

$$S_p = \frac{(3)(330)}{6} = \underline{165} \text{ ft/min}$$

Another way to calculate plunger speed is as follows:

$$S_p = \frac{49Q}{d_L^2 \, n_c \, \eta_v} \tag{6.14}$$

where

S_p = plunger (or piston) speed, ft/min
Q = total pump capacity, U.S. gal/min (gpm)
d_L = diameter of the liquid plunger or piston, in.
n_c = number of pumping chambers in the pump
η_v = volumetric efficiency of the pump

Figure 6.24 is a plot of piston or plunger speed as a function of displacement ($\eta_v = 1$) for a simplex double-acting, or duplex single-acting ($n_c = 2$) pump.

EXAMPLE 6.3

Again, using our hard-working triplex from above, $Q = 39.2$ U.S. gpm, $d_L = 2$ in., $n_c = 3$, $\eta_v = 0.97$.

$$S_p = \frac{(49)(39.2)}{(2)^2(3)(0.97)} = \underline{165} \text{ ft/min}$$

On a crank-driven power pump, the maximum plunger velocity is about 60% greater than the average calculated above. Because a direct-acting pump accelerates and decelerates quickly at both ends of the stroke, the simplex has a peak velocity which exceeds the average by a small amount. The duplex direct-acting pump is another case. Because each side is idle almost half the time, the peak velocity exceeds the average by almost 100%. For reasons unknown to the author, it is common practice to ignore this fact when comparing piston speeds and valve velocities of simplex and duplex direct-acting pumps.

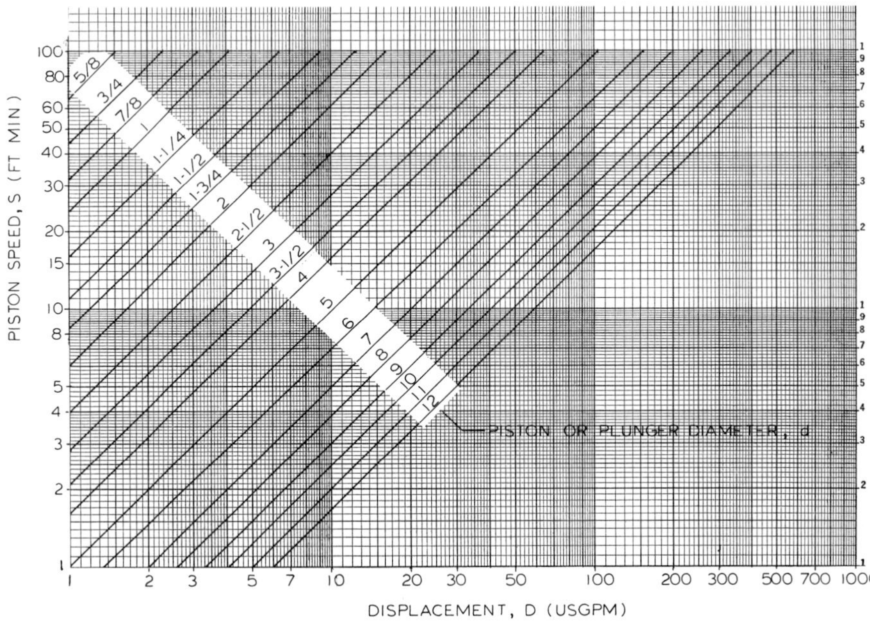

FIG. 6.24. Piston speed as a function of displacement for a simplex double-acting or duplex single-acting pump.

6.12. VALVE VELOCITY

Valve velocity is the average velocity of the liquid passing through the valve seat during the pumping stroke (except that, again, on duplex direct-acting pumps, the fact that each piston or plunger is idle almost half the time is ignored). Valve velocity can be calculated with either of the following equations:

$$V_v = \frac{S_p \, \eta_v}{\dfrac{A_v}{A_p}} \tag{6.15}$$

where

v_v = valve velocity, ft/min
S_p = plunger (or piston) speed, ft/min

η_v = volumetric efficiency of pump

$\dfrac{A_v}{A_p}$ = ratio of valve seat area to plunger area

EXAMPLE 6.4

The 2 × 3 triplex used in the above examples has a valve seat area of 3.54 in.[2] Dividing by a plunger area of 3.14 in.[2] produces an area ratio of 1.13. Therefore:

$$v_v = \frac{(165)(0.97)}{1.13} = \underline{142}\ \text{ft/min}$$

Another way to calculate valve velocity is given by the following:

$$v_v = \frac{38.5\ Q}{n_c\ A_v} \tag{6.16}$$

where

v_v = valve velocity, ft/min
Q = pump capacity, U.S. gal/min
n_c = number of pumping chambers in pump
A_v = valve seat area in each pumping chamber, in^2

EXAMPLE 6.5

Still using the 2 × 3 triplex from previous examples:

$$v_v = \frac{(38.5)(39.2)}{(3)(3.54)} = \underline{142}\ \text{ft/min}$$

As shown in Fig. 6.25, webs or wings sometimes block a portion of the flow area in the valve seat. The valve velocity through the seat should therefore be calculated using the net free area. Such is not always the case. In fact, the valve velocity calculated above for the triplex was calculated using the gross area because the manufacturer provided only that value for valve seat area.

Although the valve velocity, as defined above, does give an indication of the adequacy of the valve for the intended service, there are two other velocities in the valve assembly that are just as important. One is the velocity of the fluid as it passes between the valve and the seat of an open valve. The other is the velocity of the fluid between the outside diameter of the valve and the cylinder bore or valve retainer. Although many users

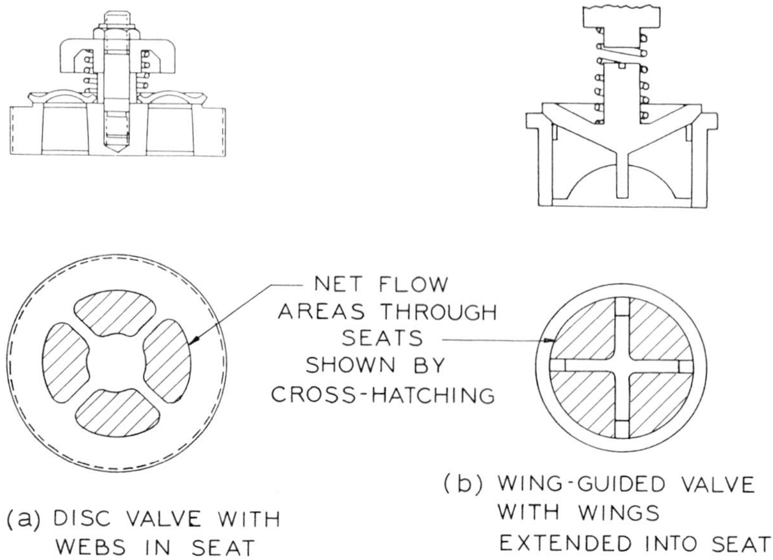

FIG. 6.25. Valve seat areas used to calculate valve velocities.

scrutinize the velocity through the seat, consideration of the other velocities seems to be left entirely to the discretion of the designer. The adequacy of the valve design, though, cannot be totally determined even by ascertaining all valve velocities. The proof is in the pump performance. If a pump gets high volumetric and mechanical efficiencies, runs quietly and smoothly, and valve life is satisfactory, then the design of the valve is adequate. An NPSH test is the best way to determine that characteristic, and must be established by the manufacturer prior to selection.

6.13. GAS DRIVE PRESSURE REQUIRED—DIRECT-ACTING PUMPS

When a direct-acting pump is sized during the selection process, seldom does the calculated required diameter of the drive piston exactly match a standard size. Therefore, the next larger standard size is selected. Because the diameter of the drive piston is then larger than required, the drive fluid (usually a gas) must be throttled to prevent the pump from running too fast. The actual chest pressure required to drive the pump may be calculated as follows.

Starting with Eq. 6.10:

$$\eta_m = \frac{(A_L)(\Delta p_L)}{(A_{dr})(\Delta p_{dr})}$$

Solving for Δp_{dr}–

$$\Delta p_{dr} = \frac{(A_L)(\Delta p_L)}{(A_{dr})(\eta_m)}$$

Also–

$$\Delta p_{dr} = p_i - p_e$$

where:

p_i = inlet drive pressure required
p_e = drive end exhaust pressure (established by the system)

Therefore

$$p_i - p_e = \frac{(A_L)(\Delta p_L)}{(A_{dr})(\eta_m)}$$

$$p_i = \frac{(A_L)(\Delta p_L)}{(A_{dr})(\eta_m)} + p_e$$

Substituting–

$$A_L = \frac{\pi}{4} d_L^2$$

and

$$A_{dr} = \frac{\pi}{4} d_{dr}^2$$

and canceling the $\pi/4$ factors,

$$p_i = \frac{(d_L^2)(\Delta p_L)}{(d_{dr}^2)(\eta_m)} + p_e$$

or

$$p_i = \left(\frac{d_L}{d_{dr}}\right)^2 \frac{\Delta p_L}{\eta_m} + p_e \qquad (6.17)$$

6.14. GAS CONSUMPTION OF DIRECT-ACTING PUMPS

Because the full chest pressure of the motive gas is maintained on the drive piston(s) throughout the entire stroke (on simple cylinders), gas consumption is calculated simply by calculating the displacement of the drive piston(s) and correcting with empirical factors. The following equations have been so determined, and are used by one manufacturer in determining gas consumption. For any gas–

$$Q_g = (C)(d_{dr})^2(S_p)\frac{520}{T_i}\frac{p_i}{14.7} \tag{6.18}$$

where

Q_g = gas consumption, standard ft^3/hr (SCFH)
C = constant taken from Table 6.4
d_{dr} = diameter of the drive piston, in.
S_p = piston speed, ft/min (as calculated with equations earlier in this chapter)
T_i = actual absolute temperature of gas entering the drive chest, in degrees Rankine (°F + 460)
p_i = actual absolute pressure of gas entering the drive chest, psia (as determined with a gage or calculated with Eq. 6.17)

It will be noted that the consumption is adjusted to standard conditions, 520°R (60°F) and 14.7 psia, by the last two fractions in the equation. If consumption at chest pressure conditions were desired, these last two fractions would not appear.

If the pump is driven by steam, the above equation will provide an accurate consumption figure, but steam is normally measured in weight/unit time, so the equation is changed to the following:

$$Q_s = (C)(d_{dr})^2(S_p)(w_i) \tag{6.19}$$

where

Q_s = steam consumption, lb/hr
w_i = specific weight of steam at chest pressure and temperature, lb/ft^3

Table 6.4. Constants for Use in Equations for Gas Consumption of Direct-Acting Pumps.

	Values of "C"	
Stroke Length (in.)	For Simplex Pumps	For Duplex Pumps
0—6	0.395	0.933
7—16	0.385	0.872
17 and up	0.372	0.839

EXAMPLE 6.6
Calculate the gas consumption of a 10 × 5 × 12 simplex pump running at 40 ft/min piston speed. Chest gas pressure is 200 psig (214.7 psia), at a temperature of 100° F (560° R). (Exhaust pressure is 75 psig.) Using Eq. 6.18:

$$Q_g = (0.385)(10)^2(40)\,\frac{520}{560}\,\frac{214.7}{14.7} = \underline{20{,}900}\ \text{SCFH}$$

This pump develops about 7 hhp on the liquid end. Gas consumption is therefore about 3000 SCFH/hhp, a fairly typical value.

EXAMPLE 6.7
If the above pump is driven by saturated steam at 135 psig inlet (10 psig exhaust), calculate steam consumption. Using Eq. 6.19:

$$Q_s = (0.385)(10)^2(40)(0.332) = \underline{511}\ \text{lb/hr}$$

As a function of the hydraulic horsepower, consumption is about 75 lb/hr/hhp, a fairly typical value.

6.15. STALL PRESSURE—DIRECT-ACTING PUMPS

The direct-acting pump has the desirable characteristic of "stalling" at a certain discharge pressure. As shown by the curve in Fig. 6.20, the mechanical efficiency can exceed 90% at zero speed (stall conditions). Since efficiency is highest at stall conditions, the maximum discharge pressure that the pump is capable of producing is at stall conditions. (See the discussion of "ram pressure" for the complete picture.) To be conservative in calculating the maximum stall pressure, the efficiency is taken as 100%, the inlet gas pressure is taken as its highest value, the exhaust gas pressure is taken as its lowest value, and the suction pressure is taken as its highest value. Figure 6.26 illustrates the concept. Solving for Δp_L from Eq. 6.10 yields:

$$\Delta p_L = \frac{\eta_m A_{dr}\,\Delta p_{dr}}{A_L} = \eta_m \left(\frac{d_{dr}}{d_L}\right)^2 \Delta p_{dr}$$

also:

$$p_d = p_s + \Delta p_L$$

combining:

$$p_d = p_s + \eta_m \left(\frac{d_{dr}}{d_L}\right)^2 \Delta p_{dr}$$

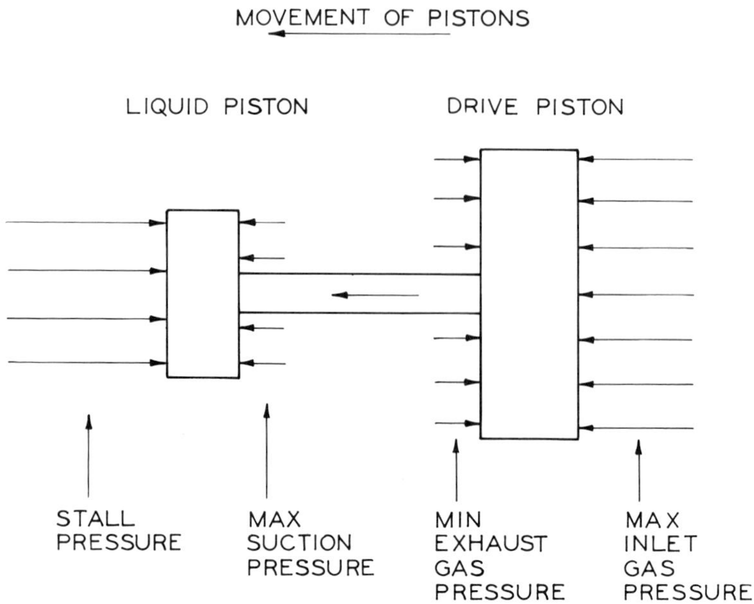

FIG. 6.26. For stall pressure calculations, mechanical efficiency is assumed to be 100%, and the other three pressures are taken at the extremes shown.

For stall conditions, we set $\eta_m = 1$, and use the maximum suction pressure and the maximum drive differential pressure. Therefore, stall pressure is calculated as follows:

$$p_{d\,\text{stall}} = p_{s\,\text{max}} + \left(\frac{d_{dr}}{d_L}\right)^2 \Delta p_{dr\,\text{max}} \tag{6.20}$$

EXAMPLE 6.8

Calculate the stall pressure for the 10 × 5 × 12 simplex pump in Example 6.6. It has been determined that the inlet gas pressure can be as high as 225 psig, that the exhaust pressure can go as low as atmospheric (0 psig), and that the suction pressure can be as high as 150 psig.

$$p_{d\,\text{stall}} = 150 + \left(\frac{10}{5}\right)^2 (225 - 0) = \underline{1050} \text{ psig}$$

Since this particular pump has a liquid end designed for a maximum pressure of only 600 psig, and the discharge system contains 300# flanges (with a maximum working pressure of approximately 720 psig), the discharge system must be equipped with a pressure-relief valve or rupture disc.

Some have attempted to apply the concept of stall pressure to power pumps. It cannot be done. If the discharge pressure were to be gradually increased on a motor-driven power pump, some portion of the pump or drive system would fail before any significant reduction in speed would occur. If motor starter heaters are sized close to the normal maximum load, the starter would disconnect the motor. Otherwise, a failure may occur in the motor, gears, belts, power end, or liquid end. The concept of stall pressure applies only to direct-acting pumps.

EXAMPLE 6.9

Now let's consider the same pump operating with the steam conditions of Example 6.7. It has been determined that the inlet steam pressure can be as high as 150 psig, the exhaust pressure can be as low as atmospheric, and the maximum suction pressure will be atmospheric. Calculate stall pressure:

$$p_{d\,\text{stall}} = 0 + \left(\frac{10}{5}\right)^2 (150 - 0) = \underline{600} \text{ psig}$$

One could logically conclude from all that's been said until now, that a pressure-relief device is not required for this system. But, alas! Life is not so simple. This particular pump has the capability of generating a discharge pressure *several times* larger than stall pressure. How? Read the next section on "ram pressure" carefully, and you shall see.

6.16. RAM PRESSURE—DIRECT-ACTING PUMPS

In the previous section we discussed stall pressure, the discharge pressure of a direct-acting pump operating at, or near, stall conditions (zero speed). Stall pressure *is* the maximum pressure possible from a pump which displaces equal volumes on both strokes (both away from and toward the drive end) such as a plunger pump (a double-acting unit with both plungers the same diameter) or a piston pump with a tail rod (a rod with a diameter equal to the piston rod, extending from the outboard face of the piston through an outboard stuffing box). But most direct-acting pumps have piston type liquid ends—and are *not* fitted w
Such a unit is capable of generating a discharge pressure seve
stall pressure.

How?

The rings or cups on a liquid piston *leak*. New parts leak less, worn parts leak more. Consider the case of the discharge block valve closed, and the pump stroking toward the liquid end, as illustrated in Fig. 6.27. Initially, stall pressure will be created on the left face of the piston, but then some of this liquid will leak past the piston to the right side. Since the displacement on the right side of the piston is less than displacement on the left side, more liquid leaks out of the left pumping chamber than can be absorbed by the right chamber. If the suction valve for the right chamber seals tightly (it will if it is in good condition), the only escape for the liquid is through the discharge valve or stuffing box. The discharge valve offers little relief because the discharge block valve is closed. If the packing holds sufficiently, the pressure in the right chamber will start to increase. As this pressure increases, the differential pressure across the liquid piston is reduced, and the piston will be pushed farther to its left, increasing the discharge pressure more, and causing more liquid to leak into the right chamber. In a "worst-case" scenario, the pressure in the right chamber will equal the pressure in the left chamber, and the only hydraulic force generated in the liquid end, to resist the force from the drive end, will be discharge pressure times the area of the piston *rod*. In fact, to calculate ram pressure, one may ignore the piston, and consider

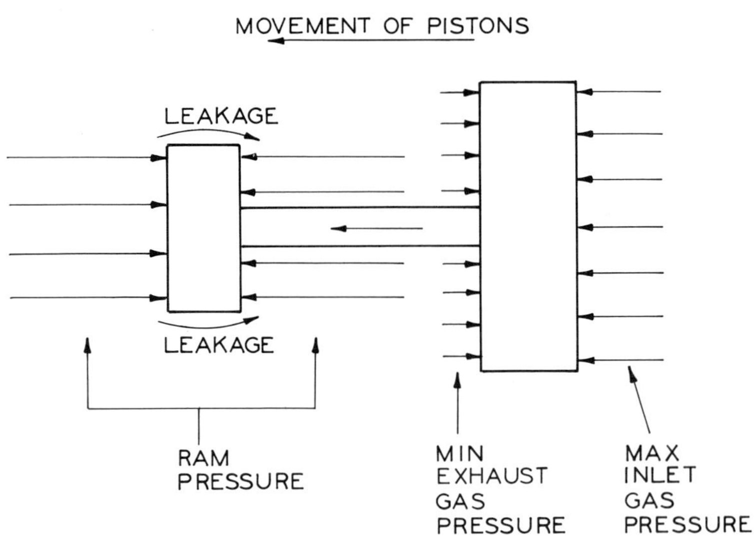

FIG. 6.27. At stall conditions, with the piston moving away from the drive end, leakage of liquid around a piston *without* a tailrod reduces the effective pumping area to that of the piston *rod*. The result is *ram pressure*.

the piston rod as a single-acting plunger. The equation to calculate ram pressure is therefore:

$$p_{d\ \mathrm{ram}} = \left(\frac{d_{dr}}{d_r}\right)^2 \Delta p_{dr\ \max} \tag{6.21}$$

where:

d_r = diameter of liquid piston rod, in.

(Suction pressure does not contribute to ram pressure.)

EXAMPLE 6.10

The pump of Ex. 6.9 has a piston rod diameter of 1.5 inches. Ram pressure is therefore calculated as follows:

$$p_{d\ \mathrm{ram}} = \left(\frac{10}{1.5}\right)^2 (150 - 0) = \underline{6670} \text{ psig}$$

Therefore, in this particular example, ram pressure exceeds stall pressure by a factor of more than ten. A pressure-relief device is an absolute requirement. Therefore, all standard direct-acting piston pumps should be fitted with a relief valve, because they are all capable of generating ram pressure. Only pumps with equal displacements on both strokes should be considered for operation without a pressure-relief device.

It should be noted that liquid compressibility could reduce the actual pressure obtainable during ram conditions. Also, various pump or system components will likely fail before stall pressure is achieved. The ram pressure calculated with Eq. 6.21 is the pressure which the drive end is *capable* of producing, not necessarily the pressure actually achieved.

6.17. SIZING THE DRIVE PISTON

The size of the liquid end of the pump is selected to satisfy the capacity requirements. The size of the gas end is selected, based on differential gas pressure available, to provide the force necessary to drive the liquid end. To select the size of the drive piston, Eq. 6.10 can be rewritten as follows:

$$A_{dr} = \frac{(A_L)(\Delta p_L)}{(\eta_m)(\Delta p_{dr})}$$

$$d_{dr} = d_L \left[\frac{\Delta p_L}{(\eta_m)(\Delta p_{dr})}\right]^{1/2} \tag{6.22}$$

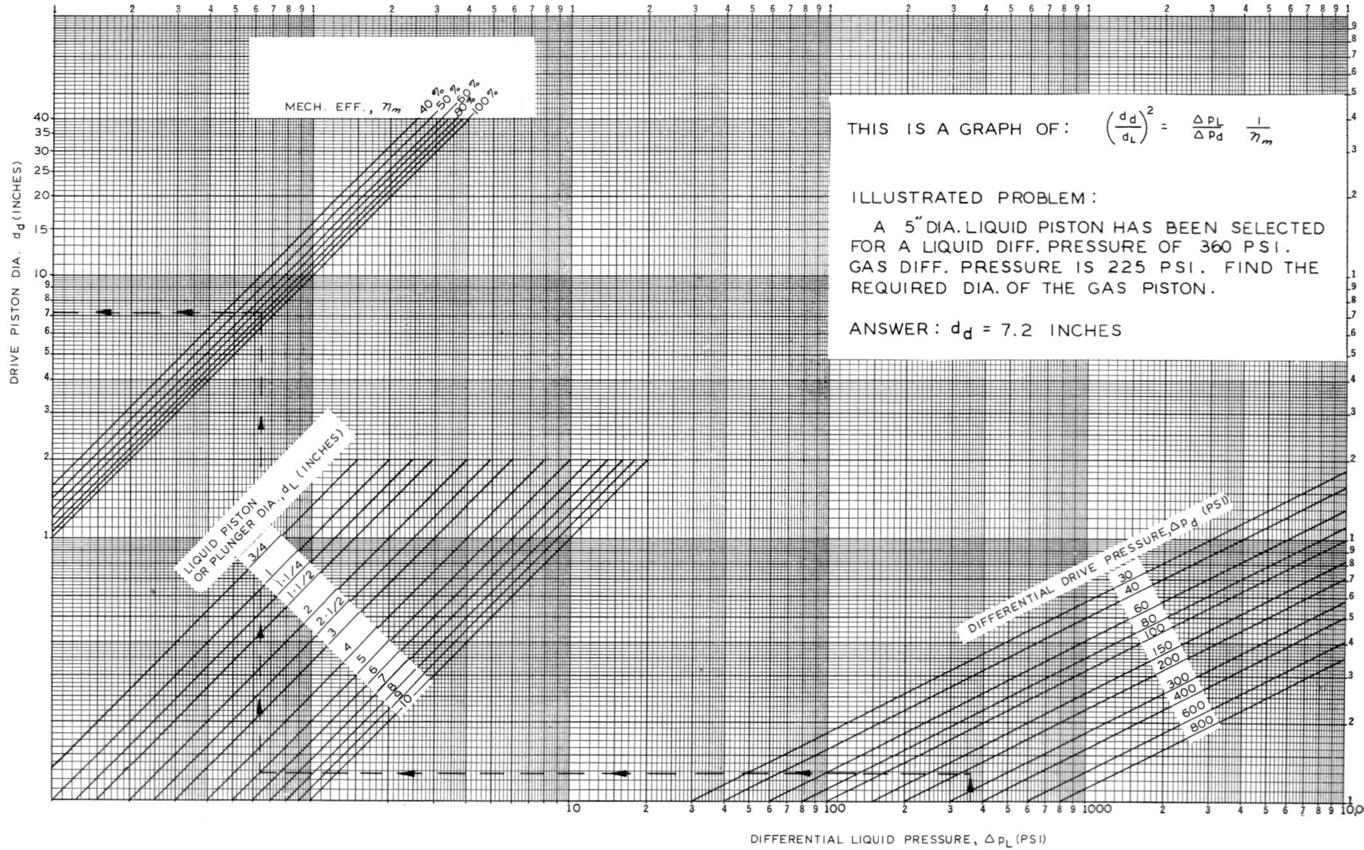

FIG. 6.28. Comprehensive sizing and performance chart for direct-acting pumps.

The solution of this equation is shown in Fig. 6.28 in a chart produced by the author. This chart can be used to select the required diameter of the drive piston, to determine the actual gas differential pressure required, to determine the stalling pressure of the pump selected, and to determine the ram pressure of a piston pump with no tail rod.

6.18. COMPREHENSIVE EXAMPLE FOR A DIRECT-ACTING PUMP (EXAMPLE 6.11)

As an example, let's assume that a 5 × 12 simplex, double-acting piston liquid end has been selected to pump 40 U.S. gpm of a light hydrocarbon from a suction pressure of 140 psig to a discharge pressure of 500 psig. Gas is available at an inlet pressure of 300 psig. Exhaust pressure is 75 psig. Using $\eta_v = 0.90$, Eq. 6.14 allows us to calculate the piston speed:

$$S_p = \frac{49Q}{d_L{}^2 n_c \eta_v} = \frac{(49)(40)}{(5)^2(2)(0.90)} = \underline{43.6} \text{ ft/min}$$

This could have been obtained from Fig. 6.24 by entering the abscissa at a displacement value of 44.4 gpm, moving up to the diagonal line for the 5 in. piston, then proceeding left to the ordinate to read about $\underline{43\frac{1}{2}}$ ft/min.

Rearranging Eq. 6.13 we can calculate the pump operating frequency in cycles per minute (cpm).

$$N = \frac{6\ S_p}{L} = \frac{(6)(43.6)}{12} = \underline{21.8} \text{ cpm}$$

By referring to Fig. 9.3 we can see that this value is below the lower line. We can therefore comfortably use an efficiency value from Fig. 6.22 for piston pumps. For a 12 in. stroke, the efficiency is 77%. We can now calculate the required diameter of the drive piston with Eq. 6.22.

$$d_{dr} = (5)\left[\frac{360}{(0.77)(225)}\right]^{1/2} = \underline{7.2} \text{ in.}$$

This same result could be obtained by entering Fig. 6.28 at $\Delta p_L = 360$ psi, moving up to $\Delta p_{dr} = 225$ psi, moving left to $d_L = 5$ in., moving up to 77%, then moving left to read $\underline{7.2}$ in.

Since a standard drive end is not available with a 7.2 in. diameter piston, it is necessary to select the next larger standard size. This drive end will therefore be equipped with an 8 in. diameter piston. Because the drive piston is larger than required, the inlet gas must be throttled. Equation 6.17 can be used to calculate the inlet pressure required.

$$p_i = \left(\frac{d_L}{d_{dr}}\right)^2 \frac{\Delta p_L}{\eta_m} + p_e$$

$$= \left(\frac{5}{8}\right)^2 \frac{360}{0.77} + 75 = 183 + 75 = \underline{258} \text{ psig}$$

The same result could have been obtained by using Fig. 6.28. Entering from the left at $d_{dr} = 8''$, move to the right to 77%, move down to $d_L = 5''$, then move to the right until you intersect $\Delta p_L = 360$ psi. The point of intersection is at about $\Delta p_{dr} = 180$ psi. To this number would be added the 75 psig exhaust pressure to obtain the inlet pressure requirement of about 255 psig.

Now let's calculate stall pressure for this pump, then use the chart to check it. Using Eq. 6.20:

$$p_{d\,\text{stall}} = 140 + \left(\frac{8}{5}\right)^2 (225) = 140 + 576 = \underline{716} \text{ psig}$$

Using the chart, enter from the left at $d_{dr} = 8$ in., move to right to 100%, move down to $d_L = 5$ in., move to right to $\Delta p_{dr} = 225$, then move down to read about 570 psi liquid end differential. To this we add the 140 psig suction pressure to obtain a stall pressure of about 710 psig—a good check with the calculated value.

Our last chore is to calculate ram pressure, then compare it to the value obtained from the chart. The piston rod diameter is 1.5 inches. Using Eq. 6.21:

$$p_{d\,\text{ram}} = \left(\frac{d_{dr}}{d_r}\right)^2 \Delta p_{dr\,\text{max}} = \left(\frac{8}{1.5}\right)^2 (225) = \underline{6400} \text{ psig}$$

We use the chart in the same fashion as for stall pressure, except we let $d_L = 1.5$ in., and we don't add in suction pressure. From the chart we read 6300 psig—a good check.

Is a relief valve required? Absolutely! Even if this pump were to be fitted with a tail rod to eliminate the possibility of ram pressure, the 716 psig stall pressure exceeds the 600 psig working pressure of the unit. If exhaust pressure would drop to 0, stall pressure would be 910 psig, even more justification for the pressure relief valve.

Chapter 7
ACCELERATION HEAD

7.1. PRESSURE PULSES

A question often asked by users and potential users of reciprocating pumps is:

"How much pressure pulsation will I get with a reciprocating pump?"

Unfortunately, this question cannot be readily answered. It is true that the flow in both the suction and discharge pipes of a reciprocating pump is of an uneven nature (as shown in Chapter 6), and that the liquid *velocity* variations can be fairly accurately predicted (more so with power pumps than with direct-acting units), but the conversion of these velocity variations into *pressure fluctuations* is a function of the system and liquid characteristics. If, for instance, the pump is discharging directly into a large vessel (a discharge line of negligible length) with a free liquid surface, such as that shown in Fig. 7.1, the pressure in the pump discharge nozzle will be steady (no variation), even if the pump is a simplex (which has a large velocity variation). On the other hand, a relatively smooth-running unit, such as a quintuplex, can experience sizable pressure pulses if the discharge pipe is long and of a small diameter, especially if the liquid is relatively incompressible, such as water.

This variation in pressure in the discharge pipe, as shown in Fig. 7.2, is caused by two factors—the friction in the pipe, and the inertia of the fluid. In low speed recips, the variation in friction can be significant, but in moderate to high-speed units, the inertial forces are usually dominant. This pressure fluctuation caused by inertial forces is called "acceleration head" because it is the acceleration (and deceleration) of the fluid which causes this portion of the pressure fluctuation.

(If each individual pumping chamber is not totally primed, or if a valve is malfunctioning due to wear or a broken or weak spring, pulsations can be severe. Before attacking acceleration head as the cause of pulsations, the user should be sure that his pump is functional and fully primed.)

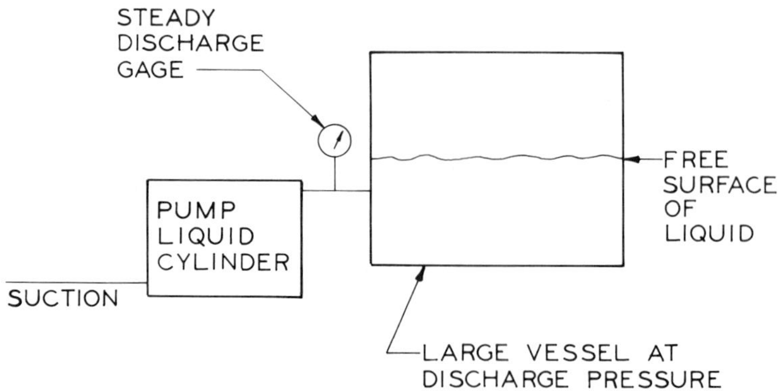

FIG. 7.1. A reciprocating pump with a very short discharge line.

7.1.1. Analyzing the Flow Curves

Let's look back at the flow curves of our popular triplex shown in Chapter 6, which, for convenience, is repeated here in Fig. 7.3. As noted in Chapter 6, the triplex actually produces *six* pulses per revolution because of the overlapping of plungers on both the suction and discharge strokes. With six pulses per revolution, a pump running 300 rpm produces 1800 pulses/min, or 30 pulses/sec (30 Hz), a relatively high frequency for fluid pulsation.

As noted on the second pair of humps in Fig. 7.3, when the curve is moving up (velocity increasing), the liquid is accelerating. When the curve is moving down (velocity decreasing), the liquid is decelerating.

Because most problems with reciprocating pumps, as with all pumps, are caused by inadequate system conditions on the *suction* side of the

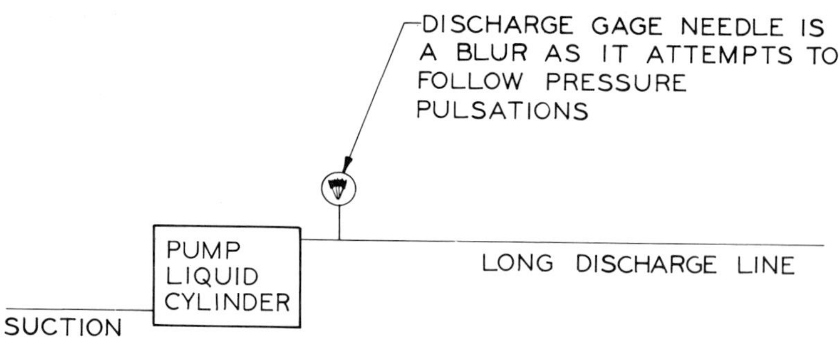

FIG. 7.2. A reciprocating pump with a long discharge line.

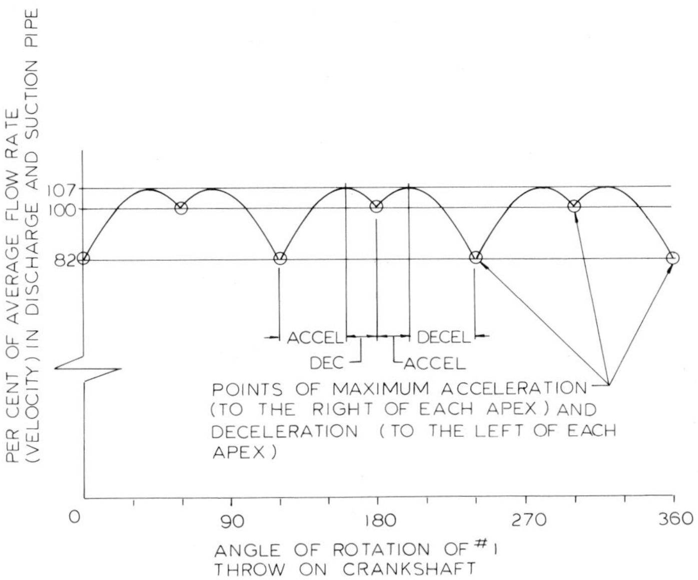

FIG. 7.3. Flow variations in the suction and discharge pipes of a triplex power pump.

pump, and because acceleration head is the major contributor to suction problems with recip pumps, the industry has concentrated on the effect that acceleration head has on suction pressure.

When liquid in the suction line is accelerating, to accommodate the pump, the energy required to produce the acceleration must come from the liquid itself. Since the liquid in the suction line has not yet reached the pump, the pump cannot add any energy to it. It is, therefore, entirely up to the suction system to provide the energy required to accelerate the entire column of liquid from the suction tank to the pump (unless the system contains some type of surge chamber, which will be discussed later). If sufficient energy is not available from the suction system, the pressure in the suction pipe (or inside the pump) will fall below the vapor pressure, causing a portion of the liquid to flash (boil) to a vapor (even if it's cold water). If vapor is ingested by the pump, it will often cause the pump to operate with a knocking sound and send shock waves into the discharge line. In mild cases, cavitation damage may occur to the piston or plunger face, and to other internal liquid end components such as valves, springs, cylinder head extensions, and internal cylinder surfaces. In moderate cases, additional problems may manifest themselves in the form of reduced life of packing, gaskets, valve seating surfaces, and power end components such as bearings. In severe cases, in addition to the above problems, liquid cylinders, crankshafts, and gears may be broken.

Some people in this field have spoken of acceleration head as being a loss, and it *is* treated as such when calculating NPSHA, but the drop in pressure caused by the acceleration is offset by the increase in pressure when the liquid decelerates. Therefore, the *average* pressure in the suction line is calculated without a consideration of acceleration head.

7.2. CALCULATION OF ACCELERATION HEAD

7.2.1. The Traditional Equation

A number of technical documents have offered methods to calculate the reduction in suction pressure caused by acceleration head, the most widely recognized of these probably being the Hydraulic Institute Standards.[1] The H.I. equation for power pumps is as follows:

$$h_a = \frac{LvNC}{Kg} \tag{7.1}$$

where

h_a = acceleration head, ft
L = actual length of suction line (not equivalent length), ft
v = velocity of liquid in suction line, ft/sec
N = rotative speed of crankshaft, rev/min
C = a constant, depending on the type of pump:
= 0.400 for a simplex single-acting
= 0.200 for a simplex double-acting
= 0.200 for a duplex single-acting
= 0.115 for a duplex double-acting
= 0.066 for a triplex, either single- or double-acting
= 0.040 for a quintuplex, either single- or double-acting
= 0.028 for a septuplex, either single- or double-acting
= 0.022 for a nonuplex, either single- or double-acting
K = a constant which compensates for compressibility of the liquid:
= 1.4 for deaerated water (relatively incompressible)
= 2.5 for hydrocarbons with high relative compressibility
g = gravitational constant (32.2 ft/sec^2)

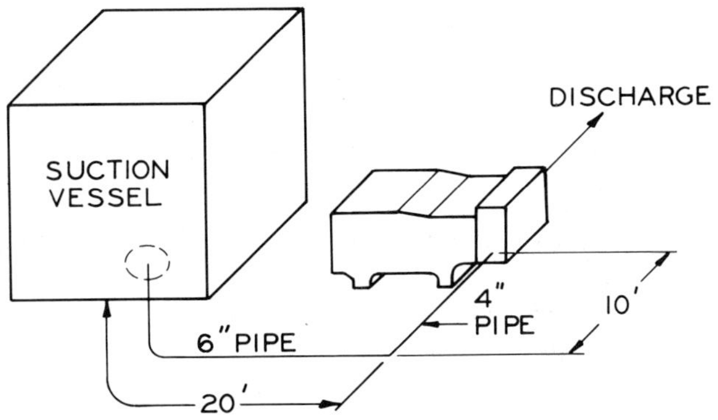

FIG. 7.4. The system analyzed for acceleration head in Example 7.1.

EXAMPLE 7.1

An example of the use of this equation is as follows (See Fig. 7.4): A $2\frac{1}{2} \times 5$ triplex pump has a crankshaft speed of 250 rpm and pumps 75 gpm of cool water. The suction line consists of 20 ft of 6-in. schedule 40 pipe and 10 ft of 4-in. schedule 40 pipe. The average velocity in the 6-in. pipe is 0.83 ft/sec, and the average velocity in the 4-in. pipe is 1.89 ft/sec. We'll calculate the acceleration head in each section of pipe, then add the two together.

For both sections of pipe, the same N, C, g, and K values apply:

$$
\begin{aligned}
N &= 250 \text{ rpm} \\
C &= 0.066 \text{ (a triplex)} \\
g &= 32.2 \text{ ft/sec}^2 \\
K &= 1.5 \text{ (a good value for most liquids)}
\end{aligned}
$$

Therefore:

$$h_a = \frac{Lv\,(250)(0.066)}{(32.2)(1.5)} = 0.341\ Lv$$

For the 6-in. pipe:

$$h_a = (0.341)(20)(0.83) = 5.66 \text{ ft}$$

For the 4-in. pipe:

$$h_a = (0.341)(10)(1.89) = 6.45 \text{ ft}$$

Total $h_a = 5.66 + 6.45 = \underline{12.11}$ ft
Converting to psi:

$$\text{Total } h_a = \frac{(12.11)(1.0)}{(2.31)} = \underline{5.25} \text{ psi}$$

This means that, at the time the fluid in the suction line is being accelerated at the maximum rate, the pressure drop, below the average suction pressure, is about 5 psi. This energy must be supplied by the suction system. The 5 psi subtracts from the amount of NPSH available (see Chapter 8 for a discussion of NPSH).

As implied above, the NPSH requirement of a power pump is normally stated in pressure units, such as psi, yet the H.I. equation provides results in feet, which must then be converted to psi. To avoid this extra step, we can alter Eq. 7.1 as follows:

$$h_a = \frac{L\,v\,N\,C\,(SG)}{K\,g\,(2.31)} \text{ answer in psi} \qquad (7.2)$$

7.2.2. The One-Step Equation

If we wish to avoid the extra step requiring the determination of velocities of liquid in the piping, we can substitute the following into Eqs. 7.1 or 7.2:

$$v = \frac{0.321\,Q}{A} = \frac{0.321\,Q}{\frac{\pi}{4}\,d_p^2} = 0.409\,\frac{Q}{d_p^2}$$

where:

v = velocity of liquid in pipe, ft/sec
Q = average flow rate, U.S. gpm
d_p = diameter of pipe for which we're calculating the acceleration head, in.

Substituting into Eq. 7.2, and also substituting 32.2 for g, yields

$$h_a = \frac{L\,0.409\,Q}{d_p^2}\;\frac{N\,C\,SG}{K(32.2)(2.31)} = \frac{L\,Q\,N\,C\,SG}{182\,d_p^2\,K} \text{ answer in psi} \qquad (7.3)$$

A couple of additional adjustments can be made to this equation to make it more "user-friendly." Since the triplex is the most popular type of power pump, the 0.066 "C" value can be rolled into the "182" in the denominator, creating a new "C" (let's call it C') which is unity for the triplex. We can also include the "1.5" value for most liquids in the denominator constant, creating a new constant K'. The equation then becomes:

$$h_a = \frac{L\,Q\,N\,C'\,SG}{4140\,d_p^2\,K'} \quad \text{answer in psi} \tag{7.4}$$

where

h_a = acceleration head, psi
L = actual length of suction pipe, ft (not equivalent length)
Q = pump capacity, U.S. gpm
N = revolutions per minute of pump crankshaft
C' = 3.00 for a duplex single-acting or simplex double-acting pump
= 1.74 for a duplex double-acting pump
= 1.00 for a triplex pump, either single- or double-acting
= 0.61 for a quintuplex pump, either single- or double-acting
= 0.42 for a septuplex pump, either single- or double-acting
= 0.33 for a nonuplex pump, either single- or double-acting
SG = specific gravity of liquid
d_p = inside diameter of suction pipe, in.
K' = 1.0 for most liquids
= 1.5 for very compressible liquids (such as ethane)

For a triplex pump, pumping most liquids, the equation reduces to the following:

$$h_a = \frac{L\,Q\,N\,SG}{4140\,d_p^2} \tag{7.5}$$

If we use this equation to calculate the acceleration head in our previous example, it is unnecessary to calculate the velocities, and two factors are eliminated.

For the 6-in. pipe, d_p = 6.065 in.

$$h_a = \frac{(20)(75)(250)(1.0)}{(4140)(6.065)^2} = \underline{2.46} \text{ psi}$$

For the 4-in. pipe, $d_p = 4.026$ in.

$$h_a = \frac{(10)(75)(250)(1.0)}{(4140)(4.026)^2} = \underline{2.79} \text{ psi}$$

Total $h_a = 2.46 + 2.79 = \underline{5.25}$ psi

EXAMPLE 7.2

Now what if we have a second identical pump running in parallel with the above pump, as illustrated in Fig. 7.5? In this case, pump "B" is not an idle standby unit. Both pumps run simultaneously. We'll divide the pipe into sections "C," "D," "E," and "F," and analyze each individually.

Pipe "C":

These two pumps will not run at exactly the same speed, so there will be times when their pulses are out of phase, and times when they'll be right in phase. When they're in phase, pipe "C" will think it is feeding one large triplex, with a capacity equal to the sum of the two pumps, running at 250 rpm. The acceleration head is therefore twice that calculated in the previous example for this pipe.

$$h_a = 2 \times 2.46 = \underline{4.92} \text{ psi}$$

Pipe "D":

The acceleration head in this pipe will be the same as in the previous example.

$$h_a = \underline{2.79} \text{ psi}$$

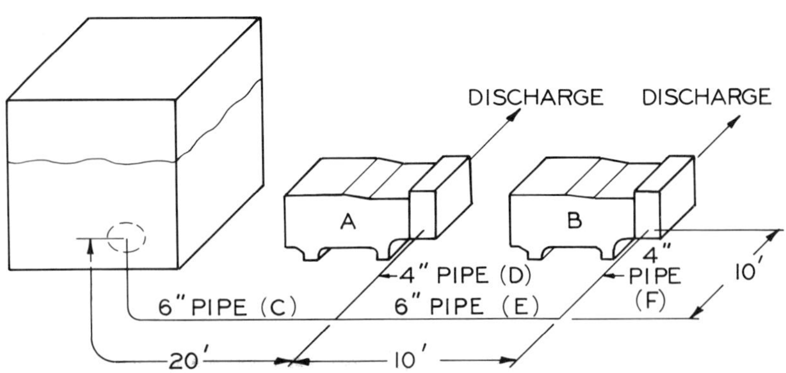

FIG. 7.5. A system containing two power pumps in parallel—Analyzed in Example 7.2.

Pipe "E":

Using Eq. 7.5—

$$h_a = \frac{L\,Q\,N\,(SG)}{4140\,d_p^2} = \frac{(10)(75)(250)(1.0)}{(4140)(6.6065)^2} = \underline{1.23}\text{ psi}$$

Pipe "F":

The acceleration head in pipe "F" is the same as in pipe "D."

$$h_a = \underline{2.79}\text{ psi}$$

For Pump "A":

The acceleration head at the inlet connection of pump "A" is the sum of the acceleration heads of pipes "C" and "D."

$$h_a = 4.92 + 2.79 = \underline{7.71}\text{ psi}$$

For Pump "B":

The acceleration head at the inlet connection of pump "B" is the sum of the acceleration heads in pipes "C," "E," and "F."

$$h_a = 4.92 + 1.23 + 2.79 = \underline{8.94}\text{ psi}$$

Therefore, pump "B" will be exposed to the higher acceleration head, which will result in a lower suction pressure. A pressure gage mounted at the end of pipe "F" (adjacent to the pump) would try to indicate a pressure that drops about 9 psi below the average suction pressure (and also rises about 9 psi above the average suction pressure) 1500 times per minute (25 times per second). The needle would be a blur. Snubbing the gage would steady the reading at the average suction pressure, masking the effect of the acceleration head. It is necessary to use a pressure transducer and an oscilloscope to get an accurate picture of suction pressure.

7.3. EFFECT OF PUMP SPEED ON ACCELERATION HEAD

It is important that we recognize the pronounced effect that pump rotative speed can have on acceleration head. Examining Eq. 7.1, 7.2, or 7.3 we can see that if the suction line length and liquid velocity are held constant, acceleration head is directly proportional to rotative speed. Often, though, speed on an existing pump is increased to increase unit through-put. The suction pipe is not changed to maintain constant velocity. Pump speed is increased by changing sheaves or gears, and/or the driver. In this case, suction line velocity is proportional to pump speed, and acceleration head therefore increases as the *square* of pump speed. For example, if pump speed is increased 50%, suction line velocity also increases 50%. The result is a 125% increase in acceleration head ($1.5 \times 1.5 = 2.25$), more than *double* the original acceleration head.

Another approach to evaluating the effect of speed on acceleration head is to rearrange Eq. 7.1 to solve for v as a function of the other parameters.

$$v = \frac{Kg}{CN}\frac{h_a}{L} \tag{7.6}$$

Now we can plug in some typical values. The value for K is typically 1.5, and $g = 32.2$. A typical value for h_a/L is 0.7. (For each foot of suction pipe length, acceleration head is 0.7 ft.)

Equation 7.6 then becomes:

$$v = \frac{(1.5)(32.2)(0.7)}{CN} = \frac{33.8}{CN} \tag{7.7}$$

We can now plot typical values of v as a function of N for different types of power pumps, as shown in Fig. 7.6. These curves graphically demonstrate that, as pump speed is increased, suction line velocity must be proportionately reduced, to retain the same acceleration head. (Some of the higher velocities shown in the low speed range would possibly be unacceptable because of high friction loss.)

7.4. SMALL HIGH-SPEED PUMP vs LARGE LOW-SPEED PUMP

A common misconception about power pumps is that a small unit, with small plungers and a short stroke-length, and running faster to obtain an equivalent capacity, will produce smaller pressure pulsations than a larger, lower-speed unit, pumping the same capacity. This *seems* reasonable because each small plunger transfers a smaller quantity of liquid from the suction pipe to the discharge pipe during each revolution. Analysis of the velocity flow curves though, reveals the fallacy of this concept. A triplex pump, for instance, has a flow variation of 25%, regardless of its size or speed. A small pump will therefore produce as much velocity variation as a large unit, and the higher speed of the small unit will produce more pulses per second, and therefore greater liquid acceleration, resulting in higher pressure pulses. A smaller high-speed unit therefore produces larger pressure pulsations than a larger low-speed unit.

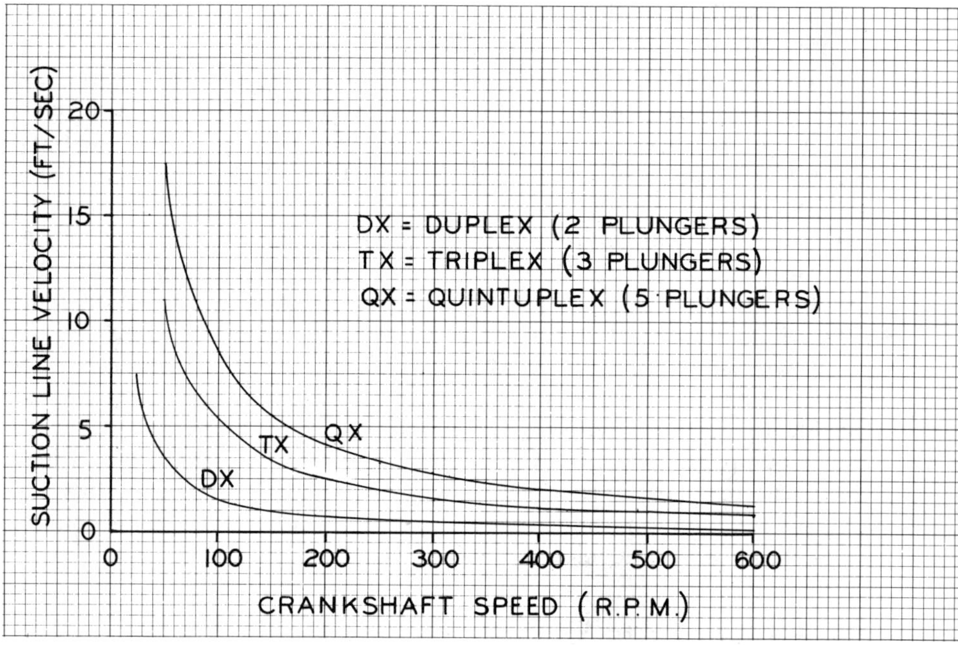

FIG. 7.6. Suction line velocity for constant acceleration head, as a function of pump type and speed.

This is confirmed by Eq. 7.1. If we have the same size pipe connected to both units, resulting in the same average liquid velocity v, the acceleration head will be proportional to pump crankshaft speed N. Let's say that the large unit speed is 200 rpm, and the small unit runs 400 rpm to achieve the same capacity. With the same size pipe, the small unit would produce twice the acceleration head. Conversely, to have the same acceleration head with the two pumps, the small one would require a pipe with twice the flow area (41% larger in diameter). If the large pump requires, say, a 4 in. suction pipe, the small pump requires a 6 in. pipe. The small pump, being a higher-speed unit, would also probably require stiffer valve springs for smooth operation, which would increase the NPSH required. (More on this in Chapter 8.)

7.5. INACCURACY OF ACCELERATION HEAD CALCULATIONS

Before we get too pleased with ourselves, thinking that we can accurately calculate acceleration head with the HI equation, we must recognize that this equation is not very accurate. It attempts to compensate for liquid compressibility (elasticity) by varying the value of "*K*," but the effect of compressibility is more complex. Computer programs are used by system consultants to more accurately predict pressure variations in piping systems. The HI equation is, at best, an approximation of the actual acceleration head. Miller[2] reported that his tests indicated acceleration head to be much less than calculated with the HI equation. Some field installations also operate satisfactorily with NPSHA considerably less than indicated necessary by this equation. The reasons for these discrepancies are not known, but could be numerous. Any gas, such as air, trapped in a high point of the suction line, will act as an accumulator, absorbing the pulsations of the pump, and reducing acceleration head. Any gas entrained in the liquid would affect the system in a similar fashion. The compressibility of the liquid itself may act favorably to absorb pulsations. It also may be that the liquid *is* actually cavitating, but not enough to significantly affect pump performance. Cavitation will affect performance less if the discharge pressure is relatively low. The author's experience has been that the HI equation is always safe—providing an answer that is equal to, or exceeds, actual acceleration head.

7.6. EFFECT OF STABILIZING THE FLOW

It should also be pointed out that an effective suction pulsation dampener (mounted in the suction pipe adjacent to the pump liquid end) will eliminate most of the velocity variation in the suction line, and therefore will eliminate most of the acceleration head. The flow-through type is usually more effective than the appendage-type dampener, but some systems successfully use no more than a simple stand-pipe filled with gas. The HI Standards state that an effective suction line dampener will reduce acceleration head to that equivalent to a suction line length of 5 to 15 pipe diameters. Using the higher number (15) and looking back at Fig. 7.5, for either pipe "D" or "F," the effective length becomes $15 \times 4 = 60$ inches (5 feet), which results in $h_a = 1.4$ psi for both pumps. To achieve such a result, each pump must be equipped with its own suction pulsation dampener. A single dampener in the common suction line would be less effective.

Some pump operators have reported that "suction stabilizers," which are designed to also separate and retain gas, have, to their surprise, required periodic venting. If the stabilizer had not been in the suction line, the vented gas would have been ingested by the pump, possibly resulting in shock operation, or, in the extreme case, causing one or more pumping chambers to become "gas-bound" or "vapor-locked." Without the stabilizer, the agitation in the suction line would have been greater, and more gas would probably have been liberated. The pressure shocks caused by ingested gas can cause failure of pump and system components.

7.7. DERIVATION OF THE ACCELERATION HEAD EQUATION

For those who wish to see how the HI acceleration head equation can be derived, we offer the following. Using Newton's Second Law ($F = ma$), we will calculate the force required to accelerate the column of liquid in the suction pipe, then convert that force to head units. As mentioned earlier, such an approach yields, at best, an approximation because it does not properly incorporate the effect of liquid compressibility.

The equation will be derived by considering the total mass of pumpage in a section of pipe (see Fig. 7.7) as moving in unison, i.e. when a particle at one end of the pipe accelerates, a particle at the other end accelerates simultaneously.

The mass of the pumpage may be calculated as follows:

$$m = \frac{W}{g} = \frac{(V)(w)}{g} = \frac{(A_L)(L_L)(w)}{g}$$

where

m = mass of pumpage in pipe length L, slugs
W = weight of liquid column in pipe, lbs
V = volume of liquid column in pipe, ft^3

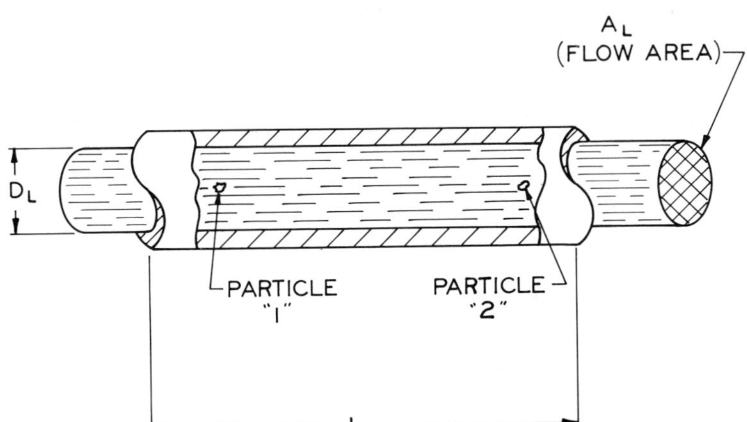

FIG. 7.7. The traditional acceleration head equation is based on all pumpage in a pipe moving in unison. (Particles "1" and "2" accelerate and decelerate in unison.)

A_L = cross-sectional area of liquid in pipe, ft^2
L_L = length of liquid column in pipe, ft
w = specific wt. of liquid, lbs/ft^3
g = acceleration of gravity, 32.2 ft/sec^2

And since: $F = ma$,

where

a = acceleration of liquid in pipe, ft/sec^2,

then:

$$F = \frac{(A_L)(L_L)(w)(a)}{g}$$

Converting to pressure units:

$$p_a = \frac{F}{144\ A_L} = \frac{(L_L)(w)(a)}{144\ g} \tag{7.8}$$

where p_a = pressure required (or created) by an acceleration (or deceleration) of the liquid, psi.

The acceleration "a" of the liquid may be determined as follows: For simplicity, we'll use a pure sine-wave flow curve, as shown in Fig. 7.8. Liquid velocity may be calculated as a function of pipe I.D., plunger diameter, and plunger velocity. Plunger velocity is a function of stroke length and pump speed.

$$v_p = \frac{\pi L_s\ N}{(60)(12)} \sin\theta = \frac{L_s\ N}{229} \sin\theta$$

where

v_p = plunger velocity, ft/sec
L_s = length of stroke, in.
N = speed of crankshaft, rpm
θ = angle of crank (starting from bottom-dead-center for discharge stroke, or top-dead-center for suction stroke)

Since liquid velocity in the pipe (due to the action of one plunger), v_L, is inversely proportional to the ratio of pipe flow area, A_L, to plunger area, A_p,

$$\frac{v_L}{v_p} = \frac{A_p}{A_L} = \frac{d_p^2}{d_L^2} = \left(\frac{d_p}{d_L}\right)^2$$

where

d_p = plunger diameter, in.
d_L = liquid column diameter, in.

Working with θ in degrees and t in seconds, $\theta = 6\ Nt$. To perform differentiation in the next operation, it is necessary to convert θ to radians, such that $\theta = \omega t = (\pi/30)\ Nt$.

Therefore, combining the last two equations:

$$v_L = \left(\frac{d_p}{d_L}\right)^2 \frac{L_s N}{229} \sin \frac{\pi}{30} Nt$$

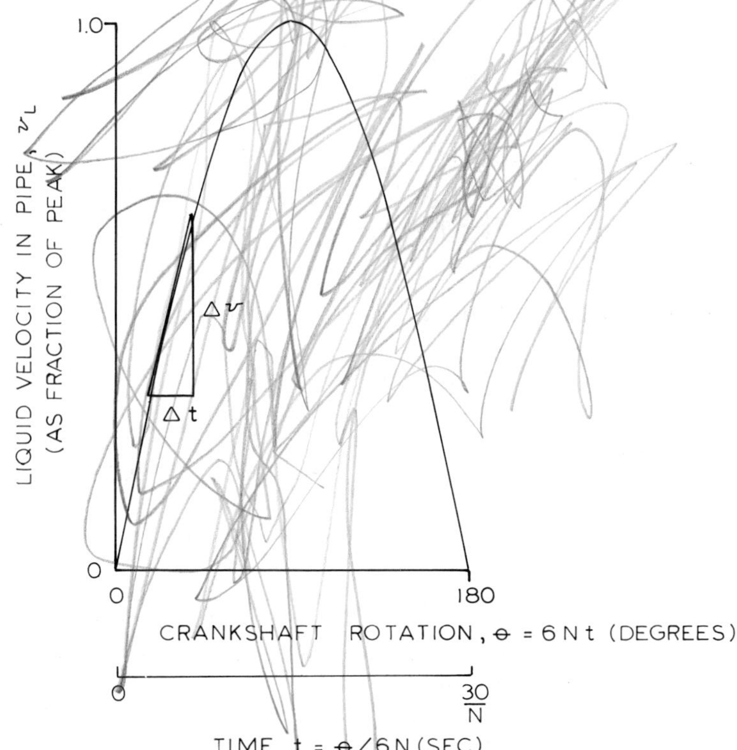

FIG. 7.8. The velocity in the suction or discharge pipe of a power pump due to the motion of one plunger (shown as a pure sine function).

Since acceleration is the time rate of change of velocity, acceleration can be determined by measuring the slope of the velocity-vs-time curve in Fig. 7.8. Figure 7.9 plots the slope of Fig. 7.8.

Using a little calculus, we can convert the above velocity equation into acceleration as follows:

$$a = \frac{dv_L}{dt} = \left(\frac{d_p}{d_L}\right)^2 \frac{L_s\, N}{229} \frac{\pi}{30} N \cos \frac{\pi}{30} Nt$$

$$a = \left(\frac{d_p}{d_L}\right)^2 \frac{L_s\, N^2}{2187} \cos \frac{\pi}{30} Nt$$

$$a = \left(\frac{d_p}{d_L}\right)^2 \frac{L_s\, N^2}{2187} \cos \theta \tag{7.9}$$

Plugging into Eq. 7.8:

$$p_a = \frac{L_L\, w\, a}{144\, g}$$

$$= \frac{L_L\, w}{144\, g} \left(\frac{d_p}{d_L}\right)^2 \frac{L_s\, N^2}{2187} \cos \theta$$

$$= \left(\frac{d_p}{d_L}\right)^2 \frac{L_L\, w\, L_s\, N^2}{315{,}000\, g} \cos \theta$$

Converting from specific weight, w, to specific gravity,

$$w = 62.3 \text{ SG}.$$

The above equation then becomes

$$p_a = \left(\frac{d_p}{d_L}\right)^2 \frac{L_L\, L_s\, N^2 \text{ SG}}{5060\, g} \cos \theta \tag{7.10a}$$

If we wish to express acceleration head as a function of the average flow-rate capacity, Q, in U.S. gpm, instead of pump dimensions, we can make the following substitution:

$$Q = \frac{\pi}{4} d_p{}^2 \frac{L_s\, N\, n}{231} = \frac{d_p{}^2\, L_s\, N\, n}{294}$$

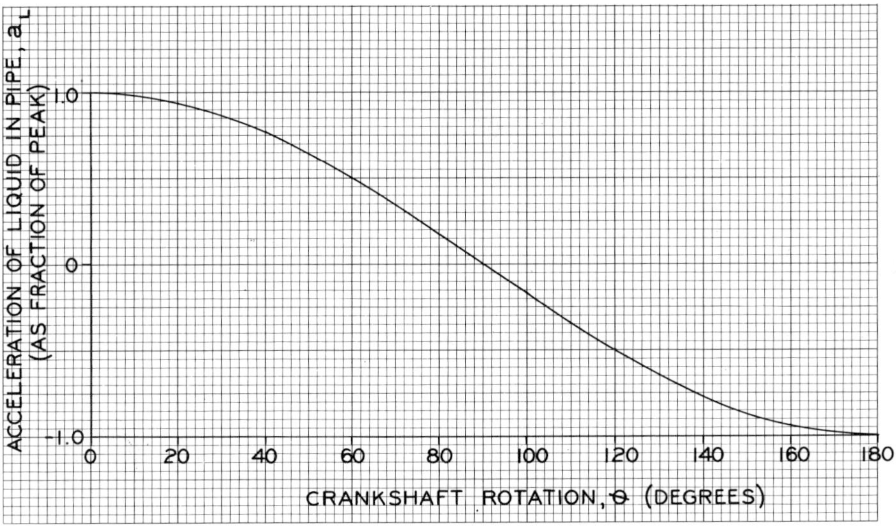

FIG. 7.9. The acceleration of the liquid in the suction or discharge pipe of a power pump due to the motion of one plunger (shown as a pure cosine function).

where:

n = number of pumping chambers

Rewriting Eq. 7.10a:

$$P_a = \left(\frac{d_p^2\ L_s\ N\ n}{294}\right)\frac{L_L\ N\ SG}{n\ d_L^2\ g\ 17.2}\cos\theta$$

$$p_a = \frac{Q\ L_L\ N\ SG}{17.2\ d_L^2\ g\ n}\cos\theta \tag{7.10b}$$

To enable us to compare this to the HI Eq. 7.1, we will first convert Q to v_L as follows:

$$Q = 3.12\ v_L\ A_L = 3.12\ v_L\ \frac{\pi}{4}\ d_L^2$$

$$Q = 2.45\ v_L\ d_L^2$$

Plugging into Eq. 7.10b yields:

$$p_a = \frac{2.45\ v_L\ d_L^2\ L_L\ N\ SG}{17.2\ d_L^2\ g\ n} \cos\theta$$

$$= \frac{v_L\ L_L\ N\ SG}{7.02\ g\ n} \cos\theta$$

Converting pressure units (psi) to head units (ft):

$$h_a = \frac{2.31}{SG} p_a = \frac{2.31}{SG} \frac{v_L\ L_L\ N\ SG}{7.02\ g\ n} \cos\theta$$

$$h_a = \frac{L_L\ v_L\ N}{g} \frac{0.329}{n} \cos\theta \tag{7.11}$$

Now let's equate Eqs. 7.1 and 7.11:

$$\frac{L\ v\ N\ C}{K\ g} = \frac{L\ v\ N}{g} \frac{0.329}{n} \cos\theta$$

We can now see that the two equations are similar, with

$$\frac{C}{K} = \frac{0.329}{n} \cos\theta \tag{7.12}$$

Solving for C:

$$C = \frac{0.329}{n} K \cos\theta$$

Letting $K = 1.4$ (HI's value for hot water):

$$C = \frac{0.461}{n} \cos\theta$$

Our equations are based on the acceleration of the liquid in the pipe being produced by only one plunger. At the points of maximum acceleration, this is true for only simplex, duplex, and triplex pumps of the single-acting configuration. We will therefore compare values for only the duplex and triplex pumps.

For the duplex pump, maximum acceleration occurs at 0°. Therefore:

$$C = \frac{0.461}{2} \cos 0° = \underline{0.231}$$

The Hydraulic Institute tabulates for the duplex a "C" value of 0.200. Our calculated value is only 16% above the HI value, a good agreement.

For the triplex pump, maximum acceleration occurs at $\theta = 60°$. Therefore:

$$C = \frac{0.461}{3} \cos 60° = \underline{0.077}$$

The HI value for a triplex is 0.066. Our value is again only 16% above HI—good consistency and good agreement.

Although possibly just a coincidence, the HI values of "C" for the quintuplex, septuplex, and nonuplex can be obtained by multiplying the 0.066 for the triplex by the inverse ratio of the numbers of plungers in the pumps, i.e. for the quintuplex $C = (\frac{3}{5})(0.066)$, for the septuplex $C = (\frac{3}{7})(0.066)$, and for the nonuplex $C = (\frac{3}{9})(0.066)$.

7.8. ACCELERATION HEAD FOR DIRECT-ACTING PUMPS

Although the low speeds of direct-acting pumps normally keep acceleration heads low enough for satisfactory operation, it is occasionally desirable to perform an acceleration head calculation for a direct-acting unit. As stated earlier, the HI equation for power pump acceleration head should be treated as an approximation. With direct-acting pumps, the calculation is even more approximate because of the variability of the adjustment of the valves on the drive end of such pumps. Valves that are improperly adjusted, or badly worn, can cause a significant increase in hydraulic shock.

The same equation used for power pumps (7.1) can be used for d-a pumps. If the pumps are operating properly, a $C = 0.200$ can be used for a simplex double-acting, and a $C = 0.060$ for a duplex double-acting.

REFERENCES

1. *Hydraulic Institute Standards, 14th ed.*, Cleveland, Ohio: Hydraulic Institute, 1983.
2. Miller, J. E., Experimental investigation of plunger pump suction conditions, *ASME Publication 64-PET-14,* New York, 1964.

Chapter 8
NPSH

8.1. THE SIGNIFICANCE OF NPSH

It has been estimated that 80% of all pump problems originate on the suction side of the pump. Most suction problems are caused by the formation of gas in the liquid. If, at any point in the suction system, or inside the pump, the static pressure of the liquid drops below the vapor pressure, the liquid will begin to boil ("flash" or "cavitate," i.e. change from a liquid to a gas). Bubbles, or "cavities," will form in the liquid. In a centrifugal pump, the bubbles normally form in the eye of the impeller and collapse somewhere inside the impeller. In a reciprocating pump, the bubbles normally form on the face of the plunger or piston during the early portion of the suction stroke, and collapse there during the latter portion of the suction stroke—or during the discharge stroke—as the local static pressure exceeds vapor pressure. Damage often occurs to these metal pump parts *when the bubbles collapse*—not when they form. As the bubbles collapse, the liquid rushes in at a high velocity to fill the void, impinging on the pump parts with a high impact (shock).

Many do not realize that cool water will boil, but it will if the pressure is reduced enough. A long suction line or one with numerous turns or restrictions, can cause the pressure to drop far enough to cause cavitation as the water enters the pump. Graber[1] stated that cool water will impose a compressive stress of 12,000 atmospheres (180,000 psi) when it "implodes" (collapses) on pump parts. The material of these parts cannot tolerate such repeated high impact loads, and the surface begins to fail in compressive fatigue.[2] Tiny particles of metal are broken from the end of the plunger (or other parts so affected), as shown in Fig. 8.1. The surface begins to look like miniature moon craters. This damage has been called "cavitation" by some observers, but it technically is not. "Cavitation" is the formation of the bubbles. The damage is caused by the collapse of the bubbles. Since the damage is the *result* of cavitation, though, it *can* correctly be termed "cavitation damage."

In addition to the metal surfaces damaged directly by the liquid implosions, the shock loads created by these implosions can damage other pump parts. Damage to reciprocating pump parts such as packing, gas-

FIG. 8.1. Cavitation damage to the end of a plunger. (*Courtesy NLB Corp.*)

kets, valves, springs, bearings, crankshafts, and gears have been attributed to severe cavitation.

"Vapor-lock" and "loss of prime" are terms used to describe the severe condition in which the pump stops pumping due to excessive gas in the pumping chamber (or chambers). This gas can come from cavitation of the pumpage and/or can be carried into the pump from the suction vessel or suction pipe.

To prevent the liquid from cavitating, it is necessary to maintain the pressure, at all points in the system and pump, and at all times, above vapor pressure. Because there is always a drop in pressure as the liquid flows from the pump suction nozzle into the liquid end, the pressure at the nozzle must exceed vapor pressure by a margin at least equal to the pressure drop, if cavitation is to be avoided. This margin is called "NPSH."

8.2 DEFINITION

The letters "NPSH" are an abbreviation of *N*et *P*ositive *S*uction *H*ead. It is defined as the difference between total (stagnation) suction pressure and vapor pressure, at the pump suction nozzle, when the pump is running. In equation form:

$$NPSH = p_s - p_v \tag{8.1}$$

where

$NPSH$ = Net Positive Suction Head, psi
p_s = Stagnation Suction Pressure, psia
p_v = Vapor Pressure of the Liquid at Suction Temperature, psia

Some authors have proposed alternate names such as "NPIH" (because "inlet" is technically more correct than "suction") and "NPIP" for reciprocating pumps (for "Net Positive Inlet Pressure"—because the units of NPSH are usually pressure units instead of head units). These concepts are intellectually appealing, but this author feels that it is sufficiently difficult for pump users, purchasers, and manufacturers to learn and discuss pump characteristics without confusing the NPSH issue by having two or more terms for the same characteristic. In this book we will stick with the most widely-used term, "NPSH."

8.3. UNITS OF NPSH

To aid in the understanding of the above definitions, "psi" units have been shown as examples. Although it is generally true that the NPSH of a reciprocating pump is expressed in pressure units such as "pounds per square inch," "kilopascals," or "bars," some recip pump manufacturers prefer to quote NPSH in specific energy units such as "feet," or "meters," such as is traditional with centrifugal pumps. Is one approach more correct than the other? Let's see.

In a reciprocating pump, NPSH is required to push the suction valve from its seat and to overcome friction losses and acceleration head within the liquid end. Because a significant portion of the NPSH required (NPSHR) is to open the valve (particularly at low pump speeds), and because this is a pressure requirement (rather than head or, more correctly, energy per unit mass requirement), NPSHR for a reciprocating pump is normally expressed in pressure units. For example, if a power pump requires 2 psi to lift the suction valve when pumping water (4.6 feet), it will also require 2 psi to lift the valve on propane (9.2 feet—since propane has one-half the density of water).

For any position of the valve between closed and fully open, a specific amount of pressure is required. Once the valve has become fully open (against its stop), though, any increase in flow results in a pressure drop which is proportional to the density of the pumpage, i.e. the *increase* in NPSHR over the fully-open NPSHR is a *head* increase.

The NPSHR of a reciprocating pump may, therefore, be a combination of pressure *and* head. Since the NPSHR of a recip is a specific pressure at each pump speed, until the speed is high enough to hold the valve against its stop during a portion of the stroke, it is usually more accurate to consider NPSHR to be a pressure requirement, independent of pumpage density.

NPSH values are neither gage pressures nor absolute pressures. The "g" in "psig" means that the pressure is measured above atmospheric pressure. The "a" in "psia" means that the pressure is measured above absolute zero, a perfect vacuum. NPSH is a measurement of a pressure above vapor pressure. If desired, one could call the units "psiv," but just "psi" is adequate.

8.4. SUCTION PRESSURE—THE FIRST HALF OF THE NPSH EQUATION

Suction pressure must be determined at the pump suction nozzle *when the pump is running*. If the suction pressure is measured with a gage, the atmospheric pressure (at the pump location) must be added to the gage reading to convert the reading to an absolute pressure. The velocity head at the gage connection must also be added to obtain the *total* (stagnation) pressure available to the pump. For a reciprocating pump, the acceleration head (as discussed in Chapter 7) must be subtracted.

In equation form:

$$p_s = p_{sg} + p_{atm} + p_{vel} - p_{acc} \tag{8.2}$$

where

p_s = total minimum absolute suction pressure, psia
p_{sg} = suction pressure measured with a gage, psig
p_{atm} = atmospheric pressure, psia
p_{vel} = velocity head, converted to psi
p_{acc} = acceleration head (applicable to a reciprocating pump), psi

If, instead of using a gage at the pump suction nozzle, suction pressure is calculated from suction tank pressure, as illustrated in Fig. 8.2, the pressure in the suction tank must be converted to absolute pressure by adding atmospheric pressure, then the liquid level above the pump suction nozzle must be added, and all friction losses subtracted. For a reciprocating pump, the acceleration head must also be subtracted.

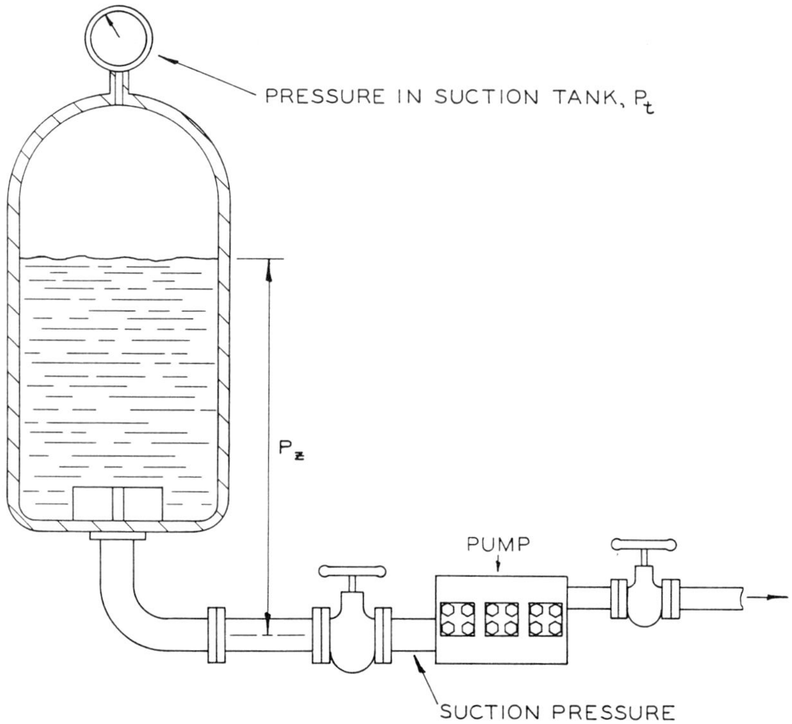

FIG. 8.2. Calculating suction pressure from tank pressure.

In equation form:

$$p_s = p_t + p_{atm} + p_z - p_f - p_{acc} \tag{8.3}$$

where

p_s = total minimum absolute suction pressure, psia
p_t = gage pressure in suction tank, above liquid, psig
p_{atm} = atmospheric pressure, psia
p_z = liquid level in suction tank, above centerline of pump suction connection, converted to pressure, psi
p_f = friction pressure drop from tank to pump, psi
p_{acc} = acceleration head (for a reciprocating pump), psi

Any other appropriate pressure units may be used in place of "psi." Equations 8.2 and 8.3 will both produce the same answer.

To calculate NPSH with this latter equation, the NPSH equation becomes:

$$\begin{aligned} \text{NPSH} &= p_s - p_v \\ &= p_t + p_{\text{atm}} + p_z - p_f - p_{\text{acc}} - p_v \qquad (8.4) \end{aligned}$$

If the liquid is at its bubble point in the tank, $p_v = p_t + p_{\text{atm}}$ and the equation reduces to:

$$\text{NPSH} = p_z - p_f - p_{\text{acc}} \qquad (8.5)$$

If, in addition, friction and acceleration heads are negligible, the NPSH available is equal to the elevation of the liquid above the pump (but the liquid must be at its bubble point for this to be true). The equation reduces to:

$$\text{NPSH} = p_z \qquad (8.6)$$

8.5. VAPOR PRESSURE—THE SECOND HALF OF THE NPSH EQUATION

Vapor pressure is more difficult to determine than suction pressure. It is a measure of the "desire" of a liquid to boil to a gas. Some liquids, such as butane and ammonia, have high vapor pressures. An open container of pure ammonia would quickly boil away, filling the area with noxious ammonia gas.

Water has a low vapor pressure. An open container of cool water, on the earth's surface, would not boil, but would evaporate slowly over a period of days. The desire of water to boil is therefore low. If we were to place that same container of water on the surface of the moon, *it would boil away,* similar to the ammonia. Why? The atmospheric pressure on the moon is zero—a perfect vacuum. The vapor pressure of *pure, air-free,* water at 80°F is about $\frac{1}{2}$ psia. This means that if the pressure on the water is reduced below $\frac{1}{2}$ psia, the water will boil.

8.5.1. A Function of Temperature

Vapor pressure is a function only of temperature. As the temperature increases, vapor pressure increases, until the critical temperature is reached. At the critical temperature, vapor pressure vanishes. Above the critical temperature, there is no distinction between a liquid and a gas. It's all just fluid.

Water at 265°F has the same vapor pressure (40 psia) as normal butane at 80°F and as propane at 0°F.

8.5.2. Boiling Reestablishes Equilibrium Conditions

Any liquid at its vapor pressure is on the verge of boiling (flashing). If the pressure is reduced slightly, the liquid will start to boil. If the temperature is held constant as the volume is increased (which requires heat input), the liquid will continue to boil until all the liquid has flashed to gas. If heat is not injected into the liquid, the portion flashing to a gas will cool and also absorb heat from the remaining liquid, causing the liquid temperature to drop. The lower temperature will result in a lower vapor pressure. The boiling will continue until the vapor pressure drops to the pressure which is imposed on the liquid. When that vapor pressure is reached and the boiling stops, the liquid-gas mixture is said to be in equilibrium. This is what happens as liquid flashes in the suction passage of a pump. Cavitation will cool the liquid and stop the cavitation.

8.6. NPSHA vs NPSHR

As seen in the previous definition and discussion, NPSH is a characteristic of the system. Each pump, though, does require the system to provide a specific amount of NPSH for "satisfactory" operation. Therefore, to distinguish between the amount of NPSH available from the system and that required by the pump, we call that which is available from the system "NPSHA" and that which is required by the pump "NPSHR."

8.7. THE EFFECT OF DISSOLVED GAS

Water, kerosene, and other liquids which interface with air or some other gas will absorb a portion of that gas. The liquid then becomes a solution of the parent liquid and dissolved gas, and has a vapor pressure higher than the pure liquid. Dissolved air causes problems during NPSH tests with cool water, often causing NPSH requirements to measure higher than they actually are, because of the assumption that the vapor pressure is that of deaerated water. Some field problems have resulted from the user neglecting to consider the dissolved air (or other gas) when calculating the NPSH available. The reciprocating pump section of the Hydraulic Institute Standards[3] recommends an extra 7 foot (3 psi) margin if the pumpage has been exposed to atmospheric air. Exposure of the pumpage to air or another gas at a pressure higher than atmospheric pressure will require an additional margin of suction pressure, to prevent the gas from flashing as the liquid moves into the pump. Some users reduce their calculated NPSHA by as much as ½ to allow for dissolved air.[4]

8.8. OTHER SOURCES OF GAS

Gas can come from sources other than cavitation of the liquid itself. One source can be the suction vessel. If the suction vessel contains no vortex breaker, and the liquid is allowed to swirl freely inside the vessel, the liquid may form a vortex as it flows into the suction line, carrying pockets of gas with it into the pump. If liquid feeds into the suction vessel above the liquid level, gas bubbles will be carried down into the liquid and may flow into the suction line.

Another source of gas can be the suction line. If the line is under a vacuum, and all connections are not absolutely tight, air will leak into the line and flow into the pump.

Still another source of gas can be pump stuffing boxes. If the pumping chamber is below atmospheric pressure on the suction stroke, air may flow through the packing and into the pump.

8.9. SOUNDS CAUSED BY CAVITATION

Cavitation of cool water and similar liquids in a reciprocating pump causes a knocking sound when the bubbles collapse on the face of the piston or plunger. In a power pump this knocking is transmitted to (and emits from) the power end, causing many mechanics and operators to think that the pump has a "loose rod" or *other* power end malfunction.

8.10. CONCEPTUAL ILLUSTRATION

Figure 8.3 may help clarify your understanding of NPSH. If a tall vertical tower is immersed in an open tank of air-free water at 80°F and at an atmospheric pressure of 14.7 psia (sea level), a column of water will rise in the tower when a vacuum is applied to the top of the tower. (Atmospheric pressure actually pushes the water up into the tower.) As the pressure in the top of the tower is reduced, the water level will rise—until the water reaches a level 33 feet above the level in the open tank. This will occur when the pressure in the top of the tower has dropped to ½ psia (about 29 in. of mercury vacuum). Any further attempt to lower the pressure, and "lift" the water higher will fail. The water column will "break." Actually the water will boil at the upper surface, pushing water vapor into the cavity above the water, and preventing the pressure from dropping below ½ psia.

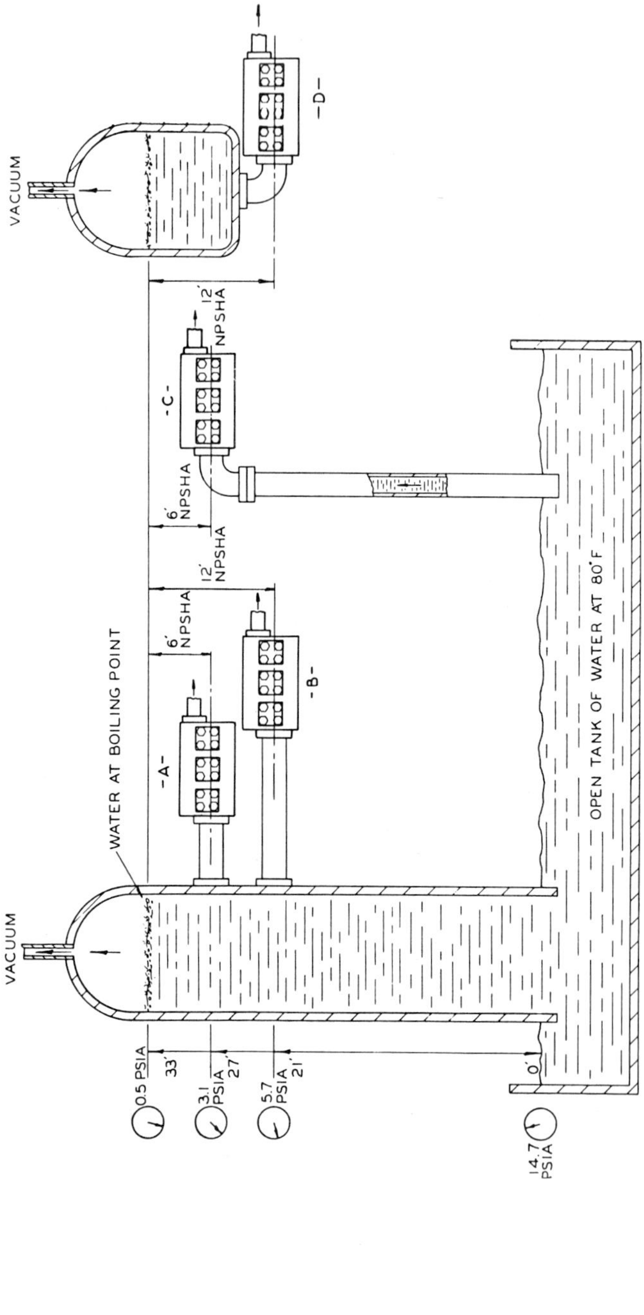

FIG. 8.3. Fundamental illustration of NPSH.

Now let's hold the column of water under these conditions and install a pump in the system. If we have a pump that requires 6 feet of NPSH, how high could we mount it on the side of the tower for satisfactory operation? Answer: 6 feet below the surface of the boiling liquid (assuming no friction loss or acceleration head) or 27 feet above the surface in the tank (position "A").

If a pump requires 12 feet of NPSH, it could be mounted no higher than position "B" for satisfactory operation.

A pump operating at position "C," with a separate suction line (with no losses or acceleration head), would be subjected to the same conditions as the pump in position "A." The water above position "A" actually has no effect on the pump. Similarly, a pump operating at position "D" would be subjected to the same conditions as position "B." (Positions "C" and "D" must have negligible friction losses and acceleration head to be equivalent to "A" and "B".)

Any system can be analyzed in a similar fashion if the actual vapor pressure of the liquid and the NPSH required by the pump are known. The point to remember is that (neglecting friction, velocity head, and acceleration head) the amount of NPSH available to a pump is equal to the distance that the liquid would rise above the pump, from the suction line, before it would start to boil.

8.11. POWER PUMP NPSH CURVES

A study of Fig. 8.4 will provide a better understanding of not only reciprocating pump NPSHR, but also of valve action, and the effect of valve spring force on operating characteristics. These curves apply to a 3 in. stroke, horizontal, triplex plunger power pump with suction valves which operate vertically. The valves are wing-guided, and are large relative to plunger diameter, resulting in a seat flow area approximately equal to the plunger area.

Because the axis of the suction valve is vertical, the valve can operate without a spring if the speed is kept low. Curves "A" and "B" represent the NPSH requirements with two different plunger diameters ($1\frac{7}{8}$ in. and $2\frac{1}{2}$ in.) tested in the same liquid end with no springs on the suction valves. Note that NPSHR with a $1\frac{7}{8}$ in. plunger at 100 rpm is only 0.7 psi (1.6 feet of water), less than most centrifugals. The NPSH required by the $2\frac{1}{2}$ in. plunger at 180 rpm is only 1.2 psi.

The speed of the pump in this configuration is limited by the ability of the suction valve to "keep up" with the plunger. Without a spring to push it back onto its seat, gravity is the only force tending to close the valve against the entering fluid. If the pump is running too fast, the valve will

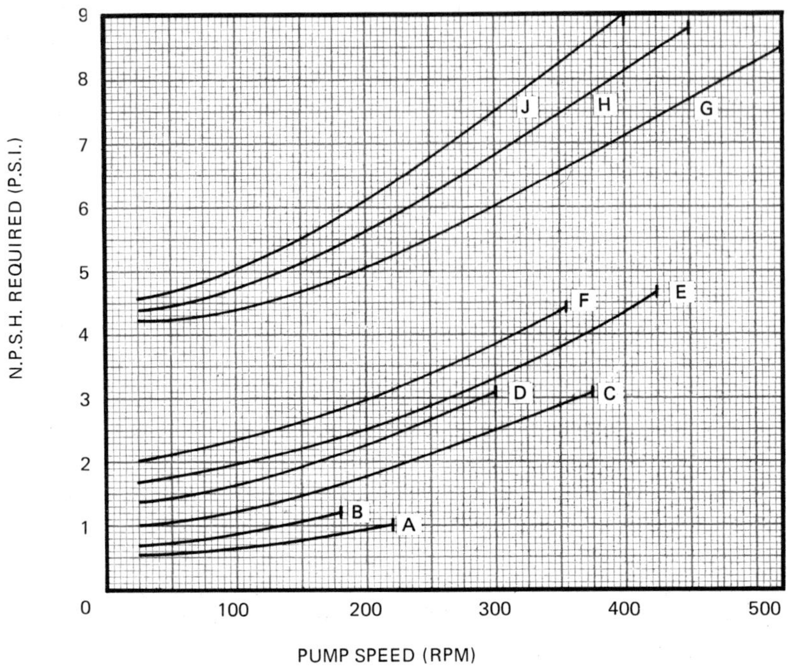

A. 1-7⁄8″ Plungers
Suction springs—None
Discharge springs—Standard

B. 2-1⁄2″ Plungers
Suction springs—None
Discharge springs—Standard

C. 1-7⁄8″ Plungers
Suction springs—Light
Discharge springs—Standard

D. 2-1⁄2″ Plungers
Suction springs—Light
Discharge springs—Standard

Maximum recommended speed is indicated by the vertical line at the end of each curve.

E. 1-7⁄8″ Plungers
Suction springs—Standard
Discharge springs—Standard

F. 2-1⁄2″ Plungers
Suction springs—Standard
Discharge springs—Standard

G. 1-7⁄8″ Plungers
Suction springs—Heavy
Discharge springs—Heavy

H. 2-1⁄4″ Plungers
Suction springs—Heavy
Discharge springs—Heavy

J. 2-1⁄2″ Plungers
Suction springs—Heavy
Discharge springs—Heavy

FIG. 8.4. NPSH requirements of a triplex plunger pump. (*Courtesy Union Pump Co.*)

still be above its seat when the plunger reverses and starts to reenter the pumping chamber. Therefore, the liquid will momentarily flow backwards through the valve, and the valve will be slammed onto its seat, sending a shock wave into the suction manifold and piping. At that instant the plunger is moving at a finite velocity, but the discharge valve is still closed. The pressure in the pumping chamber will quickly exceed the discharge pressure, and the discharge valve will be quickly driven from its seat. A shock wave will be transmitted from the pumping chamber, through the discharge manifold, and into the discharge line. The inertia of the discharge valve will carry it beyond its neutral point, compressing the spring more than normal. As the discharge valve stops rising, the spring attempts to return it to its neutral position, but again inertia carries it beyond, causing the chamber pressure to rise slightly. This discharge valve hunting is quickly dampened to zero. All of this takes place in a fraction of a second, but can be measured with a pressure transducer in the pumping chamber and displayed on an oscilloscope. The vertical lines at the ends of curves "A" and "B" indicate speeds safely within the range of proper suction valve operation.

Curves "C" and "D" are for the same two plungers, but with light (weak) springs added to the suction valves. Because the spring force, in addition to valve weight, must be overcome to open the suction valve, NPSHR has increased (about 100%) over curves "A" and "B." These springs do get the valves back onto their seats quicker, though, so that operation is smooth at higher speeds.

If speeds beyond the ends of curves "C" and "D" are desired, stronger springs are required. These springs are the standard for this pump, and allow operation in the normal speed range of 300 to 400 rpm. The NPSH requirements of curves "E" and "F," are about 3 times those of curves "A" and "B," ranging from about 2 to $4\frac{1}{2}$ psi.

Curves "A" through "F" represent a pump equipped with the same standard *discharge* valve springs. Only the suction springs have been changed.

If operation is required at speeds exceeding the limits of curves "E" and "F," heavy springs are required on *both* suction and discharge valves. The NPSH requirements are approximately double those of the standard springs, ranging from $4\frac{1}{2}$ to 9 psi.

On occasion it has been stated that a high suction pressure requires a stronger suction valve spring. Neither logic nor the author's experience support this idea. The suction valve feels only the *difference* between suction pressure and the pressure in the pumping chamber. When this difference is sufficient to lift the valve from its seat, the valve will begin to open. The pressure in the pumping chamber drops from discharge pressure as the pumpage expands due to extraction of the plunger from the pumping chamber. The rate of pressure drop is a function of the clearance volume ratio and the compressibility of the pumpage. Numerous power pumps, with *no* suction valve springs, operating at low speeds, operate satisfactorily on light hydrocarbons. The suction pressure to these pumps is 100 to 500 psi, often being only slightly above vapor pressure. Stronger valve springs *are* required for higher speeds of operation, and for denser and more viscous pumpage, but not for higher suction pressures. The reverse is usually true, though. Stronger suction valve springs do require more NPSH, and this may require an increase in suction pressure.

The term "strong" or "heavy" as used above does not necessarily mean an increase in stiffness. Spring stiffness is the rate of change of spring force as a function of displacement, i.e. a stiff spring will increase its resistance with a small displacement. As used above, a "stronger" or "heavier" spring is one which pushes with greater force to push the valve closed. This can be accomplished by increasing spring stiffness, *or* by increasing the initial (preload) force.

How is required suction pressure established from a known NPSHR? If the pumpage is, for example, deaerated water at 70°F (0.4 psia vapor pressure), the above pump, with $2\frac{1}{2}$ in. diameter plungers, no suction valve springs, and operating at 150 rpm, would require a suction pressure of only 1.5 psia (1.1 psi of NPSHR + 0.4 psia vapor pressure), or about 27 in. of mercury vacuum at sea level. Please note that the water must be totally free of air to have such a low vapor pressure. As stated earlier, water almost always contains dissolved air. The dissolved air actually increases the vapor pressure of the solution, but is often overlooked when NPSH calculations are performed. If we add in the 3 psi margin recommended by the Hydraulic Institute for water exposed to air, the suction pressure should be at least 4.5 psia. A liquid, such as propane, though, which is at its bubble point in the suction vessel, requires no such margin.

To minimize the problem of dissolved air, the NPSH tests which produced the above curves were performed with water at, or near, its boiling point in the suction vessel.

As shown by Fig. 8.4, a reciprocating pump, under the proper conditions, is capable of operating with a suction pressure below atmospheric pressure. Such a situation, though, can lead to air being "drawn" through the stuffing box packing and into the pumping chamber on the suction stroke. This air will cause as many problems as air entrained in the pumpage. Capacity will drop, the pump may operate noisily, the system may vibrate, and damage may occur to pump and system components. This inward air leakage can be reduced by an external sealing liquid, such as lubricating oil, being directed onto the plunger surface or into the packing. One very effective method is to install an endless lip seal packing ring on each side of a lantern ring, with both lips pointing toward the lantern ring, and inject oil with a mechanical lubricator into the lantern ring. Care must be exercised to prevent overpressuring the lantern ring, though, because excessive pressure can damage the packing, lubricator, and/or stuffing box.

8.12. NPSH TEST CRITERIA

Power pump NPSH tests are performed by holding pump speed and discharge pressure constant, varying the NPSH available (NPSHA) in the system, and measuring pump capacity. Capacity remains constant for all NPSHA values above a certain point, but as NPSHA is reduced below this value, capacity begins to fall. Figure 8.5 plots data from such a test. Similar to the *3% head* drop used in centrifugal pump NPSH testing, the reciprocating pump industry has established a *3% capacity* drop as the criteria for defining NPSHR. (Head or pressure cannot be used because, with a positive-displacement pump, the system establishes discharge pressure—not the pump.) Therefore NPSHA is reduced until the capacity drop has exceeded 3%, then the NPSH that was available at the 3% reduction is established, by definition, as the NPSHR for that speed.

The NPSH required by the triplex pump represented in Fig. 8.5 was 0.85 psi, and this is the value at which the pump will be quoted and sold. But, just as with a centrifugal pump, if this unit is operated with 0.85 psi of NPSHA, the unit will experience cavitation. Recognition of this fact raises two questions: 1) Will operation of the pump with cavitation cause damage to the pump or system? and 2) How much additional NPSHA is required to eliminate all cavitation?

Unfortunately, neither question is easy to answer.

If this unit were to be operated on a light hydrocarbon with a low differential pressure (200–300 psi), the effect of cavitation would be difficult to detect. With 0.85 psi of NPSHA, the ratio of vapor volume to liquid volume would be less than with water, and the compression ratio (discharge pressure to suction pressure) would be less than the 100+ which existed during the hot water test, both of which would reduce the distance which the plunger would travel on its discharge stroke before compressing the pumpage to discharge pressure. The result would be a higher capacity. Also, the collapse of the vapor during the discharge stroke is much ''softer'' with the light hydrocarbon, causing less damage to metal surfaces, and less shock transmitted to power end bearings. The

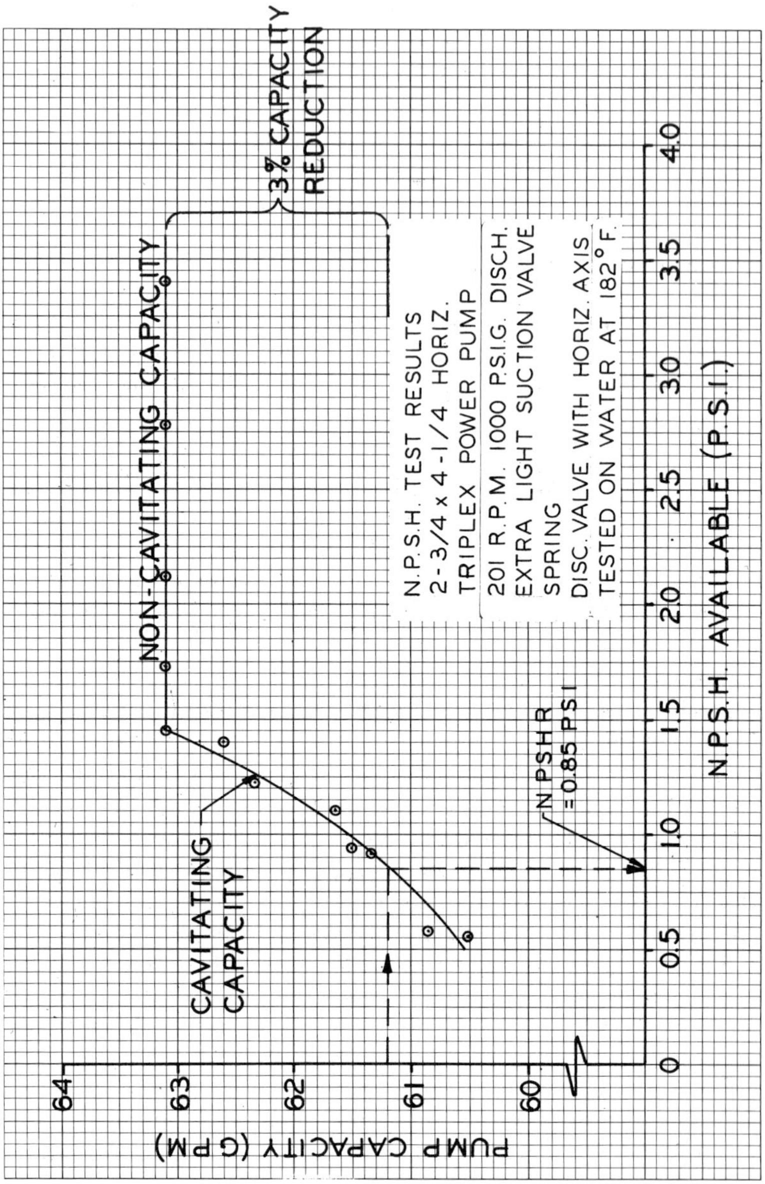

FIG. 8.5. NPSH test results of a triplex plunger pump. (*Courtesy Union Pump Co.*)

power end would also be lightly loaded, and the components much better able to absorb shock loads. The result would be a quiet, smooth-running pump with long bearing life.

On the other hand, if this unit were pumping water at 1000 psig discharge, the effect of cavitation would be much more pronounced. With 0.85 psi of NPSHA, the capacity would be less; and the more severe collapse of the vapor bubbles would be more damaging to metal surfaces—and would create larger shocks. Power end bearings (and other components), being more heavily loaded initially (due to the higher discharge pressure) would be less able to tolerate the shock without damage.

How much additional NPSHA is required to eliminate all cavitation with this pump? Examination of Fig. 8.5 might lead us to conclude that, since full capacity is achieved with 1.5 psi of NPSHA, that amount would eliminate all cavitation. But, alas—we are foiled again. As pointed out by Collier,[5] cavitation can occur in a reciprocating pump without causing any reduction in capacity. This can happen when incipient cavitation occurs during the initial portion of the suction stroke, because the plunger or piston is accelerating, then disappears during the latter portion because the plunger is decelerating. The pumping chamber is therefore completely filled with liquid at the end of the suction stroke, and pump capacity is not affected. Capacity is reduced only when cavitation is so severe that complete bubble collapse does not occur until the plunger is on its discharge stroke. But even when the bubbles completely collapse on the suction stroke, damage can occur. During collapse, the liquid impinges on the face of the plunger or piston, causing damage similar to that seen in the eye of a centrifugal pump impeller exposed to similar conditions. The resulting shock is transmitted through the open suction valve and into the suction line, sometimes causing vibration and noise. No known attempt has been made at quantifying, for reciprocating pumps, the extra margin of NPSHA required to preclude all cavitation.

8.13. NPSHA VS NPSHR (A SUMMARY)

NPSH *available* is a characteristic of the system, and must be calculated by the system designer or operator. NPSH *required* is a characteristic of the pump, and must be determined by the pump manufacturer, normally by test. NPSHA must exceed NPSHR for smooth, quiet operation. The margin by which NPSHA should exceed NPSHR is, unfortunately, a function of the application (service).

REFERENCES

1. Graber, P., Solids handling pumps—Part 2, *World Pumps Magazine,* pp. 83–87, 1983.
2. Stepanoff, A. J., *Centrifugal and Axial Flow Pumps,* New York: John Wiley & Sons, Inc., 1957.
3. *Hydraulic Institute Standards, 14th ed.,* Cleveland, OH: Hydraulic Institute, 1983.
4. Taylor, Irving, What NPSH for dissolved gas?, *Hydrocarbon Processing Magazine,* pp. 133–134, Aug. 1967.
5. Collier, S. E., Know your mud pump—Part 5: Knocking, *World Oil Magazine,* Gulf Publishing Co., Houston, 1958/1959.

Chapter 9
SELECTION

9.1. DEFINITION

For the purposes of this book, "selection" is defined as that function, normally performed by the seller, when determining the pump which is most suitable for the intended service. This function is sometimes called "application," but the term application will herein mean designing the system and installing the pump, and is covered in Chapter 11.

Included among the items which must be determined in selection are the pump model, plunger diameter, and speed; the effect of clearance volume and fluid compressibility on capacity; the effect of suction pressure—if significantly above atmospheric; materials of construction; packing and its lubrication; driver type, size, speed, and construction; type and size of the speed reducer; possible additional auxiliary equipment; and any special features or construction required by environmental or service conditions. If not covered elsewhere in this book, these subjects will be discussed in this chapter.

9.2. SELECTION GRAPH

For a reciprocating power pump, the equivalent to the centrifugal pump performance curve is the selection graph which plots, primarily, displacement as a function of plunger diameter and speed. This graph also usually tabulates power requirements at maximum plunger load. Different manufacturers produce these graphs in different forms, but they are all basically straight lines, since displacement for any plunger is directly proportional to speed. Figure 9.1 shows one such graph for a 3 in. stroke triplex pump. Crankshaft speed is plotted against displacement for the different plunger diameters available in the pump. The maximum discharge pressures listed across the top of the chart are the pressures at which each plunger would operate to load the power end to its rated load (4460 lbs). (The pressures shown for the $1\frac{1}{4}$ in. and $1\frac{7}{8}$ in. plungers are fluid end working pressures, and are less than allowed by the power end rating.)

The break in the displacement scale at 10 gpm (which causes the straight lines to be bent) allows the lower displacement values to be ex-

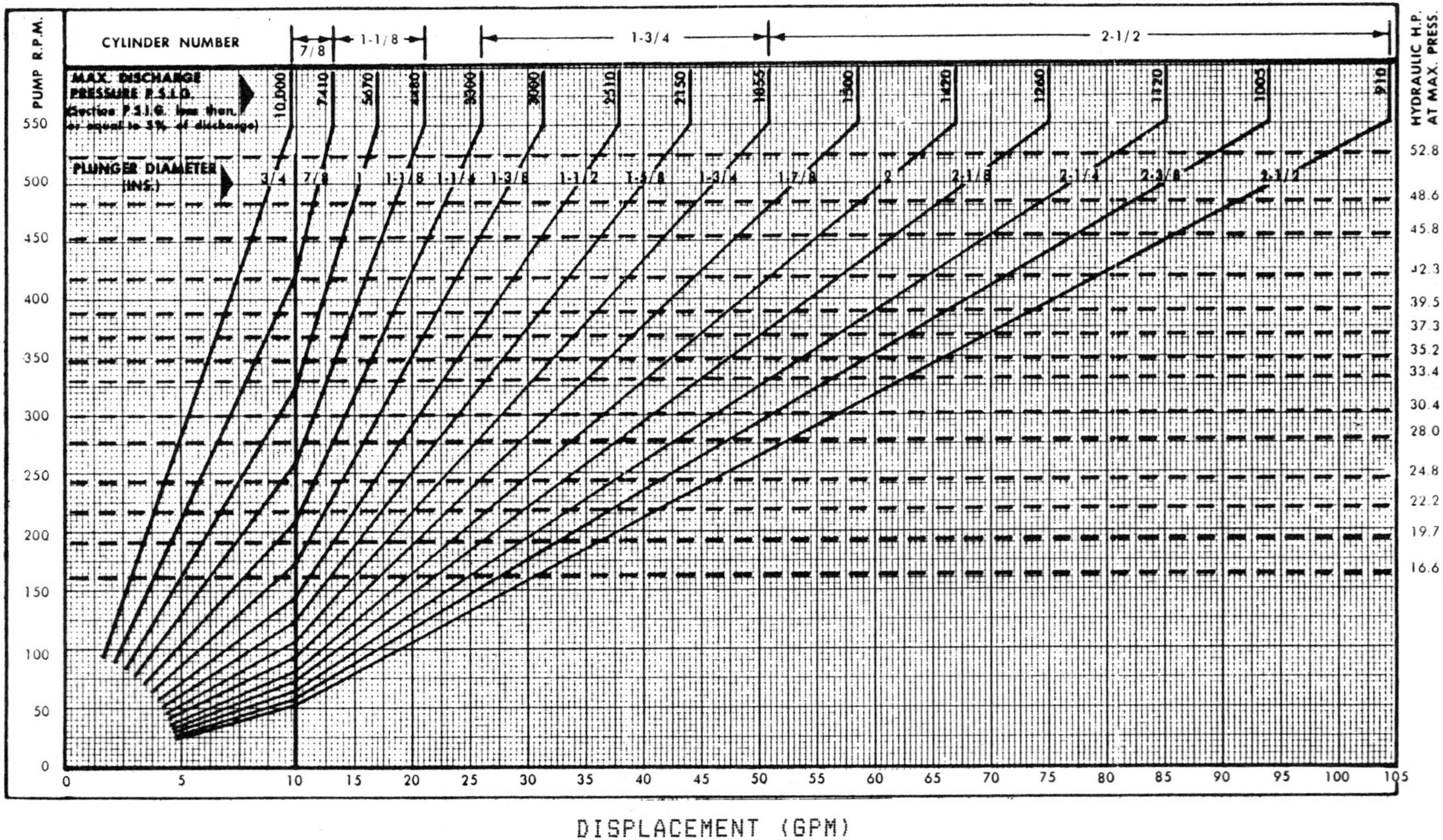

FIG. 9.1. Selection graph for a 3 inch-stroke triplex plunger pump. (*Courtesy Union Pump Co.*)

panded for easier reading. Actual pump capacity is determined by multiplying the displacement by the volumetric efficiency, η_v. For most applications, volumetric efficiency will be about 95%, so 0.95 is normally used as the multiplier to determine pump capacity.

The left vertical scale (ordinate) is the rotative speed of the pump crankshaft. The horizontal dashed lines indicate speeds which result from the combination of 60 Hz electric motors and standard v-belt drives. The right vertical scale tabulates the hydraulic power for 100% volumetric efficiency, atmospheric suction pressure, and maximum plunger load. The four brackets across the top of the chart indicate the four different cylinders available for this pump.

Although generally not done, direct-acting pump performance can be represented in a similar fashion, with displacement shown as a straight-line function of speed.

9.3. OPERATING SPEEDS

9.3.1. Power Pumps

The subject of operating speeds is one which finds a wide divergence of opinion among power pump manufacturers. The mechanical efficiency of a power pump varies little as speed is increased, so no power consumption penalty is paid for running at a higher speed. To minimize cost, some manufacturers naturally tend to offer smaller units at higher speeds, while other manufacturers traditionally offer their units at more conservative speeds. The two-decade period from about 1940 to 1960 saw a number of manufacturers increase the operating speeds of their units by a factor of about three. As an example, one manufacturer gradually increased the top speed of his 3-in. stroke triplexes from 150 to 520 rpm. The Hydraulic Institute lists 530 rpm as being the "basic speed" for a 3-in. stroke pump. These increases were made possible by improvements in liquid end and power end components, but packing and stuffing box development lagged behind the speed increase, and the difficulty in providing systems which could tolerate the increased acceleration head resulted in maintenance requirements for pump and system components beyond those considered acceptable by many users of continuous-duty pumps.

The opinions of users tend to be more uniform than those of manufacturers, with the preferred maximum speeds, for continuous duty, being somewhat below the average of those recommended by the manufacturers.

Through field observations and with input from numerous users, the author developed the set of maximum speeds, for continuous-duty operation, shown in Fig. 9.2. These speeds are lower than offered by most manufacturers, but agree with those preferred by many users (although too fast for a few). The top solid line represents the Hydraulic Institute "basic speeds." Because different purchasers have specified 50%, 70%, and 80% of the HI basic speeds as their allowable maximum speeds, these values are also shown, as dashed lines. The next lower solid curve represents the author's maximum recommended speeds for continuous-duty (24 hr/day, 7 day/wk) operation. Units in intermittant and cyclic duty can usually be operated satisfactorily at higher speeds. If a lower-speed unit is available for little or no cost premium, it is generally preferred.

For satisfactory operation, even at the author's relatively moderate speeds, all operating conditions must be good, i.e. the system must be properly designed, adequate NPSH must be available, and the liquid must be cool, clean, and of a low viscosity and density (equivalent to water). Conversely, conditions which may dictate a reduction in speed include a poorly-designed system; low NPSHA; pumpage that is hot, contains gases or solids, has a high viscosity, has a high density, and/or is corrosive; pumps with large plungers or with heavy valves (such as balls); a requirement for low sound level; and a requirement for extended life of "expendables" (pump parts, such as packing, which require periodic replacement).

The *minimum* speed of a power pump is determined by its ability to provide sufficient lubrication to all bearing surfaces in the power end. Some units can run satisfactorily at speeds as low as 10 rpm. Others must be maintained at 100 rpm or more to provide oil to the bearing surfaces and/or to generate the required lubrication "wedge" at the sliding surfaces. Unlike rotary or centrifugal pumps, the *liquid end* of a reciprocating pump has no minimum speed for satisfactory operation (unless its valves are leaking). A reciprocating pump liquid end creeping along at one cycle/min. will still develop full pressure and will have a capacity proportional to its speed.

9.3.2. Direct-Acting Pumps

Direct-acting pumps operate at speeds that normally range from 0 to 50 cycles/min, the upper limit decreasing as the stroke length increases, the same as with power pumps—but at much lower speeds than power pumps. The maximum operating speeds of direct-acting pumps are normally only 15–20% of the power pump maximums for pumps of equal stroke length.

The ability of these machines to operate at speeds down to 0 (stall) is desirable in some applications, such as hydrostatic testing, where, after

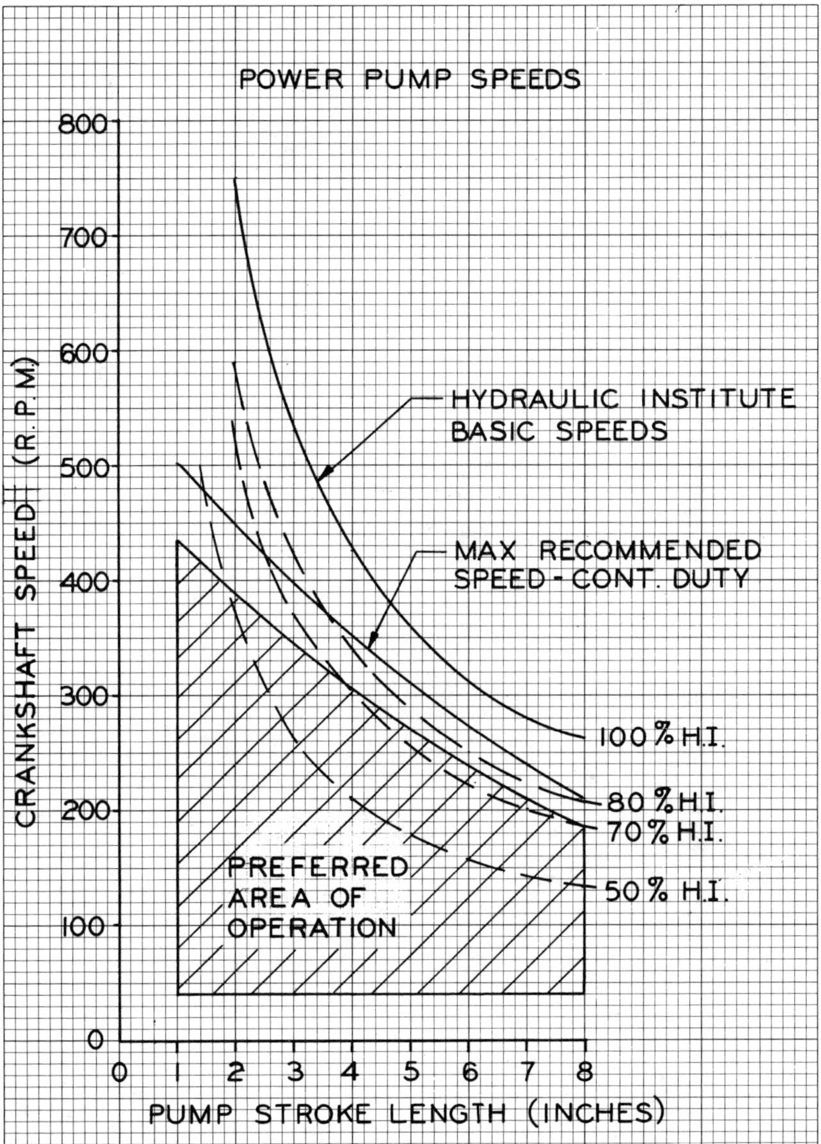

FIG. 9.2. Operating speeds of power pumps. The author's recommended continuous-duty speeds are compared with various percentages of the Hydraulic Institute "basic" speeds.

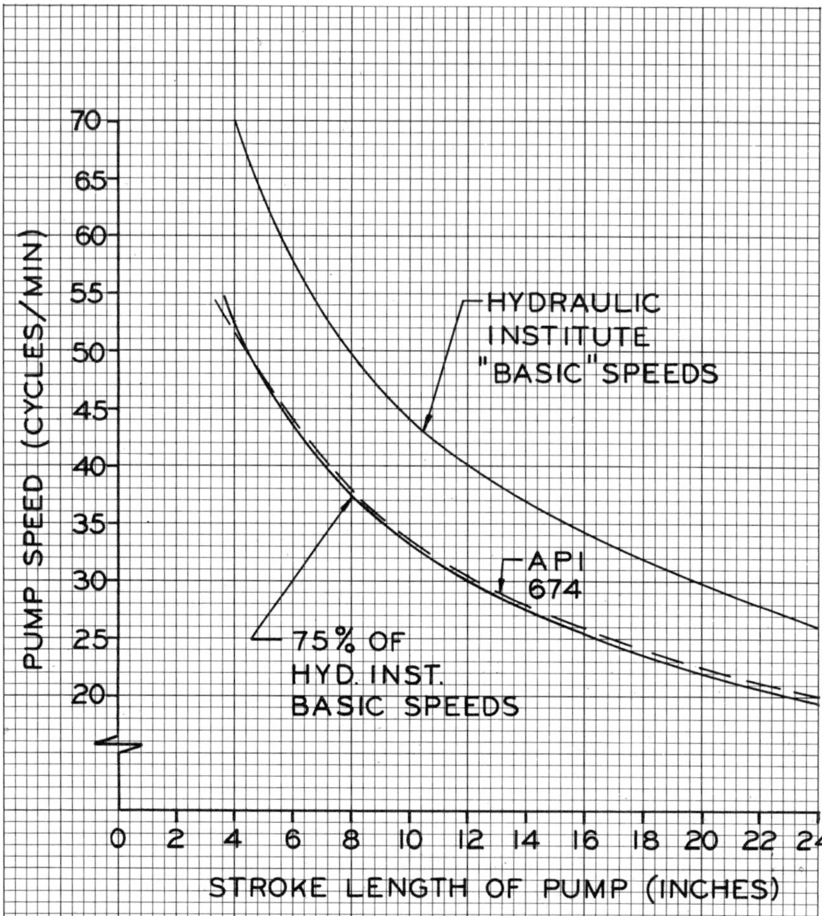

FIG. 9.3. Operating speeds of direct-acting pumps: Hydraulic Institute vs API 674.

the system is filled with liquid and up to pressure, it is necessary only to maintain that pressure—to add liquid only if some is lost from the system.

Unlike power pumps, the mechanical efficiency of a direct-acting pump falls significantly as the speed increases, as shown in Fig. 6.20. The steepness of the efficiency curve is determined, to a large extent, by the type and size of valve in the drive end. As discussed in Chapter 6, it is the slope of the efficiency curve that allows the speed of the pump to be controlled by throttling the motive gas. This characteristic of higher efficiency at lower speeds provides additional incentive, beyond smooth operation and long life, to select these units at low speeds. The higher efficiency will reduce gas consumption per unit power output, and allow the use of smaller drive ends. Oversizing the drive end not only increases initial cost, but increases operating cost, and makes speed control more difficult.

Figure 9.3 plots the speeds of direct-acting pumps as a function of stroke length. The upper line is a plot of the Hydraulic Institute[1] published "Basic" Speeds. The lower line is 75% of the upper line and coincides closely to the maximum continuous speeds allowed by API 674.[2]

If one compares the mechanical efficiency of the 6 × 4 × 6 duplex piston pump (as shown in Fig. 6.20) with the top η_m curve in Fig. 6.22, it can be seen that the approximate efficiency of this 6 in. stroke piston pump (64%) is obtained at about 48 cpm. This speed falls between the two curves on Fig. 9.3. If operated at the HI speed of 58 cpm, the efficiency would drop to 54%. If operated at the API 674 speed of 44 cpm, the efficiency would rise to 68%.

9.4. HIGH SUCTION PRESSURE CONSIDERATIONS

High suction pressure requires special considerations for reciprocating pumps. The suction flange and manifold must often be upgraded to be suitable for the increased pressure. The effect of high suction pressure on the total force imposed on a power pump power end varies with the type of liquid end. As illustrated in Fig. 9.4, for constant discharge pressure, double-acting liquid ends transmit reduced forces to their power ends as suction pressure is increased.

A double-acting piston pump sometimes requires a tail-rod to equalize the forces imposed on the piston during both strokes (See Fig. 9.5). Otherwise, the force of the suction pressure acting on the larger outboard face of the piston may exceed the force of the discharge pressure acting over the smaller area (because of the piston rod) of the inboard face of the piston. If this occurs, the drive rod will be in compression at all times, an undesirable condition for both direct-acting and power pumps.

Conventional single-acting power pumps are designed for operation with the suction pressure near atmospheric. As suction pressure exceeds atmospheric, the load imposed on the bearings, crankshaft, and other power end components increases, even though discharge pressure remains constant. With single-acting power pumps, adding a tail-rod is not possible. A crankshaft bending moment diagram for a triplex pump reveals that a 3 psi increase in suction pressure imposes the same bending moment on the crankshaft as a 1 psi increase in discharge pressure.

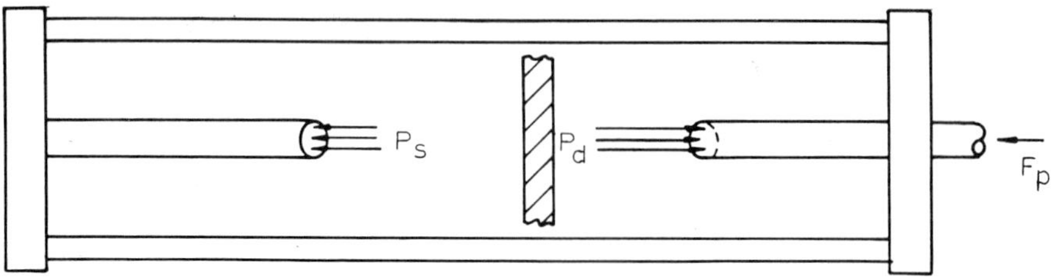

(a) On a double-acting plunger pump, an increase in suction pressure *reduces* the forces imposed on the drive end.

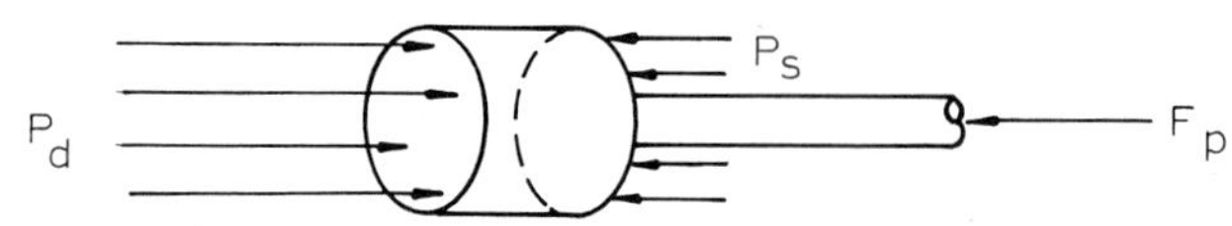

(b) On a double-acting piston pump, an increase in suction pressure *reduces* the forces imposed on the drive end.

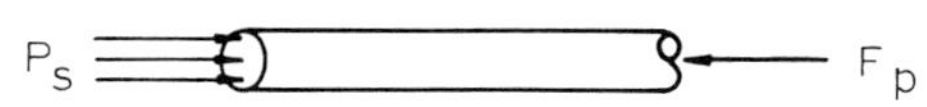

(c) On a single-acting plunger or piston pump, an increase in suction pressure *increases* the forces imposed on the drive end.

FIG. 9.4. Effect of suction pressure on loads imposed on the drive end.

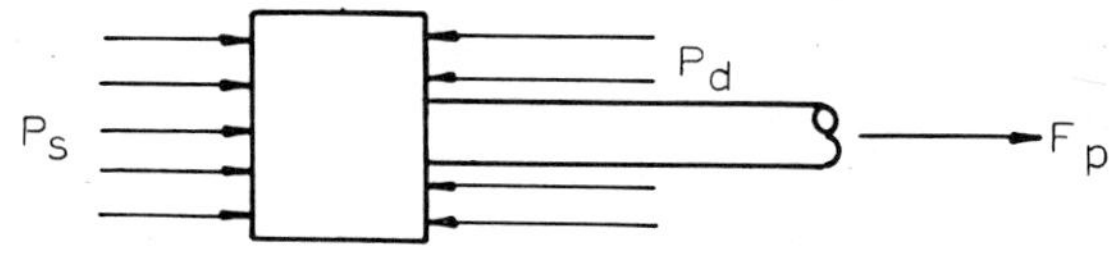

(a) This double-acting piston, with no tail rod, exposes more area to pressure on the left side than on the right.

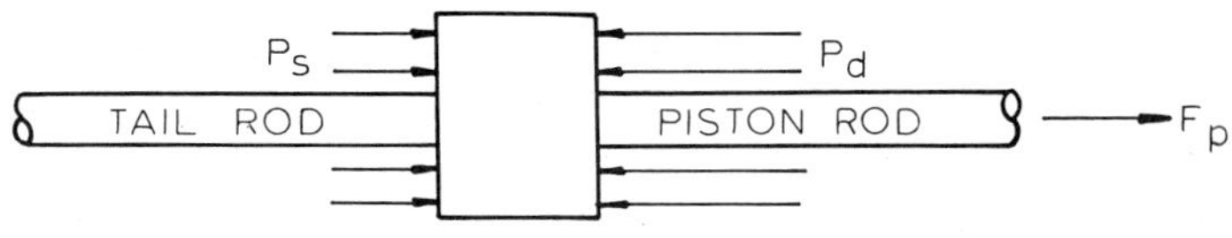

(b) This piston, equipped with a tail rod of the same diameter as the piston rod, has equal areas on both sides.

FIG. 9.5. The addition of a tail rod to a piston equalizes the areas exposed to fluid pressures, thereby equalizing the rod load during both strokes.

Therefore, to determine the "effective discharge pressure," from the standpoint of crankshaft bending moment, $\frac{1}{3}$ of the suction pressure is added to discharge pressure.

Another factor, usually more important than crankshaft stress in a single-acting pump, is the effect of high suction pressure on bearings. If suction pressure exceeds approximately 5% of maximum allowable discharge pressure, the connecting rod remains in compression (or tension) on the suction stroke. Because of this continuous load, the bearings at both ends of the rod are continually loaded in the same area. If oil is fed into either bearing through a hole in the center of this area, a fairly common design, the oil flow will be blocked. An additional problem exists at the crosshead pin because the rod bearing *oscillates* on the pin, swinging through a total arc of only about 22°. This small arc of oscillation, and the low velocity of the bearing surface, is not sufficient to carry oil into the pressure area. Table 9.1 summarizes these effects.

Pressurized lubrication systems contribute little to the solution of this problem because bearing pressures are approximately 1000 psi, and lubrication systems seldom operate over 60 psi.

Where single-acting power pumps have been applied to high-suction-frame-load conditions without adequate provisions in their construction and/or selection, high power end temperatures and premature bearing wear have resulted. The bearing experiencing the most wear is usually at the crosshead pin, but because it remains loaded in the same direction at all times, there is no knocking. The pump runs quietly.

The problems of high suction pressure can sometimes be resolved by relocating oil entry points and/or by special bearings, and by making allowance for the high suction pressure in the selection process. To allow for the effects of high suction pressure, it is necessary to select a pump power end suitable for a discharge pressure higher than the operating

Table 9.1. Effects of Increased Suction Pressure.

Single-Acting Pump	Double-Acting Pump
Reduced power requirement.	Reduced power requirement.
Increased frame load.	Reduced frame load.
Increased friction in power end.	Reduced friction in power end.
Connecting rod stays in compression (or tension) during both strokes.	
Bearings stay loaded on same side during both strokes.	
Increased temperature of power end.	
Reduced overall mechanical efficiency.	

discharge pressure. Stated differently, any pump must have its maximum allowable discharge pressure reduced when exposed to a high suction pressure. If crankshaft stress were the only concern, maximum discharge pressure would be reduced by $\frac{1}{3}$ of the suction pressure. Tests have shown, though, that power end temperatures, even with properly designed bearings, are often significantly increased by high suction pressures, indicating that bearing friction is increased. With one pump it was found that an increase in suction pressure of 1 psi was equivalent, in power end temperature rise (over ambient) to about $1\frac{1}{2}$ psi of discharge pressure increase.

A high suction pressure also plays havoc with mechanical efficiency. Most plunger power pumps are single-acting machines. The loads imposed on the power end are therefore *not* proportional to differential pressure (discharge minus suction), as is the case with a double-acting pump, but are more nearly equal to the *sum* of discharge pressure plus suction pressure. In other words, the power end feels the effect of full discharge pressure for each plunger on the discharge stroke, and, for each plunger on the suction stroke, it feels a load almost equal to that which would be experienced if it were discharging at suction pressure. This increase in bearing loads and in packing friction cause the friction power to increase. Simultaneously, the increased suction pressure causes the output (hydraulic) power to drop. This combination causes mechanical efficiency to drop, sometimes significantly.

It has been found that power consumption can be fairly accurately predicted for such conditions by calculating the power which the pump would require with the actual discharge pressure and atmospheric suction pressure, then subtract the amount of power derived by the pump from the suction pressure. The only trick is the determination of the "suction efficiency," and this has been found to be 3 to 5 points less than "discharge efficiency." To be on the conservative side, we will subtract 5 points. The equation is therefore:

$$P_i = \frac{(Q)(p_d)}{(1714)(\eta_m)} - \frac{(Q)(p_s)(\eta_m - 0.05)}{1714} \tag{9.1}$$

where

P_i = Power input required by pump, hp
Q = Pump capacity, U.S. gpm
p_d = Pump discharge pressure, psig
p_s = Pump suction pressure, psig
η_m = Mechanical efficiency of the pump for rated p_d if $p_s = 0$

Note that suction efficiency, since it is a power-recovery efficiency, appears in the *numerator* of the second term.

The following example will illustrate the calculation of the power requirement with a high suction pressure.

EXAMPLE

A light hydrocarbon, suction pressure = 500 psig, discharge pressure = 1000 psig, capacity = 38 gpm. The selection is a 2 × 3 triplex, with a maximum plunger load rating of 4500 lbs, running 330 rpm. Discharge plunger load = $(\pi/4)(2)^2(1000)$ = 3140 lbs. Fraction of maximum load = 3140/4500 = 0.70. From the efficiency curve (not shown) we read η_m = 0.89.

$$P_i = \frac{(Q)(p_d)}{(1714)(\eta_m)} - \frac{(Q)(p_s)(\eta_m - 0.05)}{1714}$$

$$= \frac{(38)(1000)}{(1714)(0.89)} - \frac{(38)(500)(0.89 - 0.05)}{1714}$$

$$= 24.9 - 9.3 = \underline{15.6} \text{ hp}$$

What is the overall pump mechanical efficiency?

$$P_o = \frac{(Q)(\Delta p)}{1714} = \frac{(38)(500)}{1714} = 11.1 \text{ hp}$$

$$\eta_m = \frac{P_o}{P_i} = \frac{11.1}{15.6} = \underline{0.71}$$

Therefore, a pump which has a mechanical efficiency of 89% when the suction pressure is 0 psig and the discharge pressure is 1000 psig, experiences a reduction in efficiency to 71% if the suction pressure is increased to 500 psig. Even though the "discharge efficiency" used in the above equation is 89%, and the "suction efficiency" is 84%, the overall efficiency is 71%. To avoid confusion in communicating efficiencies for pumps with high suction pressures, the above phenomenon needs to be understood by vendors and users, and care must be exercised in identifying the efficiency in question.

Although the above equation is not normally used until suction pressure exceeds 5% of discharge pressure, it yields accurate results for all suction pressures.

Although it has been contended that high suction pressure requires strong suction valve springs, such is not the case. Stronger valve springs are required by higher operating speeds, more viscous pumpage, and denser pumpage; but not by higher suction pressure. For a more complete discussion, see Chapter 8.

9.5. CLEARANCE VOLUME AND LIQUID COMPRESSIBILITY

The clearance volume of a reciprocating pump is the "dead-space" volume remaining in the pumping chamber when the plunger (or piston) is at the end of its discharge stroke (see Fig. 9.6). Clearance volume is normally expressed as a percent or fraction of displacement, and is identified as the "C/D (Clearance/Displacement) ratio" or "percent clearance."

Most reciprocating pumps have a C/D ratio of 2 to 4 (200% to 400%), with plunger pumps normally having larger values than piston pumps. Some plunger pumps, though, have been designed with C/D ratios as low as 0.5 and "minimum-clearance" piston pumps, designed specifically for volatile liquids and liquids containing gases, have C/D ratios of approximately 0.15.

All liquids compress as pressure increases—some more than others. From 15 psia to 10,000 psia, water compresses about 3%. From 150 psia to 3000 psia, propane compresses about 5%. As the plunger withdraws from the pumping chamber, the liquid trapped in the clearance volume at discharge pressure must expand to suction pressure before the suction valve can open. If a pump with a C/D of 3 is pumping a liquid which compresses 5%, the plunger must move 15% of its suction stroke, in addition to the 1–2% required for discharge valve slip, before the pressure in the chamber drops to suction pressure, thereby resulting in a 15% capacity reduction (as measured in the suction pipe) compared to a totally-incompressible liquid. When a pump is originally sized for this service, it must be sized about 15% larger than for water. When selecting a drive for this unit, it is best to size it for the power which would be required for water because the pumpage may change to a less compressible mixture, or may not be as compressible as originally anticipated.

The C/D ratio also has an effect on the ability of a reciprocating pump to prime itself, or to clear itself of vapors if it becomes "vapor-locked." With a 15% clearance, a pump can produce a pressure ratio of about 15 on air; but with a 300% clearance, a pressure ratio of only 1.5 is possible. With atmospheric air initially in the chamber, the 15% C/D unit can discharge at about 200 psig; the 300% unit at only 7 psig.

It can therefore be seen that a plunger-type power pump is generally not capable of ingesting any quantity of gas without a significant degradation of performance, and is not capable of priming itself unless the discharge pressure is kept low during the priming process. Free gas affects the reciprocating pump in much the same way as cavitation, i.e. shock loads which cause the shortened life of liquid end and power end components.

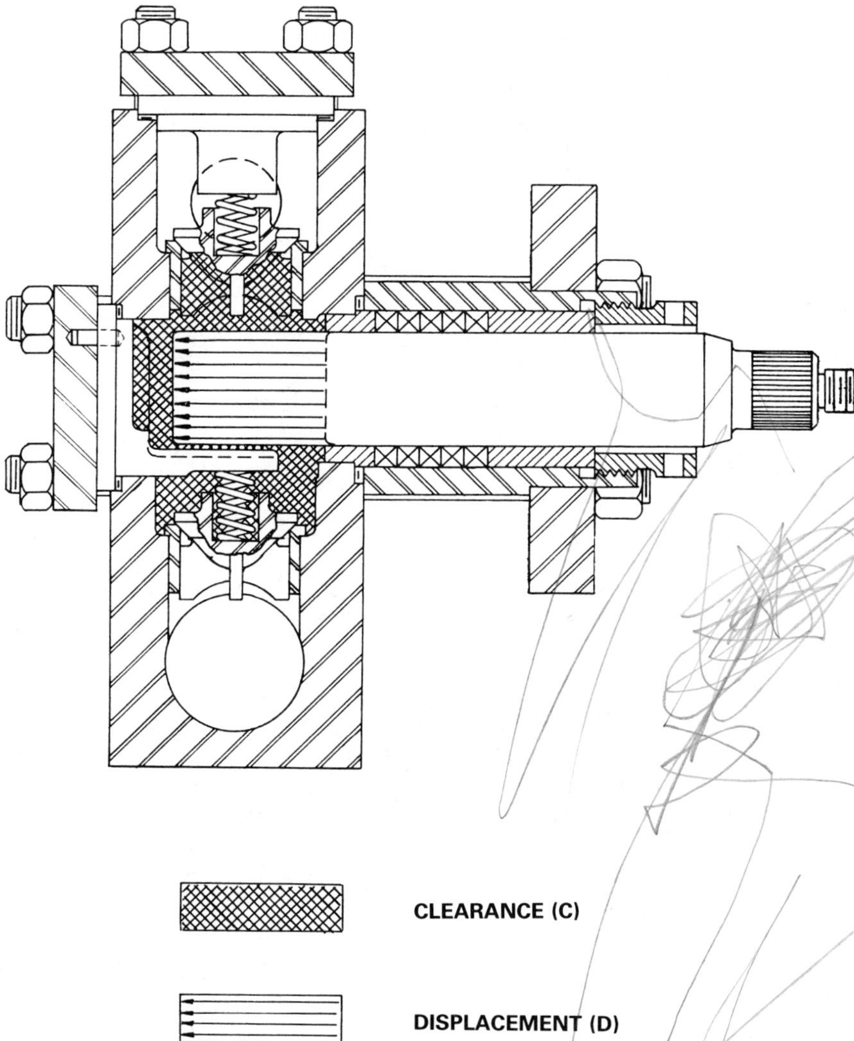

FIG. 9.6. Illustration of pump clearance volume.

A more detailed explanation of the compression process follows.

Figure 9.7 illustrates the pressure-volume curve for the complete cycle of a single pumping chamber. The cycle has been divided into seven distinct events. Event #1 has the plunger at the end of its discharge stroke and the beginning of its suction stroke. The discharge valve is closing, but has not completely seated. After the plunger moves a short distance on its suction stroke, the discharge valve seats (but a small quantity of fluid

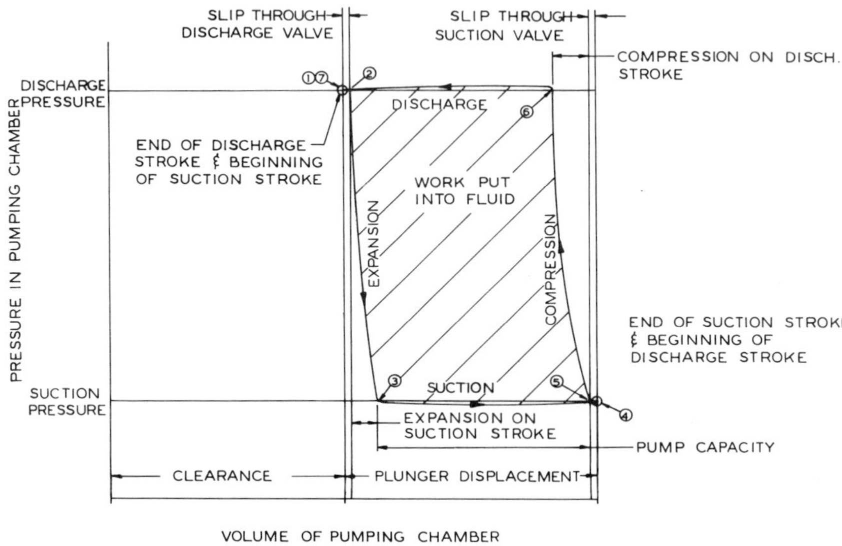

FIG. 9.7. Pressure-volume diagram of the pumping cycle of a single-acting plunger, showing effects of compression and expansion of the fluid, and valve slip.

returns to the chamber from the discharge manifold prior to full closing), and the fluid trapped in the clearance volume begins to expand (event #2). As the plunger moves further on the suction stroke, the trapped fluid expands until chamber pressure drops to slightly below suction pressure, and the suction valve begins to open (event #3). Continued retraction of the plunger from the pumping chamber further reduces chamber pressure, allowing the suction pressure to push the suction valve open more, and push liquid into the chamber. As the plunger passes midstroke and begins to decelerate, the suction valve begins to close. At the end of the suction stroke, "bottom-dead-center" (event #4), the plunger reverses and starts to reenter the chamber. At event #5, the suction valve becomes fully seated, but not before some fluid has returned to the suction manifold. From event #5 to event #6 the fluid is compressed to a pressure slightly above discharge pressure, a pressure just high enough to push the discharge valve open (event #6) and force fluid into the discharge manifold. The plunger continues on its discharge stroke until it reaches "top-dead-center" (event #7), at which time the cycle begins anew.

Because there is a larger volume of fluid in the pumping chamber at the end of the suction stroke than at the end of the discharge stroke, the amount of compression on the discharge stroke, exceeds the amount of expansion on the suction stroke. The difference between these two values is the amount that the fluid compresses from suction conditions to discharge conditions. For example, propane that compresses 5% from suction to discharge would have only 95% of its initial volume at discharge pressure. If the pump ingested 10 $in.^3$ of propane on the suction stroke, it would discharge only 9.5 in^3. If the flow rate on the suction side is 100 gpm, the discharge flow rate will be 95 gpm, even though the mass flow rate is unchanged. Because the fluid properties and flow rate are usually determined by establishing fluid properties at suction conditions, pump capacity is defined at suction conditions.

The cross-hatched area bounded by the expansion-suction-compression-discharge lines represents the amount of work absorbed by the fluid.

Allowing for fluid compressibility and valve slip, pump capacity may be calculated as follows. (Refer to Fig. 9.7.)

Let:

D = Displacement of one plunger during one stroke, $in.^3$
C = Clearance volume of one pumping chamber, $in.^3$
Q = Capacity of one pumping chamber during one cycle, $in.^3$
S_d = Slip through a discharge valve during one cycle, $in.^3$
S_s = Slip through a suction valve during one cycle, $in.^3$
S = Total slip = $S_d + S_s$ = total backflow through suction and discharge valves during one cycle, $in.^3$
w_d = Specific weight of pumpage at discharge conditions, $lb/in.^3$
w_s = Specific weight of pumpage at suction conditions, $lb/in.^3$
E = Expansion of fluid on suction stroke, $in.^3$
η_v = Volumetric efficiency = Q/D
v_s = Specific volume of fluid at suction conditions, ft^3/lb
v_d = Specific volume of fluid at discharge conditions, ft^3/lb

The mass of fluid in pumping chamber at end of discharge stroke, point 1 = the mass of fluid in pumping chamber at end of expansion, point 3. (Slip through the discharge valve can be ignored here because it is negligible relative to the clearance volume.)

$$C\,w_d = (C + E)\,w_s = C\,w_s + E\,w_s$$

$$E = \frac{C\,w_d - C\,w_s}{w_s} = C\left(\frac{w_d}{w_s} - 1\right)$$

and:

$$Q = D - S - E$$

Therefore:

$$Q = D - S - C\left(\frac{w_d}{w_s} - 1\right)$$

$$= D\left[1 - \frac{S}{D} - \frac{C}{D}\left(\frac{w_d}{w_s} - 1\right)\right]$$

$$\eta_v = \frac{Q}{D} = 1 - \frac{S}{D} - \frac{C}{D}\left(\frac{w_d}{w_s} - 1\right) \tag{9.2}$$

Since fluid density is often expressed as specific volume, the inverse of specific weight, the following substitution may be made:

$$\frac{w_d}{w_s} = \frac{v_s}{v_d}$$

Therefore:

$$\eta_v = 1 - \frac{S}{D} - \frac{C}{D}\left(\frac{v_s}{v_d} - 1\right) \tag{9.3}$$

Slip is usually expressed as a ratio, as shown in Eq. 9.3, or as a percent. A typical value of 0.03 (3%) can be used for most applications.

It should be noted that, although units were defined for C, D, w_d, and w_s for the derivation of Eq. 9.2, since these values are expressed in the equation as ratios, any suitable units may be used. In other words, C and D may be expressed in any units of volume, such as gallons, cubic feet, cubic centimeters, or cubic meters, so long as both C and D are in the same units. Similarly, specific weights can be expressed in units of kilograms/cubic meter or liter or cubic centimeter, pounds/cubic foot, or specific gravities, if both densities are in the same units. To calculate the volumetric efficiency with Eq. 9.2, using 0.03 for S/D, it is necessary to determine only C/D (a pump characteristic) and w_d/w_s (a fluid characteristic) ratios, but establishing firm values for these ratios is not always easily accomplished.

9.5.1. Determining the C/D Ratio

If the manufacturer has determined the clearance volume ratio for the specific plunger-liquid end combination, this number is used in the calculation. If this value has not been established, it is necessary to calculate this volume, or, better yet, to measure it. Measurement requires removal of the discharge valve and, with the plunger at top-dead-center, carefully measuring the volume of water, with a graduated beaker, required to fill the clearance volume. The correct volume of water is that which just starts to seep past the discharge valve, into the suction manifold, as the valve is returned to its normal closed position on its seat.

9.5.2. Determining the w_d/w_s Ratio

The pump specifier seldom provides the discharge density on his data sheet or in his inquiry, so it becomes the responsibility of the manufacturer or his representative to attempt to establish this number. The reduced-pressure reduced-temperature method of determining w_d, as provided in the Hydraulic Institute Standards,[1] requires the knowledge of discharge temperature, but discharge temperature cannot be determined, prior to running the pump, without a pressure-enthalpy (or similar) diagram. Since the pressure-enthalpy diagram also establishes specific weight, the reduced-pressure reduced-temperature method is not needed at all.

To best determine the fluid density ratio, a pressure-enthalpy diagram, for the specific fluid being pumped, should be used. Figure 9.8 is the left portion (liquid region) of such a diagram.

The following example will illustrate the use of this graph:

Fluid	propane
Suction temperature	80°F
Suction pressure	140 psig (155 psia)
Specific gravity	0.49
Discharge pressure	2000 psig (2015 psia)

We enter this chart from the left at 155 psia, proceeding to the right until we reach the line for 80°F. At this intersection, we can read a specific volume of approximately 0.0331 ft³/lb. Dividing this number into 0.0161, the specific volume of water at 60°F, confirms the specific gravity of 0.49. From this intersection, we proceed up a line of constant entropy (constant S) until we reach a pressure of 2015 psia. This establishes the discharge conditions of about 95°F and 0.0315 ft³/lb. Note that the propane has increased in temperature by 15°F—not because of any inefficiency, but because of compression. The specific volume at discharge is about 5% less than that at suction.

The compression process in a reciprocating pump is not totally isentropic (constant entropy), but it is very close. Any attempt to compensate for the small amount of inefficiency in the liquid end changes the results less than the error inherent in the calculated result. Such would *not* be the

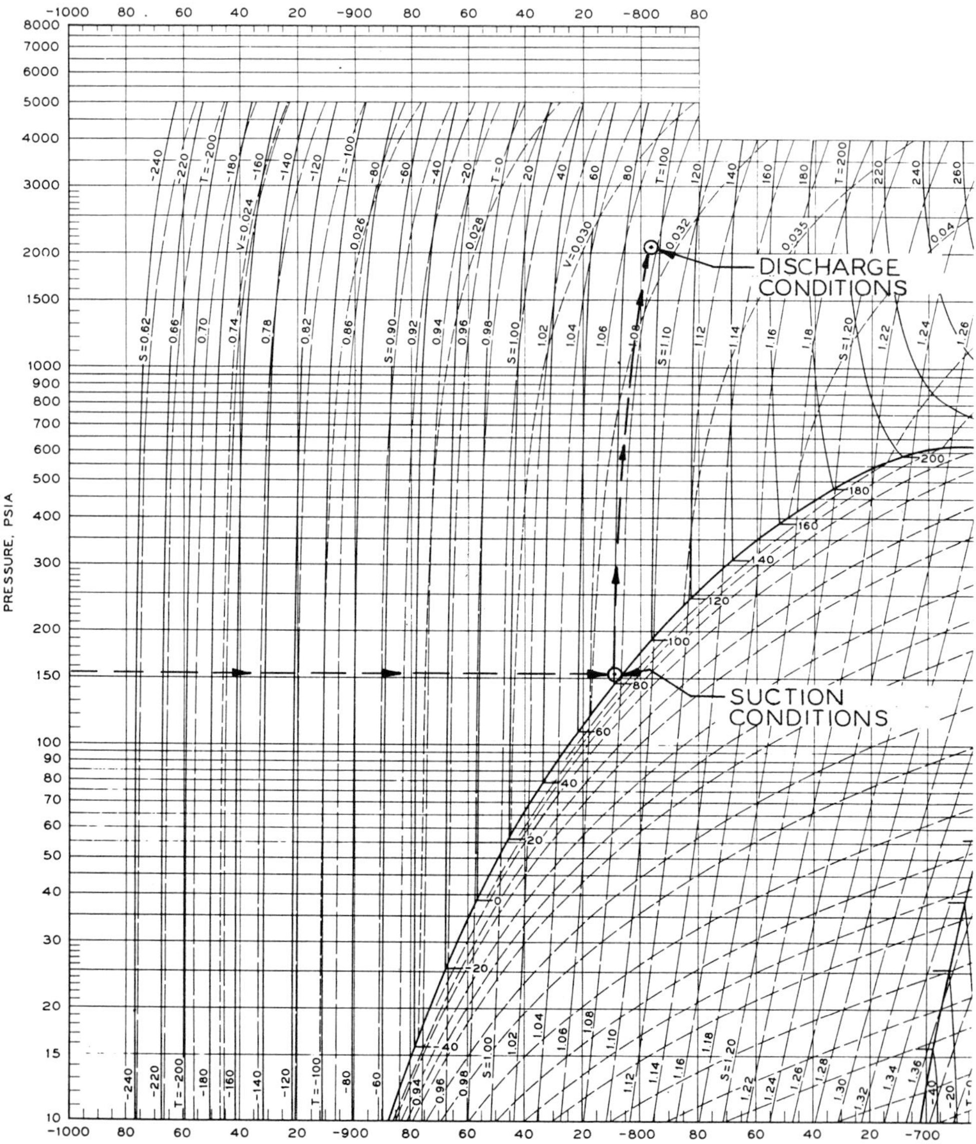

FIG. 9.8. Pressure-enthalpy diagram for propane. (Only the left portion of chart is shown.) (From *Fluid Thermodynamic Properties for Light Petroleum Systems* by Kenneth E. Starling. Copyright © 1973 by Gulf Publishing Company, Houston, Texas. Used with permission. All rights reserved).

case for a centrifugal pump, whose inefficiency usually has an appreciable effect on discharge temperature and specific volume.

With the discharge specific volume now established at 0.0315 ft³/lb, and assuming we have determined that the pump has a C/D ratio of 3.0, we can proceed to calculate volumetric efficiency. Repeating Eq. 9.3, and plugging in the values:

$$\eta_v = 1 - \frac{S}{D} - \frac{C}{D}\left(\frac{v_s}{v_d} - 1\right)$$

$$= 1 - 0.03 - 3.0\left(\frac{0.0331}{0.0315} - 1\right)$$

$$= 0.97 - 3.0\,(1.05 - 1) = 0.97 - 3.0\,(0.05) = 0.97 - 0.15 = \underline{0.82}$$

The volumetric efficiency of this pump, pumping propane under the above conditions, will be about 82%. For example, a pump displacing 100 gpm would have a capacity (in the suction line) of 82 gpm. If it is desired to have a capacity of 100 gpm, pump displacement must be 122 gpm (100/0.82).

When the pumpage is a mixture of different fluids, this procedure should be followed for each component, then a weighted average of all specific volume ratios should be calculated for use in Eq. 9.3.

9.6. POWER REQUIREMENT WITH COMPRESSIBLE FLUID

As stated earlier, the cross-hatched area of Fig. 9.7 represents the work absorbed by the pumpage. It is proportional to the hydraulic power of the unit. A close approximation of the hydraulic power can be calculated with the traditional Eq. 6.6 by using the average capacity at suction and discharge conditions. For example, if the pump in the above example were sized for a capacity (at suction conditions) of 100 U.S. gpm, the discharge capacity would be about 95 U.S. gpm. With a suction pressure of 140 psig and a discharge pressure of 2000 psig, the differential pressure is 1860 psi. Hydraulic power may then be calculated as follows:

$$P_o = \frac{Q\,\Delta p}{1714} = \frac{(97.5)(1860)}{1714} = \underline{106}\text{ hp}$$

To calculate the required power input we will use Eq. 9.1 for high suction pressure. Restated:

$$P_i = \frac{Q\,p_d}{1714\,\eta_m} - \frac{Q\,p_s}{1714}\,(\eta_m - 0.05)$$

For a compressible fluid, Q is the average of suction and discharge capacities. Continuing with the above example, $Q = 97.5$ U.S. gpm. Let's assume that the pump we've selected has a mechanical efficiency, for 0 psig suction and 2000 psig discharge, of 90%. Plugging these values into the equation:

$$P_i = \frac{(97.5)(2000)}{(1714)(0.90)} - \frac{(97.5)(140)}{1714}(0.85) = 126 - 7 = \underline{119} \text{ bhp}$$

The overall pump mechanical efficiency is therefore:

$$\eta_m = \frac{P_o}{P_i} = \frac{106}{119} = 0.89 = \underline{89\%}$$

If this pump were to pump water, the volumetric efficiency would be about 95%, and the capacity would be $0.95 \times 122 = 116$ U.S. gpm. For the same pressure conditions as above:

$$P_i = \frac{(116)(2000)}{(1714)(0.90)} - \frac{(116)(140)(0.85)}{1714} = 150 - 8 = \underline{142} \text{ bhp}$$

If we used the 119 bhp calculated previously to select an electric motor, a 125 hp motor would be selected. It is possible, though, to overload the motor if the design fluid does not compress as much as calculated, or if the fluid changes to one of less compressibility. It is therefore recommended that the driver be sized for a less compressible fluid, such as water. On water, this pump would require 142 bhp. A 150 hp motor is therefore recommended, unless the 125 hp motor has a service factor adequate to allow continuous operation at the 142 bhp.

REFERENCES

1. *Hydraulic Institute Standards, 14th ed.,* Cleveland, OH: Hydraulic Institute, 1983.
2. *Positive Displacement Pumps—Reciprocating (API Standard 674), 1st ed.,* Washington, DC: American Petroleum Institute, 1980.

Chapter 10
AUXILIARY EQUIPMENT AND LUBRICANTS

10.1. DRIVERS FOR POWER PUMPS

10.1.1. Electric Motors

Popularity and Speeds. The most popular driver for power pumps is the electric motor. To minimize the total unit cost, motor speeds are normally selected at 1800 rpm (4-pole) or 1200 rpm (6-pole) in 60 Hertz areas, and 1500 or 1000 rpm in 50 Hertz areas. A lower speed (900 or 600 RPM) is sometimes selected, but the premium price for the lower-speed motor is usually more than any premium required for a higher ratio speed reducer.

Starting Torque. The most popular electric motor for power pumps is the normal-starting-torque NEMA[1] design B. Table 10.1 lists the starting (locked-rotor) torques for these four-pole and six-pole motors. As shown in Fig. 6.17, the break-away torque required by a typical triplex pump, started against full discharge pressure, is about 150% of full-load torque. As can be seen in the table, NEMA B motors 40 hp and larger if four-pole, and 15 hp and larger if six-pole, have starting torques less than 150% of full load, and therefore may have difficulty starting a power pump against full discharge pressure. Because of this potential starting problem, some users have purchased high-starting-torque NEMA design C motors as power pump drivers—at a premium price. As explained in Chapters 11 and 13, this is not the preferred solution to this problem. The pump should be started against negligible discharge pressure. This preferred procedure protects the pump and drive system against overload conditions, and allows the use of a standard normal-starting-torque NEMA B motor. This procedure also keeps inrush motor current to a minimum amount of time, thereby minimally impacting the plant electrical system.

Long vs Short Shaft. Many electric motors are available in either a "long shaft" or "short shaft" configuration. Examples of long-shaft motors are those built in frames 405U and 405T. Motors of exactly the same power and speed ratings, but made with short shafts, are designated as

Table 10.1. NEMA Minimum Locked-Rotor Torques of Design B, 4 and 6 Pole, 60 and 50 Hertz, Single-Speed, Polyphase, Squirrel-Cage, Integral-Horsepower Induction Motors (When Supplied with Rated Voltage and Frequency), Expressed as a Percent of Full Load Torque.

Horsepower	1800 and 1500 rpm	1200 and 1000 rpm
1	275%	170%
$1\frac{1}{2}$	250%	165%
2	235%	160%
3	215%	155%
5	185%	150%
$7\frac{1}{2}$	175%	150%
10	165%	150%
15	160%	140%
20	150%	135%
25	150%	135%
30	150%	135%
40	140%	135%
50	140%	135%
60	140%	135%
75	140%	135%
100	125%	125%
125	110%	125%
150	110%	120%
200	100%	120%
250	80%	100%
300	80%	100%
350	80%	100%
400	80%	
450	80%	
500	80%	

frames 405US and 405TS, the "S" standing for "short shaft." The long shaft unit has not only a longer shaft, but also a larger diameter shaft and larger bearings. The long-shaft motor is intended for side-load applications such as v-belt and chain drives, where the sheave or sprocket is mounted directly on the motor shaft. The short-shaft motor is not recommended for such side-load applications, but is intended for direct-connection though a flexible coupling. The price for both style motors is usually the same. Two-pole motors are normally supplied only with short shafts because these high-speed units are rarely used with belt or chain drives.

To increase motor interchangeability within a plant, a user will sometimes specify that, when available, long shaft motors be supplied for all applications, whether side-loaded or direct-connected. When used for a direct-connected application, the longer shaft will require slightly more installation space, and the larger shaft diameter may require a larger coupling.

Conduit Box Location and Rotation of Box. As illustrated in Fig. 10.1, the standard electric motor, when facing the motor shaft extension, has its conduit box located to the left. This is designated by NEMA as an "F-1" assembly. This box location can sometimes be an inconvenience to the user, such as when the box falls between the motor and pump power end in a back-mounted motor arrangement, and hampers access to the power

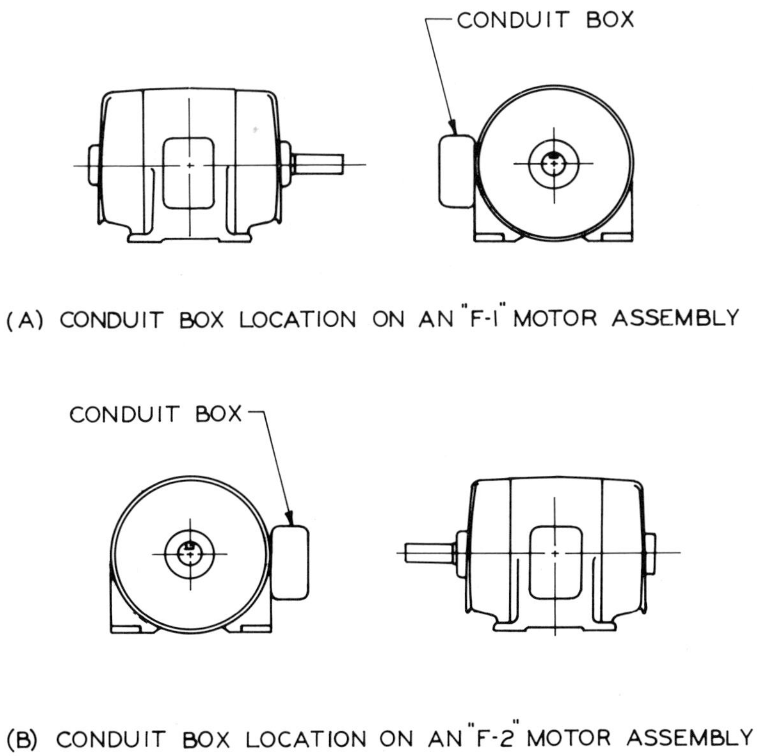

Fig. 10.1. Electric motors with alternate conduit box locations.

end for service. It can also be an inconvenient location for the routing of the electrical conduit to the motor. If desired, the motor can be supplied in an "F-2" assembly, which locates the box on the opposite side of the motor. (The motor is actually assembled with the same components, but with the frame reversed, end-for-end.) This special handling may cost a premium, though, so the user needs to determine if the extra cost is justified.

It is sometimes helpful to rotate the conduit box about its axis, to one of four positions, as shown in Fig. 10.2. This rotation is accomplished in the

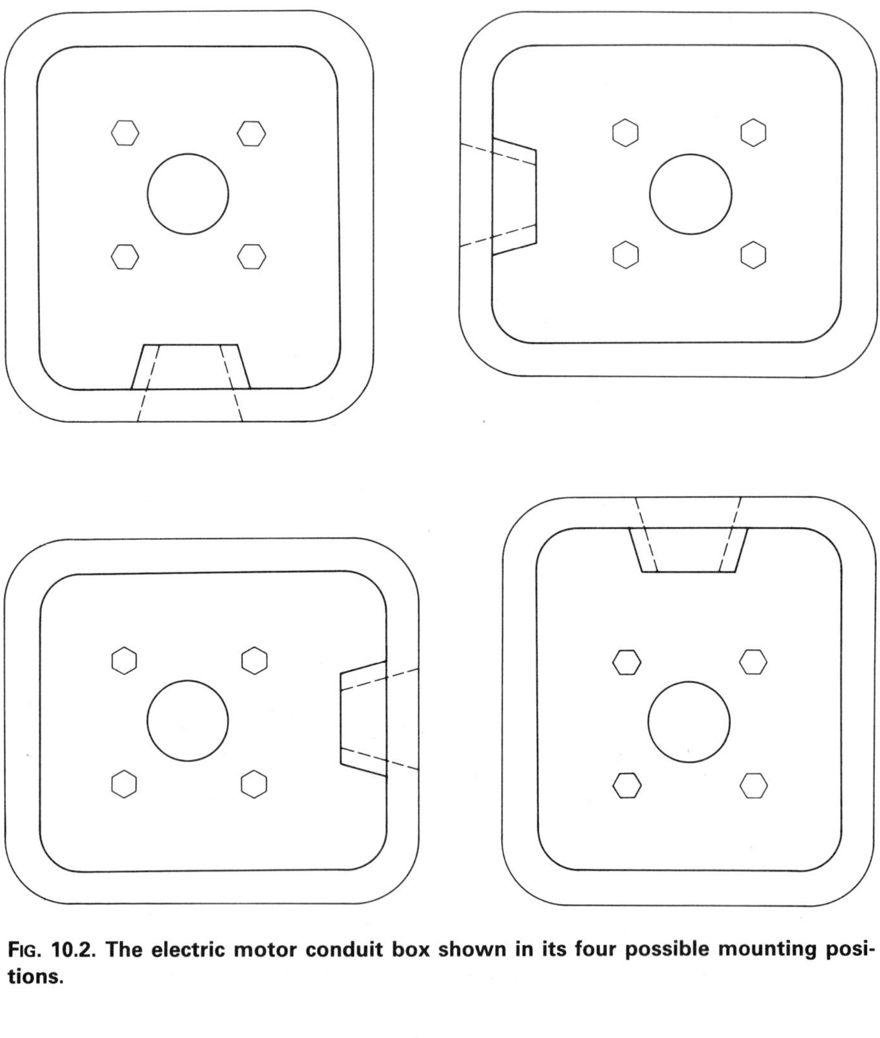

FIG. 10.2. The electric motor conduit box shown in its four possible mounting positions.

field, usually just prior to connecting the conduit. The box is normally fastened to the motor frame with four cap screws, located on 45 degree diagonals, about the box centerline. The motor is supplied with the conduit opening facing down, but by removing the four internal screws, the box can be reinstalled with the opening facing horizontally either direction, or up.

10.1.2. Engines

Internal combustion engines are used to drive power pumps when electricity is unavailable, unreliable, or a more expensive power source. Engines are fueled by natural gas, gasoline, or diesel fuel. A clutch should be supplied on the engine to allow it to be started and warmed up prior to the application of any load from the pump. Figure 10.3 shows an engine driving a triplex pump through a clutch, transmission, coupling, and gear unit. During the selection process (unless the unit is too small to warrant the expense), a torsional analysis of the complete system should be performed by the engine manufacturer to determine if any torsional natural

FIG. 10.3. A triplex power pump driven by an engine through a clutch, transmission, coupling, and gear. (*Courtesy of GASO Pumps, Inc., a Goulds Pumps Company*)

frequencies occur within the anticipated operating speed range. If any such dangerous frequencies are determined, it is necessary to change the stiffness or mass of the drive train, normally by selecting a different coupling.

It is essential that the rotation of pump and speed reduction system be carefully selected to match the engine. Engines usually rotate counter-clockwise, facing the engine shaft extension, and the rotation cannot readily be changed in the field.

Engines are available with a long list of optional features. The following is offered, not as recommendations, but as a check list to assist as reminders for those preparing an engine specification. The explosion-proof features are utilized if the installation is in a potentially explosive environment.

1. Standard compression heads
2. Oil bath air cleaner
3. Barring device
4. Box type base mounting
5. Natural gas carburetor, regulator, and gage
6. Engine-mounted instrument panel with gages
7. Oil temperature thermometer
8. Water-cooled exhaust manifold
9. Oil cooler
10. Friction type throttle control
11. Water pump and by-pass assembly
12. Low tension shielded (low fire hazard) ignition
13. Hydraulic governor with low fire hazard overspeed mechanism
14. Air-powered actuator to control governor setting
15. Radiator with pusher-type fan
16. Heavy-duty spark-arresting silencer
17. Air or gas starter
18. Tachometer with clutch
19. Special flywheel, carburetor, and lubrication system to allow engine speed to be varied from _ to _ rpm
20. Clutch and power take-off
21. Sentinel safety shut-down system to cut off fuel supply when actuated by low oil pressure or high water temperature
22. Low fire hazard enclosure for engine start-stop toggle switch
23. Crankcase oil heaters and motor-driven oil circulating pump, all explosion-proof, all thermostatically-controlled
24. Sole plates for engine—to be grouted to foundation

25. Torsional analysis
26. Start-up service

10.1.3. Steam Turbines

The starting torque of a steam turbine is typically 140% of full load torque, not too unlike a typical electric motor. A turbine-driven power pump should therefore be started against negligible discharge pressure.

To get optimum efficiency from a steam turbine, it is necessary that it be operated at a high speed, typically from 3600 to 4000 rpm. Such a high speed will require a speed reducer with a speed ratio higher than required for an electric motor.

As with an engine, the rotation of a steam turbine cannot be changed in the field. Therefore, during the selection process, it is necessary to carefully match the turbine, pump, and speed-reducer so that the rotation of all equipment is correct.

If a steam-turbine-driven power pump is being considered for purchase, with all of its speed-reduction and control requirements, it may be well to compare it with a steam-driven direct-acting pump, a much simpler and lower-speed machine. Especially in such services as slurry and hot oil, the low-speed d-a pump has developed a reputation for less maintenance than its higher-speed cousin.

10.1.4. Air Motors

An unusual driver for a power pump is the rotary air motor, as illustrated in Fig. 10.4. This air motor is mounted over the pump power end on adjustable legs and connected to the pump shaft with a v-belt drive. This unit is mounted on a trailer for portability, and can be used where electricity is not available or is considered a safety hazard. Here, again, the d-a pump may be a better selection.

10.2. MECHANICAL DRIVE SYSTEMS

Because power pump crankshaft speeds range from about 20 to 500 rpm, and their drivers typically run at speeds of 1000 to 1800 rpm, it is necessary to connect the two shafts with some type of speed-reduction power transmission. For units of 100 hp and below, the v-belt drive is popular, although cog belts, chains, and gears are also used in this range. Above 100 hp the gear is more popular, with belts and chains used with less frequency.

FIG. 10.4. An air motor driver top-mounted on a triplex pump. (*Courtesy Union Pump Co.*)

10.2.1. V-Belts

V-belts are economical, efficient (less than 2% power loss), quiet, normally furnished of static-conducting materials (to minimize accumulation of static electricity), offer a wide range of speed ratios, and provide maximum flexibility in the relative mounting of driver and pump. On the negative side, though, belts do wear more readily than gears, must be kept properly tensioned and aligned, and impose a side load on both the driver and pump shafts. Drivers connected to power pumps with v-belts are normally mounted either behind the power pump power end (back-mounted) or above the power end (top-mounted).

To uniformly share the load, all v-belts in a set must be very nearly the same length. As produced, belts vary slightly in length. The manufacturer therefore measures and marks each belt with a number which indicates its measured length. Each belt will be marked with a "46," "48," "50," "52," or "54." A "50" represents a belt of average length. A "46" or "48" represents a belt that is slightly undersize. A "52" or "54" represents a belt that is slightly oversize. All belts in a set should be the same number.

This match number should not be confused with the number indicating the nominal belt length. For example, a "710" belt has a 71-inch nominal circumferential length. A "1400" belt has a 140-inch nominal circumferential length. Each of these belts could, for example, be marked with a "50" match number, indicating minimal variation from the nominal length.

Back-Mounted Driver. To provide a method for adjusting the belt tension, the driver is mounted on an adjustable base or on two separate slide rails. These devices are furnished by motor suppliers. When using the separate rails, the rails should be installed with the adjusting screws pointing opposite directions (see Fig. 10.5), so that the driver shaft can readily be kept parallel to the pump shaft.

A pump and back-mounted driver are often mounted on a common baseplate or skid, the skid providing portability for the entire pumping unit. When the unit is to be permanently installed, such as in a processing plant, the baseplate may be omitted; with the pump, adjustable motor base, and belt guard bolted directly to the concrete foundation. Such an arrangement not only costs less, but is more rigid and runs quieter. It does, though, require more care during installation to assure proper positioning of the driver to obtain proper drive alignment, and the v-belt drive must be field-installed.

Top-Mounted Driver. To save space and simplify installation, a pump with a v-belt drive is often furnished with the motor mounted above, and supported by, the pump power end, as shown in Fig. 10.6. Motors up to about 150 hp are readily top mounted. The motor plate is supported by adjustable legs so that belt tension can be maintained. The belt guard is also supported by the power end.

10.2.2. Gears

Gears for power pumps are furnished in four basic types:

1. Internal— internal to the pump power end as standard equipment. Figure 10.7 shows a unit with internal gears and an integral motor driver.
2. Bolt-on— external to the pump power end, but the gear housing bolts to the power frame and the main gear mounts on the crankshaft—this unit contains its own independent

lubrication system—such a unit is optional equipment. Figures 10.3 and 10.8 show such units.

3. External— a unit separate from both the pump and driver—supported independently—must be coupled with flexible couplings to both pump and driver shafts.
4. Gearmotor—the gear and motor are built together—the pinion gear is mounted on the motor shaft—a special motor is required.

A single gear unit is usually adequate, but when pump speed is low (below about 150 rpm), pumps with built-in gears sometimes also require an external gear, belt, or chain drive.

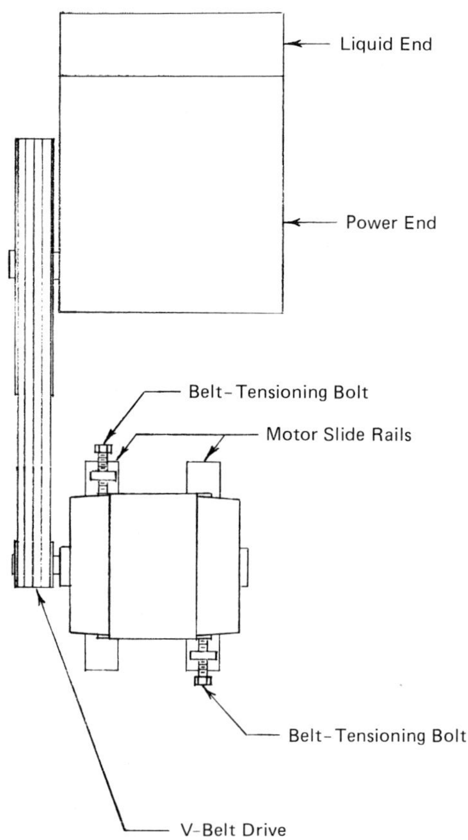

FIG. 10.5. Top view of a back-mounted motor connected to a pump with a V-belt drive.

FIG. 10.6. Horizontal triplex pump with top-mounted motor driver. (*Courtesy Union Pump Co.*)

10.2.3. Couplings

When a driver shaft is coupled to a gear unit, or an external gear unit to a pump, perfect alignment of the two shafts is not possible. It is therefore necessary that a flexible coupling be used. The coupling allows for some initial shaft misalignment, and for inevitable additional shaft movement as both driver and driven equipment warm up and are subjected to operating loads. Metal gear and grid type couplings can tolerate less operating misalignment than some couplings made with elastomeric (rubber) elements. (For a discussion of coupling alignment, see Chapter 12—Maintenance.)

10.2.4. Variable-Speed Drives

Variable-speed drives are used for power pumps when it is desired to accurately meter the pumpage into the system—and/or to save power during requirements of reduced capacity.

FIG. 10.7. A triplex power pump with internal gears and an integral motor. (*Copyright Kobe, Inc. Used with permission.*)

Variable-Pitch Sheaves. Small variable-speed drivers often utilize variable-pitch sheaves, as seen in Fig. 10.9. This type of unit has occasionally been a maintenance problem when a sufficient service factor was not used in its selection. Although manufacturers of such transmissions typically recommend a 1.5 service factor for driving a triplex pump, it has been common for such units to experience unsatisfactory belt and spline life unless a service factor of 2.0 is used in their selection. One reason for the unsatisfactory experience with these units is the requirement that they be run through the entire speed range daily. This procedure reduces the possibility that the mating splines will take a "set" at the common running position and that the belt will wear grooves in the sheave faces. Such a procedure is impractical, though, in processing plants which require that a unit operate continuously for days or even weeks at one setting. The higher service factor reduces the load on the parts, thereby reducing the

FIG. 10.8. Installation of a triplex pump with a bolt-on gear reducer. (*Courtesy Union Pump Co.*)

possibility that these parts will become set or locked at their operating positions. Since the largest variable-pitch transmission is rated at about 120 hp, the 2.0 service factor reduces its applicability to 60 hp. These units are usually made integral with electric motors, and sometimes with a speed-reduction gear on the output shaft. The 2.0 service factor does not apply to the motor or gear. A 1.0 SF is usually adequate for the motor. The gear should be selected in accordance with Table 10.2.

Hydraulic Couplings and Eddy-Current Clutches. Above 60 hp, variable speed of a power pump is commonly achieved with a hydraulic coupling or eddy-current clutch. Figure 10.10 shows a hydraulic coupling driving a quintuplex pump. Although effective at controlling pump speed, and thereby capacity, no energy is saved when operating at reduced speeds if pump discharge pressure does not drop as capacity is reduced. In process plants, constant discharge pressure is a common condition. The vessel into which the pump is discharging is usually at a constant pressure, regardless of the process flow rate. As shown in Chapter 6, for a constant discharge pressure, a power pump requires a constant torque, regardless of speed. The hydraulic coupling and eddy-current clutch both

FIG. 10.9. Variable-pitch sheave drive with integral gear driving a small simplex pump. (*Courtesy Union Pump Co.*)

FIG. 10.10. Hydraulic coupling driving a quintuplex plunger pump. (*Courtesy Union Pump Co.*)

require a torque input equal to torque output. If torque output is constant, then torque input must be constant. Since input torque *and* speed are constant, input power is constant. Therefore, reducing pump speed, although reducing the power required by the pump, does not reduce the power required by the coupling or clutch. Where does all the extra energy go? Into heat—inside the coupling or clutch. That's why these units require coolers. The lower the ratio between output and input speeds, the larger the cooler required. The large cooler can be seen in the foreground of the unit in Fig. 10.10. As a matter of fact, these drives are not 100% efficient even at top speed because they slip about 3%. Their maximum efficiency is therefore about 97%. Their efficiency is approximately equal to the speed ratio (output to input). (When used to drive a centrifugal pump, these units can save energy at reduced-speed operation because the torque requirement of a centrifugal unit varies approximately as the square of speed.)

An erroneous assumption is also occasionally made, during the selection process, that such a drive can be considered as a speed-reduction gear. The power requirement of the pump is calculated at a speed less than top speed for the clutch, and this power requirement is incorrectly used to select the clutch and motor. Such a process leads to a driver and clutch loaded higher than calculated. The correct procedure is to calculate the power requirement of the pump at the speed it will operate when the clutch is operating at its top output speed. The driver must be selected for a power output about 3% above this value.

Because an energy premium is required by such a clutch, when driving a power pump at a constant discharge pressure, many users simply drive the pump with a constant-speed driver, and bypass excess capacity back to the suction vessel.

Steel and aluminum rolling mills save energy with power pumps in their large hydraulic systems by having them charge a large accumulator (gas-filled or weighted), then switching them off to a negligible discharge pressure, or shutting them down altogether. The requirement for hydraulic fluid is intermittant, so the pumps may be in the idle condition for a long time, all the time saving energy. As fluid is required by the system, the accumulator pressure, or weight, drops; and at a predetermined level, the pump is brought back on line. If the level continues to drop, a second, and sometimes a third, pump will be brought on stream. These pumps will then be sequencially dropped out of service as the accumulator becomes recharged.

Multi-Speed Motors. The multi-speed motor is a compromise between the infinitely-variable-speed drive and the standard single-speed motor. Such a motor is able to run at more than one speed. For example, it could be built to run at 1800 or 900 rpm; at 1800 or 1200 rpm; or at any one of 1800, 1200, or 900 rpm. The greater the number of speeds, the larger and more expensive the motor. The starter also becomes larger and more complex, or more than one starter can be used in a less flexible arrangement. A multi-speed motor is used in conjunction with a bypass valve to establish the capacity delivered to the system. Let's say, for example, that an 1800/1200 rpm motor is selected to drive the pump. Full capacity is delivered by the pump at 1800 rpm. Two-thirds capacity is delivered at 1200 rpm. If the system requires 75% capacity, the motor would run 1800 rpm, and 25% would be bypassed. If the system requires 60% capacity, the motor would be operated at 1200 rpm, and only 7% of total rated capacity would be bypassed. The power requirement of the pump would then be only 67% of that at full speed.

Other Variable-Speed Drives. Although used sparingly to date, it is felt that, because of their reduced energy consumption at reduced pump speeds, variable-frequency-driven induction motors and hydrostatic transmissions could find increased application as power pump drivers.

10.2.5. Service Factors for V-belts, Gears, and Couplings

As discussed in Chapter 6, if suction and discharge pressures stay constant throughout the pumping cycle, the input torque required by a power pump varies with the same amplitude as the flow velocity. Because all liquid systems have some friction losses, the discharge pressure rises, and the suction pressure falls, as the liquid velocities from and to the pump increase. Therefore, even in the best of systems the input torque required by the pump varies by a greater magnitude than flow. The degree of this

Table 10.2. Service Factors for V-Belts, Gears, and Couplings.

Type of Pump	Maximum Torque/ Average Torque	Service Factor
Simplex single-acting	3.20	4.15
Simplex double-acting	1.60	2.08
Duplex single-acting	1.60	2.08
Duplex double-acting	1.27	1.65
Triplex single-acting	1.07	1.39
Quintuplex single-acting	1.02	1.33
Septuplex single-acting	1.01	1.31
Nonuplex single-acting	1.01	1.31

excess torque is totally a function of the variation of pressures imposed on the pump by the system, i.e., how the system reacts to the velocity variations produced by the pump.

To allow for the torque variations, it is necessary to oversize mechanical drive components. This oversizing is achieved by applying a "service factor" when selecting the component. The service factor is multiplied times the average pump power requirement, and the drive component is selected for this peak power. Table 10.2 lists the ratios of the maximum torque to the average torque, for constant pressures, and the recommended service factors, which are the torque ratios increased by 30% to allow for pressure variations in the system. If the pump will be subjected to high pressure variations or shock conditions, the service factor must be increased even more.

10.3. LUBRICATORS

In the language of reciprocating pumps, the term "lubricator" describes the device which provides oil to packing on either a power or direct-acting pump. A direct-acting pump can also use a lubricator to provide oil to the drive end, valve gear bearings, side rod bearings, and/or crosshead support shoes. This lubrication is a once-through process, and the system must be periodically refilled. It should not be confused with the lubrication system on the power end of a power pump. The two systems are independent. Although each power pump must have a lubrication system for its power end, it is common for a power pump to have no lubricator for the packing. In such a case, the pumpage provides the lubrication for the packing. As an intermediate arrangement (between continuous external lubrication and no external lubrication) some stuffing boxes are equipped with grease fittings which allow grease to be periodically injected, manually, for packing lubrication.

10.3.1. The Mechanical Lubricator

Figure 10.11 shows a mechanical lubricator mounted on top of a power pump power end, driven from the end of the crankshaft. The three tubes provide oil to the packing in the three stuffing boxes.

The mechanical lubricator is actually a small controlled-volume type power pump, with its own built-in sump, as illustrated in Fig. 10.12. A typical individual pumping unit is shown in Fig. 10.13. Instead of a conventional crankshaft and connecting rods, the plungers are driven by a cam shaft. On the suction stroke of each plunger, oil flows from the sump, through a sight-flow indicator, through a suction valve, and into the

FIG. 10.11. A mechanical lubricator mounted on the power end provides oil to the stuffing boxes on the liquid end. (*Courtesy of GASO Pumps, Inc., a Goulds Pumps Company*)

pumping chamber. On the discharge stroke, oil flows through the discharge valve, through the lubrication tubing, and to the point being lubricated. The stroke length of the plunger is adjustable, and can be adjusted while the pump is running. The seal between the plunger and its pumping chamber is accomplished by only a close clearance, so some of the oil which flows through the sight-flow indicator leaks back into the sump on the discharge stroke. It is therefore possible, with a worn lubricator, and/or with a low-viscosity oil, to have a flow indicated, but with no oil actually flowing through the discharge tubing.

Mechanical lubricators are built with some method of manually operating the plungers so that the pumping chambers can all be primed, and the lubrication lines filled with oil, prior to initial start-up of the main pump. If the pump is started with the lubricator lines dry, the packing will run dry during the first few minutes of pump operation, and will fail prematurely. The lubricator in Fig. 10.12 is manually operated by turning the small hand crank on the shaft extension opposite the input shaft. Some units provide a small mushroom cap on each little pump, which can be pumped by hand. On a new pump it is best to purge air from the system prior to

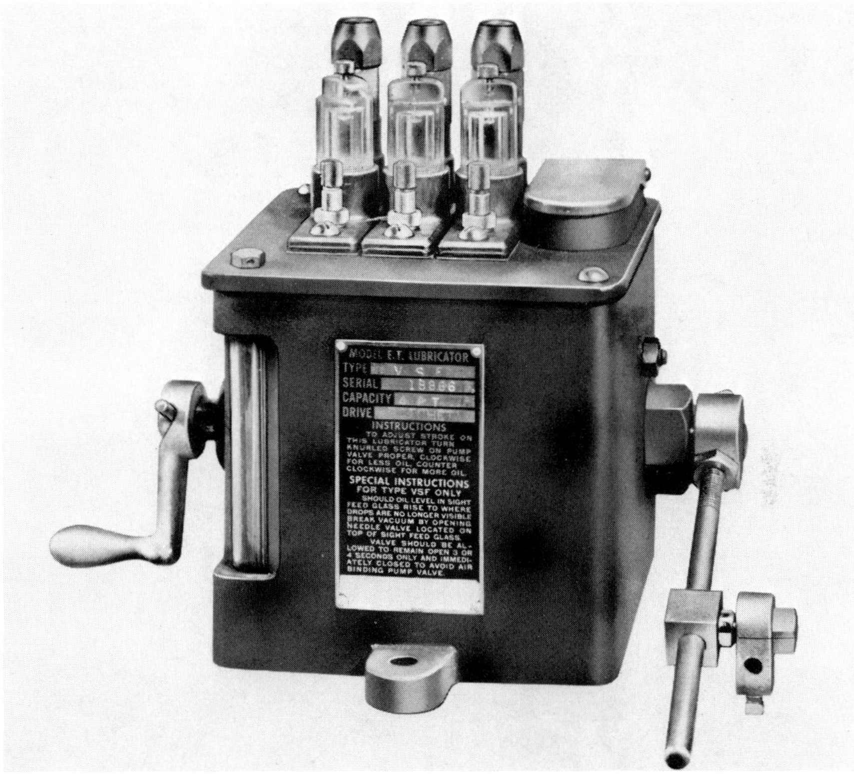

FIG. 10.12. A mechanical lubricator. (*Courtesy The Duriron Company/Pressure Products Industries Div.*)

installation of the packing. This allows the entire line to be purged. Any solids which are carried into the box by the oil can be wiped out prior to packing installation. If the packing is already installed, the lubricator connection at each box should be loosened to allow all air to escape.

For outdoor operation in cold climates, mechanical lubricators can be provided with sump heaters to keep the oil warm enough to flow through the lines. Both electric and steam heaters are available.

The sump capacity of a mechanical lubricator is usually $\frac{1}{2}$ to 1 gallon. Therefore, the lubricator normally needs to be refilled about every 24 to 48 hours. If it is desired to extend the period between refills, the lubricator can be ordered with a constant-level float valve. When connected to an elevated 55-gallon drum of oil, the lubricator will operate unattended for 1 to 6 months, avoiding the sometimes daily chore of refilling.

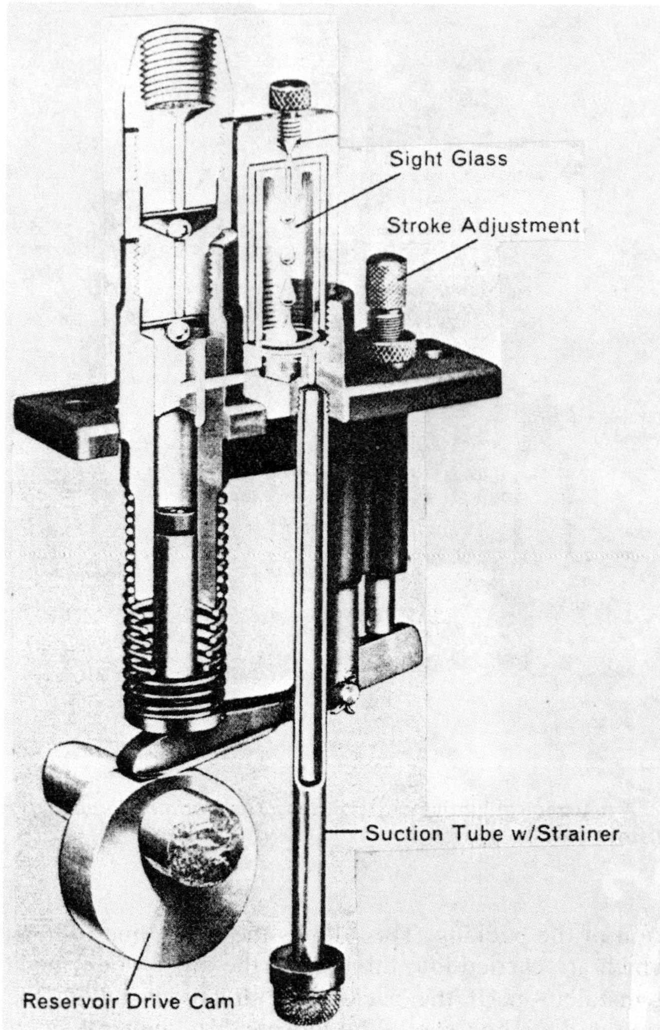

FIG. 10.13. A pumping unit of a mechanical lubricator. (*Courtesy The Duriron Company/Pressure Products Industries Div.*)

The mechanical lubricator, like any power pump, will discharge at the pressure imposed on it by the system. If injecting into a lantern ring bracketed by packing, the maximum pressure it will see is normally the main pump discharge pressure. The lubrication system is rarely fitted with relief valves, although it is possible to overpressure the lubricator if the lubrication rate is too high, or if the discharge tubing is restricted (pinched). Such overpressuring, although detrimental to the lubricator and pump stuffing box and packing, is normally not considered a hazard to personnel because of the small flow rate of the lubricator.

When mounted on a power pump, the lubricator shaft revolves at a speed only about 10% that of the main pump. This speed reduction is achieved in different ways, depending on the drive mechanism. The lubricator can be driven through a rod, as seen in Fig. 10.12, connected to the crosshead extension; it can be driven with a belt from the main pump shaft, as in Fig. 10.11; or can be directly connected to its own electric motor driver. When the lubricator is rod-driven, it is built with a rachet on the input shaft. The number of teeth "clicked through" during each revolution of the main pump is established by the manner in which the drive linkage is assembled. A click-through of two teeth is about right for many applications. Setting the linkage to click through more teeth causes the lubricator shaft to rotate at a higher speed. The plunger stroke length must then be reduced to prevent over-lubrication. The resulting short stroke sometimes causes the lubricator to lose prime and quit pumping. It is better to keep the stroke length up, and the speed down.

Some rachet-driven lubricators are not designed for the high oscillating frequencies common on current-day power pumps. Frequencies higher than 300 cycles/min cause some rachet mechanisms to wear rapidly. For satisfactory life at higher speeds, the lubricator must be properly selected.

Another problem common to mechanical lubricators is contamination of the lubricant in the sump. Contaminated oil is a poor lubricant, and sometimes causes the lubricator to loose prime (stop pumping). One source of contaminants can be the container used to fill the lubricator. Any bucket or can used for this purpose must be spotlessly clean prior to use. Another contaminant is rain. Some lubricators are not suited for outdoor installation, and a portion of any liquid falling onto the lubricator will find its way into the sump (sometimes through the top of the oil level sight glass). Yet another source of contamination is just moisture condensing from the atmosphere. Condensation can be reduced by keeping the lubricator full of oil, but the only way to avoid all condensation is to keep the lubricator warm.

10.3.2. The Drip-Type Lubricator

As discussed in Chapter 4, it has been found that oil dripped onto the surface of the plunger on the *atmospheric side* of the packing is often more effective than pumping oil into a lantern ring in the center of the packing. This arrangement not only extends the life of a mechanical lubricator (because of its reduced discharge pressure), but allows, in its place, the use of an atmospheric drip-type lubricator, as shown in Fig. 10.14. This system is totally gravity-fed. The lubrication rate for each box is established by the setting of a little metering valve. The drip rate is visible in a sight glass below each valve. A toggle lever controls a shut-off valve in the reservoir outlet. If desired, a solenoid valve can be supplied at the reservoir outlet to stop oil flow when the pump is shut down.

Drip lubricators are also available with an adjustable thermostatically-controlled electric heating element to maintain the temperature, and thereby the viscosity, and feed rate, of the oil.

Advantages of the drip-type lubricator over the mechanical lubricator are:

1. Simplicity—no moving parts
   —installation and adjustment are easy
2. Cannot lose prime
3. Lubrication rate is monitored on discharge side
4. Not affected by speed of power pump.

FIG. 10.14. A drip-type lubricator. (*Courtesy Oil-Rite Corp.*)

10.4. PULSATION DAMPENERS AND STABILIZERS

A pulsation dampener is a device which reduces the pressure pulsations created by the interaction of a pump and a system. The dampener performs this function by smoothing the flow in the suction or discharge line.

As shown in Chapter 6, the flow into and out of a reciprocating pump is not smooth, i.e., the flow pulses. Unless the suction and discharge vessels are mounted very close to the pump, these flow variations will be converted by the system into pressure variations. To reduce these pulses, a small vessel can be installed in the suction and/or discharge line. Such a vessel mounted in the discharge line is called a "pulsation dampener." If the suction pressure is high enough, this same type device can be used in the suction line, but it is more desirable to use a *special* dampener there which also separates any free gas from the liquid. Such a device is called a "suction stabilizer," as is shown in Fig. 10.15.

10.4.1. Types of Pulsation Dampeners

Pulsation dampeners are provided in a variety of configurations. Three basic types are (1) the gas-charged dampener, (2) the acoustic type, and (3) the spring-loaded type. Figure 10.16 shows a gas-charged dampener without a bladder (an air bottle), and Fig. 10.17 a gas-charged dampener with a bladder. Figure 10.18 shows a variation of the conventional appendage-mounted bladder-type unit in which a baffle deflects the fluid into the pulse chamber to increase effectiveness. Figures 10.19 and 10.20 show acoustic types, and Fig. 10.21 a spring-loaded type. The gas-charged dampener functions by the gas compressing and expanding to reduce the velocity variations in the pipe. The acoustic dampener functions by using the compressibility of the pumpage itself to absorb the variations. The spring-loaded dampener functions by the compression and expansion of the spring. Figure 10.22 shows a quintuplex pump in water reinjection service equipped with both a suction stabilizer and a discharge dampener (acoustic type). Some installations use a section of hose to connect the suction line to the pump. The flexing of the hose during operation absorbs much of the pulsation from the pump. The hose does have the shortcoming of being unable to separate free gas.

In the 1960s, when power pumps were first used in quantity for high-pressure water cleaning, water was fed from the water supply system, at about 50 psig, directly into the pump. No suction stabilizer was used. A high frequency of cylinder failures lead to a test during which the incoming water was observed to contain free air (even at 50 psig). The suction system was changed to include an atmospheric tank, adjacent to the

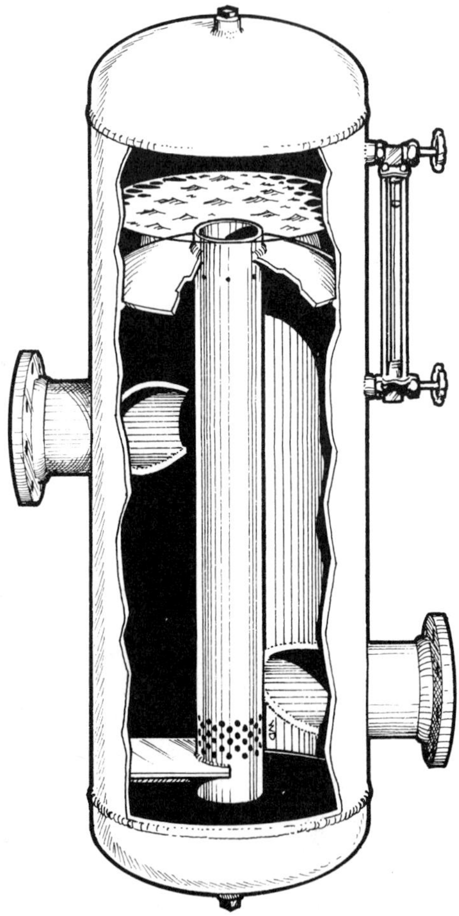

FIG. 10.15. A suction stabilizer. (*Courtesy Fluid Kinetics Corp.*)

pump, which allowed the air to escape from the water. Even though suction pressure was reduced to atmospheric, the free air was eliminated and acceleration head was mostly eliminated, and cylinder failures dropped markedly. This tank is nothing more than an atmospheric suction stabilizer.

10.4.2. How a Pulsation Dampener Smooths the Flow

The dampener absorbs a small volume of liquid from the pipe when the pressure in the pipe rises above the average, then it discharges the liquid back into the pipe when the pressure in the pipe falls below the average.

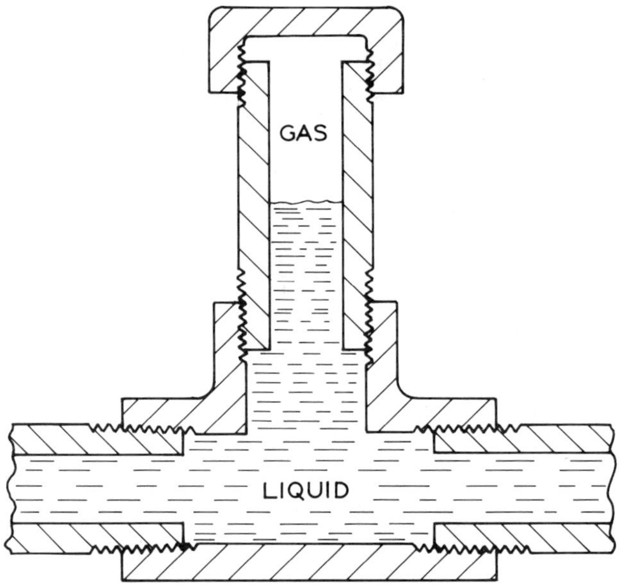

FIG. 10.16. A simple stand-pipe dampener (air bottle) installed in a "T."

In effect, it chops off the peaks and fills in the valleys. The *flow* variations in the pipe between the dampener and the pump are not affected by the dampener, but the flow outboard of the dampener is made more uniform. Because the flow in the line is more uniform, the *pressure* variation is reduced. The dampener therefore reduces pressure variations by reducing the velocity variations. The dampener benefits not just the system, but also the pump, by reducing the pressure variations imposed on the pump.

As an example, let's consider a triplex pump with a capacity of 100 gpm. As shown in Chapter 6, this unit has an instantaneous maximum flow of 107 gpm and an instantaneous minimum flow of 82 gpm. The task of the dampener is to reduce this 25 gpm variation. If the dampener reduces the flow variation to 10 gpm, then the pressure variation may drop 70% (say from 200 psi to 60 psi), although the degree of attenuation will be a function of the system. The effectiveness of a discharge pulsation dampener is shown in the oscilloscope traces of Figs. 10.23 and 10.24.

10.4.3 How a Suction Stabilizer can Reduce Discharge Pulsations

Often severe discharge pulsations are caused by poor suction conditions rather than by the normal pumping action of the pump. Cavitation due to acceleration head, and/or free gas in the fluid, can cause shock conditions

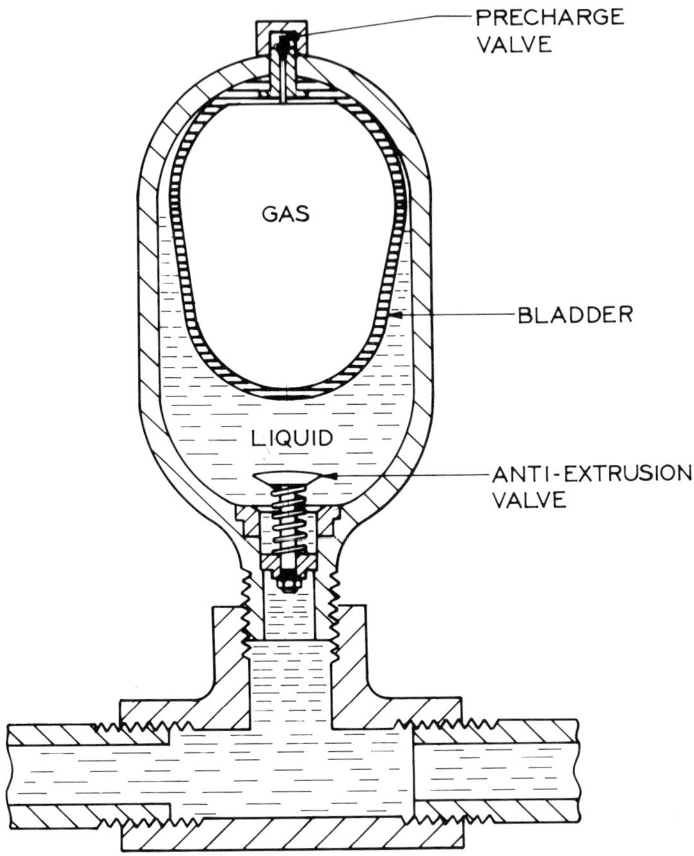

FIG. 10.17. A bladder-type dampener installed in a "T."

in the pump that are transmitted into the discharge line. Such conditions can be alleviated by an effective suction stabilizer. Installation of a discharge dampener will reduce these discharge pulses, but such action is actually just masking the suction problem. A suction stabilizer can reduce pulsations in both the suction and discharge piping, while significantly adding to the life of pump components. In summary, although a discharge dampener may reduce system pulsation to an acceptable level, it may be masking poor suction conditions. In such a case, a suction stabilizer would be a better investment than a discharge dampener.

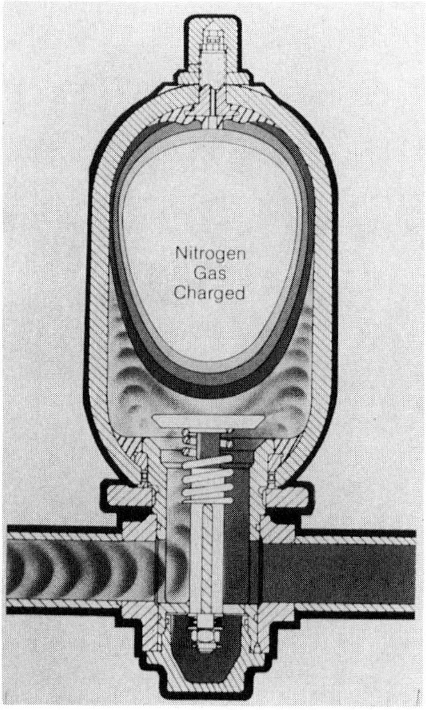

Fig. 10.18. A flow-through bladder-type dampener with a deflector baffle. (*Courtesy Greer Hydraulics, Inc.*)

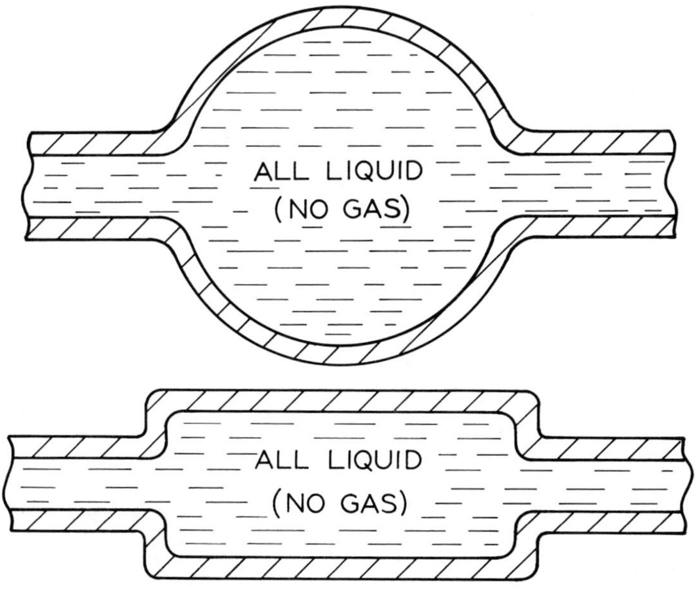

Fig. 10.19. Acoustic type dampeners.

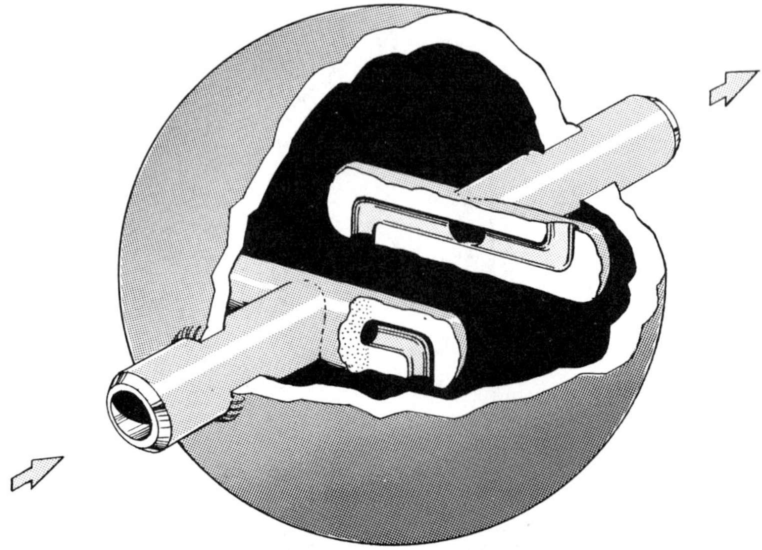

FIG. 10.20. An acoustic-type discharge pulsation dampener. (*Courtesy Fluid Kinetics Corp.*)

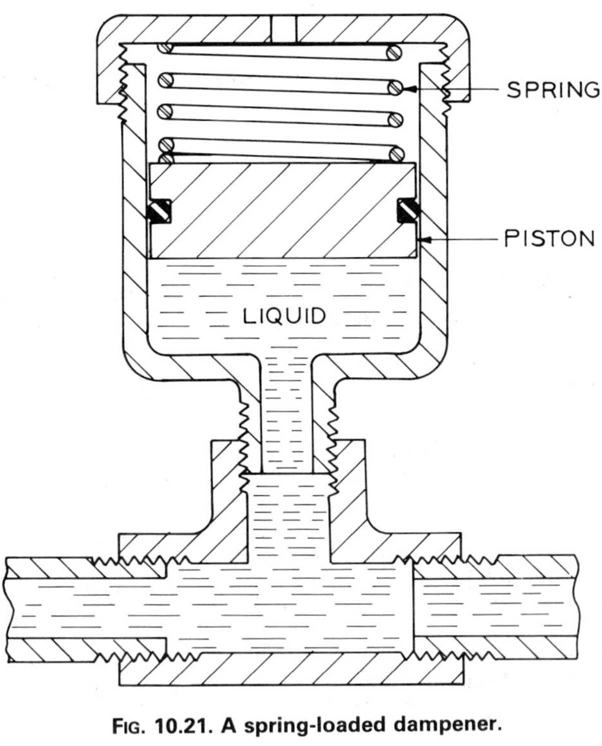

FIG. 10.21. A spring-loaded dampener.

FIG. 10.22. A pumping system with suction stabilizer and acoustic discharge dampener. (*Courtesy Fluid Kinetics Corp.*)

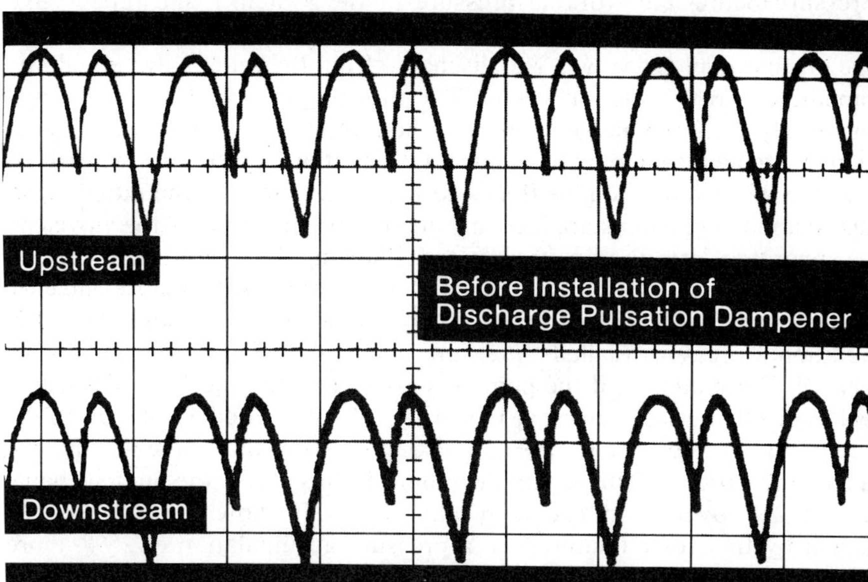

FIG. 10.23. Traces of discharge pressure before installation of pulsation dampener. (*Courtesy Greer Hydraulics, Inc.*)

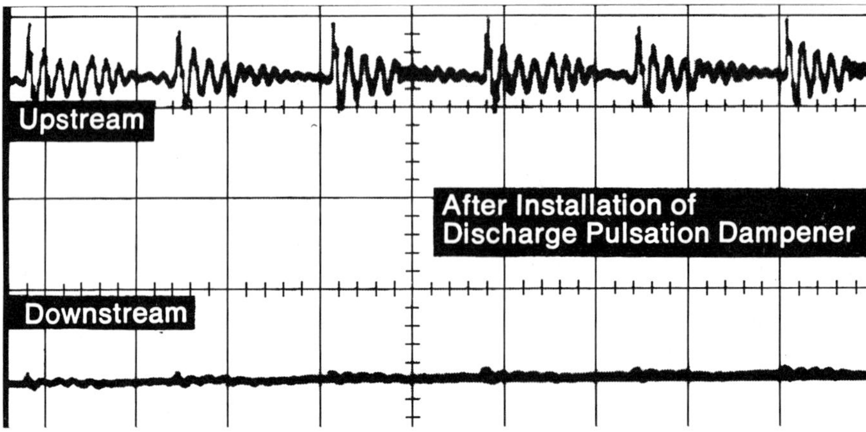

FIG. 10.24. Traces of discharge pressure after installation of pulsation dampener. (*Courtesy Greer Hydraulics, Inc.*)

10.5. PRESSURE-RELIEF VALVE

Because it is a positive displacement pump, the discharge line of a reciprocating pump should be fitted with a pressure-relief valve. (The only exception to this rule is the direct-acting plunger pump which has a ram pressure below the working pressure of the system.) The relief valve should be installed in the piping between the pump discharge nozzle and the discharge block valve. The discharge from the relief valve should be piped back to the suction vessel. The throttling action of the valve will inevitably create some gas, and this should not be dumped into the suction line to be ingested by the pump. If the suction vessel is not accessible, and the valve must be piped back to the suction line, it should be piped upstream of a suction stabilizer so that any gas generated by the valve can be separated from the liquid prior to its flowing back into the pump.

The relief valve should be adequately sized to pass the entire capacity of the pump, and while passing that capacity, to create no more than 10% additional pressure (called "pressure accumulation") on the pump discharge. For example, if the pump discharge is 1000 psig, the relief valve will typically be set to crack at 1100 psig (10% over rated discharge). If the accumulation pressure is 10% of cracking pressure, the valve will impose a backpressure of 1210 psig on the pump. In this mode, the pump is being loaded 21% over rated load. Typically relief valve flow ratings are established by their manufacturers at a pressure accumulation of 25%, more than desired for pump application. Therefore, to obtain the desired 10% accumulation, it is necessary to select a relief valve larger than one rated at a 25% accumulation.

It is also necessary to size the discharge line of the relief valve for the total flow of the pump. An undersized line will increase the back pressure on the valve, and therefore also on the pump. Don't scrimp.

10.6. LUBRICANTS

Lubricants are required for the power end of power pumps, usually for the drive end of direct-acting pumps, and often for the packing in the liquid ends of both types of pumps. Occasionally, other wear points on d-a pumps will be provided with mechanically-fed lubrication. These points include the valve gear mechanism (levers and shafts), side rods on plunger pumps, support shoes under large plungers, and the packing in the drive end.

10.6.1. Packing Lubricants

Liquid end packing that receives inadequate lubrication from the pumpage should be lubricated from an external source. The lubricant found to be most effective in extending packing life has been steam cylinder oil (the same oil used to lubricate the steam end of a steam pump). Steam cylinder oil is compounded with a tallow which gives it quite a tenacity for metal surfaces. This property, which makes it suitable for clinging to a cylinder wall under the scraping action of piston rings and washing conditions of a hot and wet steam, also makes it suitable for clinging to the surface of a plunger sliding through packing while being washed by the pumpage.

If the pump is unprotected from winter temperature extremes, and the lubricator is not equipped with a heater (or if the pumpage is cold), steam cylinder oil may become too viscous to flow and/or lubricate properly. In this case, rock drill oil is usually the choice. It, too, is compounded for the scraping action of a piston, and therefore performs well as a packing lubricant. Rock drill lubricants are available with pour points as low as 35°F below zero.

If the pumpage cannot tolerate even the slightest trace of one of the above lubricants, a special lubricant must be selected. Its lubricating properties should be as close as practical to those of steam cylinder or rock drill oil.

10.6.2. Drive End Oils for Direct-Acting Pumps

The drive end of a direct-acting pump, unless constructed for nonlubricated operation, must be lubricated. The lubricant is required primarily for the piston rings and "D" slide valve. If the pump is driven by steam, the oil should be a steam cylinder oil suitable for the temperature of the steam. If driven by a cool gas, the lubricant is usually an industrial oil.

10.6.3. Power End Oils

Enclosed power ends require some type of oil for lubricating all sliding and rolling parts. Smaller, higher-speed pumps, with smaller running clearances, use lower viscosity oils equivalent to SAE 10W, 20, or 30. Recent tests, based on operating temperature, show that a multiviscosity motor oil, such as 10W–40, outperforms the single-grade industrial and motor oils. Larger, lower-speed pumps, with larger running clearances, require more viscous oils. This is particularly true for power ends with internal gears, which usually require an oil equivalent to an SAE 80 or 90 gear oil with "extreme pressure" properties to meet the total lubrication requirements of the power end components.

If the pump is to be operated in a cold environment, the oil should be warmed to some minimum temperature before a load is applied to the bearings. Otherwise, lubrication will be inadequate, and bearing life will be reduced. This minimum recommended temperature is as high as 70°F for some of the heavier oils. Power end oil may be warmed by operating the pump unloaded (with low discharge pressure), or with electric heaters installed in the power end.

10.7. POWER END OIL PUMPS

The power ends of larger power pumps are often provided with a small pump to circulate oil to the bearings, gears, and crossheads. Some pumps utilize the internal speed-reduction gears to circulate the oil, some drive a small oil pump from the crankshaft, and some utilize a totally-external motor-driven oil pump. Figure 10.25 shows a power pump with a motor-driven oil pump and filter.

10.8. POWER END HEATERS

For outdoor installations, especially in colder climates, the power end of a power pump is often equipped with an electric heater to prevent the oil from becoming too viscous. Care must be exercised in selecting a heater so that the power density (watts/sq in.) of the heating surface does not exceed that recommended for the particular oil. A more viscous oil requires a lower power density because it is more sluggish and unable to carry heat away from the heater as quickly as a lighter oil. A heater with a power density too high for the oil will cause the oil to locally overheat and

FIG. 10.25. A power pump with external motor-driven lube-oil pump and filter. (*Courtesy of GASO Pumps, Inc., a Goulds Pumps Company*)

carburize. A typical maximum recommended power density is 15 watts/sq in. These heaters are often controlled by thermostats which are also immersed in the oil.

10.9. POWER END OIL FILTERS

A power pump with a power end pressure lubrication system can be equipped with an oil filter. The filter must be carefully selected to be suitable for the maximum viscosity and pressure expected during a cold start, and must be capable of bypassing the oil if it loads up with solids.

10.10. POWER END OIL LEVEL INDICATORS

The level of the oil in the power end should be readily determinable, and preferably while the pump is running. An oil level indicator can be a simple dip stick, a transparent bullseye, an elbow, or a sight-level glass. The bullseye and elbow may be limited to checking only when the pump is idle. Some elbows give a false oil level indication by trapping a small volume of oil in the bend of the elbow.

The sight-level glass can be a problem if the pressure in the power end is not exactly atmospheric. The top of most sight glasses is vented to the atmosphere. Some power end breathers contain a check valve which creates a slight vacuum in the power end for the purpose of limiting the leakage of oil through seals. Such a differential pressure will cause a sight-glass to give a false indication of a low oil level. In corrosive atmospheres, some power ends are provided with a small positive pressure of dry gas to exclude the environmental air. This slight differential pressure will cause a sight glass to give a false indication of a high oil level. In such cases it is necessary to vent the top of the sight glass back to the top of the power end to get a correct oil level indication.

10.11. POWER END OIL PRESSURE SWITCH

If the power end is pressure-lubricated, a pressure switch may be provided to indicate low oil pressure. The switch may be connected to an alarm to sound a warning, or to the driver (to the starter if motor-driven) to shut the unit down. Such a switch must be selected with the proper enclosure and materials for the environment, such as explosion-proof in an explosive atmosphere and/or copper-free in an ammonia atmosphere. The electrical circuit must provide a time-delay provision for start-up, to allow oil pressure to be established prior to the switch becoming effective.

10.12. POWER END OIL LEVEL SWITCH

A switch may be provided which signals a low oil level. This switch is subject to the same considerations stated above for the oil-pressure switch and the oil-level indicator.

10.13. POWER END OIL COOLER

If the power end generates more heat than can be dissipated by the normal conduction and convection of air around the power end, a cooler must be provided. If the unit is pressure-lubricated, the oil can be circulated through a heat exchanger prior to feeding to the points of lubrication. Otherwise a cooling jacket may be provided in the cover, below the oil level; or in the sump area.

10.14. SPEED-CONTROL DEVICES FOR DIRECT-ACTING PUMPS

The speed of the direct-acting pump is normally controlled by throttling the inlet gas. (See Chapter 6 for more details.) The throttling valve may be controlled manually, by the speed of the pump, by a liquid level in a vessel, or by the pump discharge pressure. The constant-speed governor shown in Fig. 10.26 contains a small reciprocating oil pump whose dis-

FIG. 10.26. A constant-speed governor on a direct-acting pump. (*Courtesy Union Pump Co.*)

placement is proportional to the main pump speed. The oil pump discharge pressure controls the throttle valve.

REFERENCE

1. NEMA—National Electrical Manufacturers Association—Motor and Generator Section, 2101 L Street NW, Washington, DC 20037

Chapter 11
APPLICATION

11.1. DEFINITION

For the purposes of this book, "application" will be defined as the function performed by the purchaser, or his contractor, to design the system and install the pump.

11.2. SYSTEM DESIGN

System design is equal in importance to pump speed in producing a satisfactory installation. Improper design has often resulted in systems which vibrate and are noisy. Pulsations may be severe enough to damage instrumentation, system components, and pump components.

Field experience and information from the Hydraulic Institute Standards are condensed in the following list of system design guidelines, as illustrated in Fig. 11.1:

A. The suction vessel should:
   1. Be large enough to provide sufficient retention time to allow entrained gas bubbles to rise to the liquid surface.
   2. Contain a weir plate, to force gas bubbles toward the surface, with its top sufficiently below minimum tank level to avoid turbulence.
   3. Have feed and return lines enter below the minimum liquid level.
   4. Include a vortex breaker over the outlet (pump suction) line. (Fig. 11.2 shows a simple vortex breaker.)

B. The suction piping should:
   1. Be as short and as direct as possible.
   2. Be one or two pipe sizes larger than the pump suction connection.
   3. Have an average liquid velocity less than the value from the curves in Fig. 11.3. (See Chapter 7 for a discussion of Fig. 7.6.)
   4. Contain a minimum number of turns. Necessary turns should be accomplished with long-radius elbows or laterals.
   5. Be designed to preclude the collection of vapor in the piping. (There should be no high points, unless vented.)

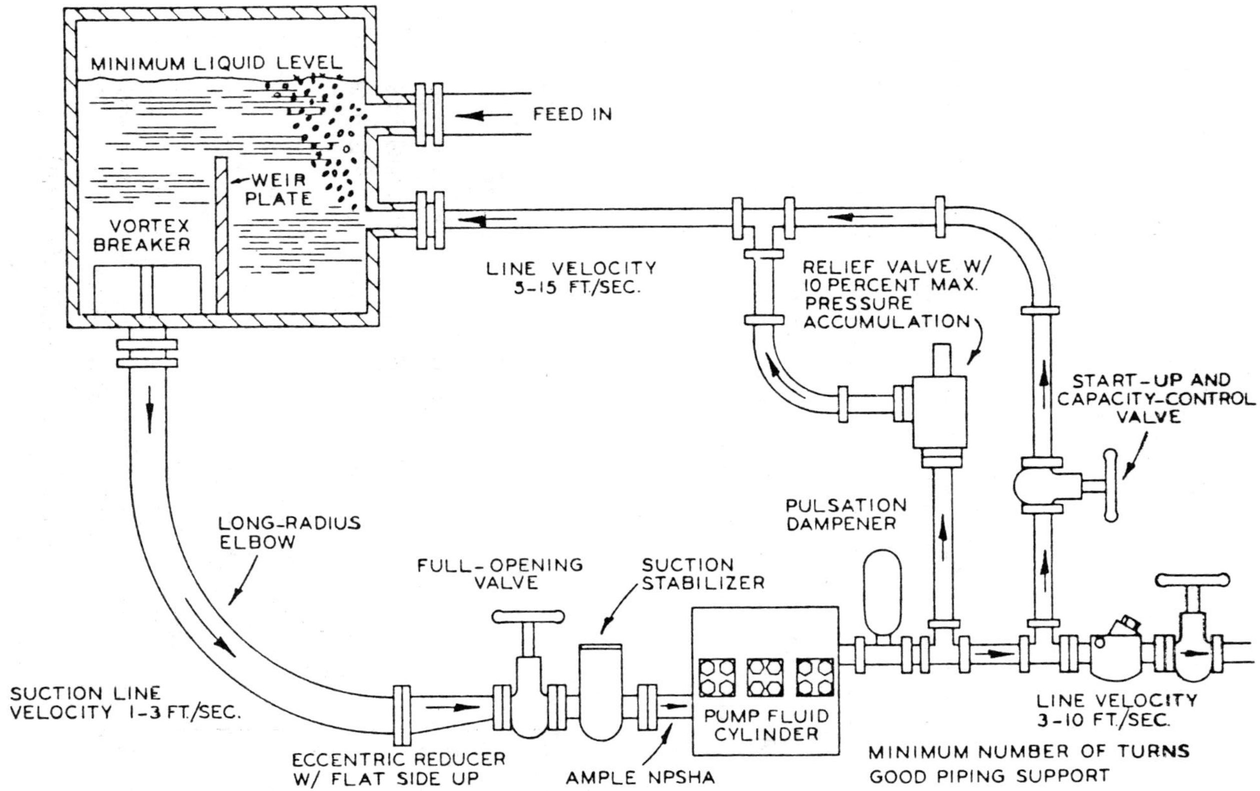

Fig. 11.1. Reciprocating pump system—Good design.

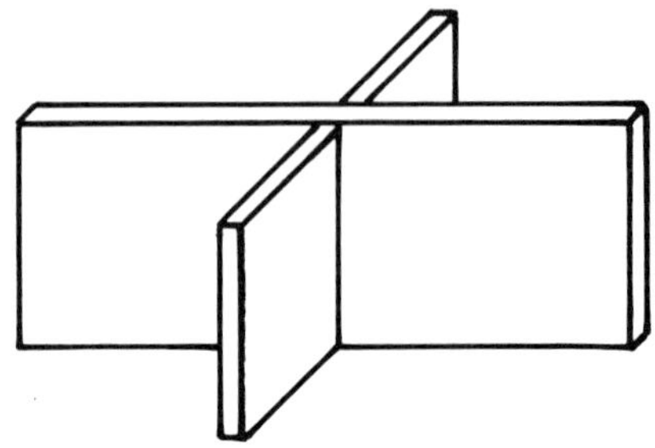

Fig. 11.2. A simple cross, fabricated from plate and placed over the suction vessel outlet connection, will reduce the tendency of the liquid to vortex.

6. Be designed so that NPSHA, allowing for acceleration head, exceeds NPSHR.
7. If acceleration head is excessive, include a suitable suction stabilizer, bottle, or pulsation dampener located in the suction pipe, adjacent to the liquid end.

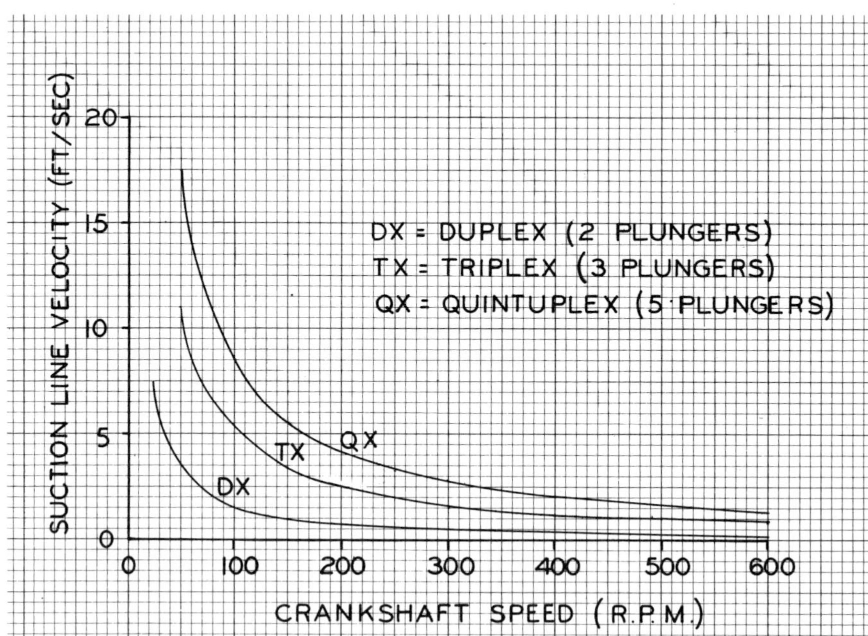

Fig. 11.3. Suction line velocity for constant acceleration head, as a function of pump type and speed.

8. Contain a full-opening block valve so that flow to the pump is not restricted.
9. Not contain a strainer or filter unless regular maintenance is assured. (The starved condition resulting from a plugged strainer can cause more serious damage to the pump and system than is caused by solids.)

C. Discharge piping should:
1. Be one or two pipe sizes larger than the pump discharge connection.
2. Have an average velocity less than three times the maximum suction line velocity.
3. Contain a minimum of turns, utilizing long-radius elbows and laterals where practical.
4. Include a suitable pulsation dampener, or provisions for adding one, adjacent to the pump liquid end. The maximum discharge pressure pulse should be limited to 10% of the average discharge pressure, to minimize the load on the pump and the setting of the relief valve cracking pressure.
5. Contain a relief valve, sized to pass full pump capacity at a pressure that does not exceed 110% of cracking pressure. The discharge from the relief valve should be piped back to the suction vessel so that gases liberated through the valve are not fed back into the pump.
6. Contain a bypass line and valve so that the pump may be started against negligible discharge pressure.
7. Contain a check valve to prevent the imposition of system pressure on the pump during start-up.

Figure 11.4 illustrates a system with features that often lead to serious problems.

In addition to compliance with these guidelines, calculations must be performed to establish pipe friction losses and piping natural frequencies.

11.3. LOCATION OF DAMPENERS AND STABILIZERS

To be most effective, a dampener (or stabilizer) must be mounted close to the fluid cylinder (see Fig. 11.5). Location just a few feet away will reduce its effectiveness. Because the dampener does not reduce the velocity variations between itself and the pump, it is important to minimize the length of pipe between the pump and dampener.

As shown in Fig. 11.5, a discharge dampener (an appendage-mounted type) is often mounted on a "T" adjacent to the fluid cylinder. In such an

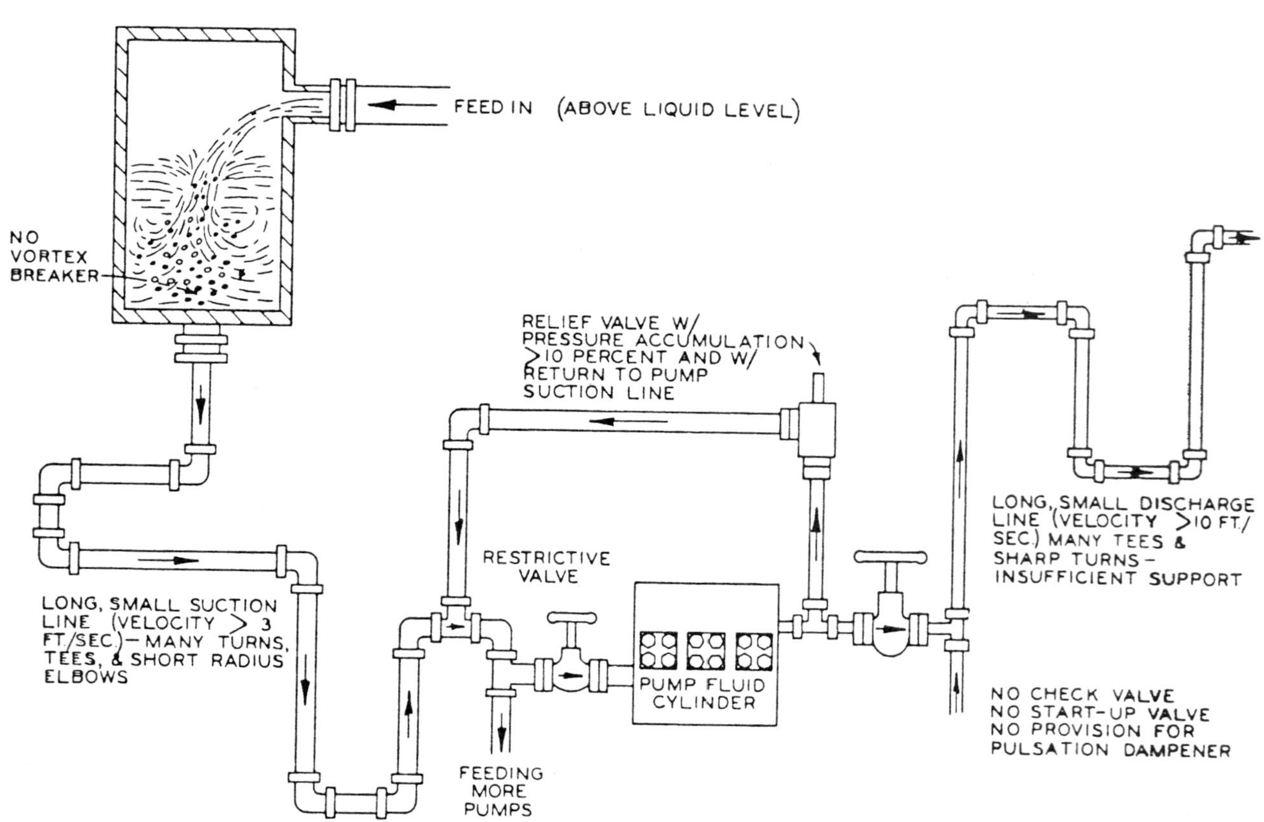

FIG. 11.4. Reciprocating pump system—Poor design.

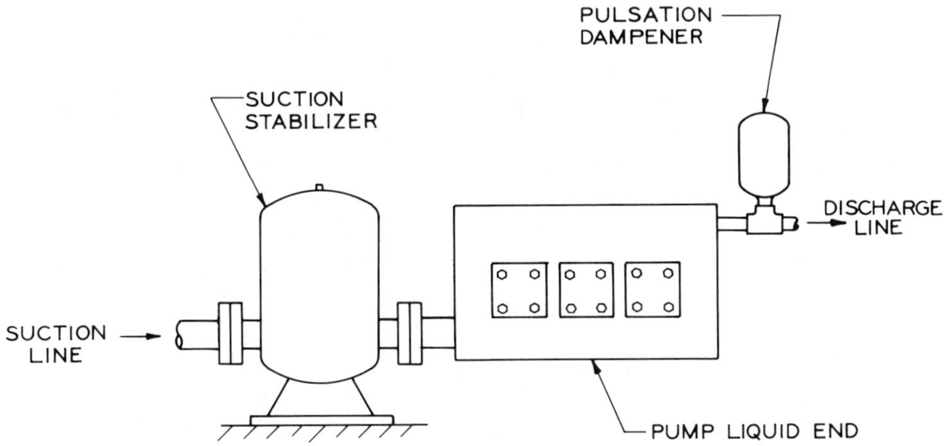

FIG. 11.5. Dampeners are most effective when mounted adjacent to the pump liquid end.

arrangement, the dampener should be right on top of the "T," with no restriction between the "T" and dampener. There should be no valve between the "T" and dampener, because the extra liquid in the valve would increase the inertia of the liquid leg. Also, the valve would increase the friction drop between the "T" and dampener. Both would reduce the effectiveness of the dampener. An alternate arrangement, illustrated in Fig. 11.6, puts the dampener in the straight run of the "T," and the discharge line off the leg of the "T." This provides more direct communi-

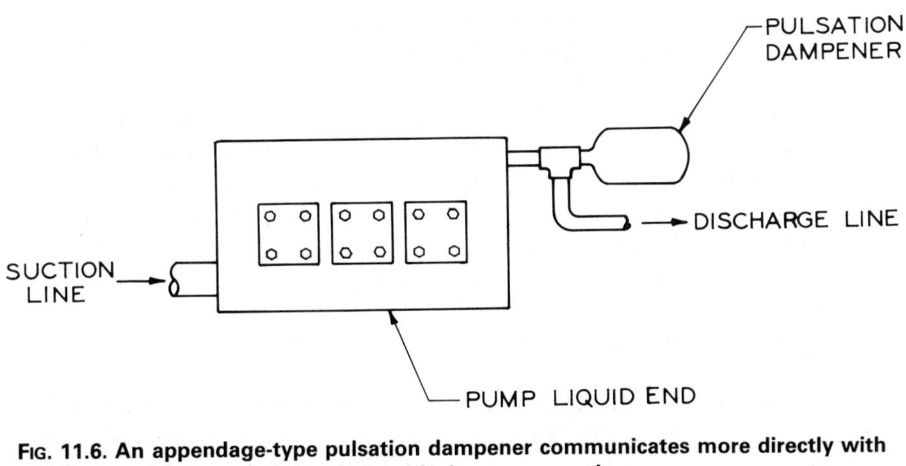

FIG. 11.6. An appendage-type pulsation dampener communicates more directly with the liquid end if mounted coaxially with the pump nozzle.

cation between the pump and the dampener. (Flow-through type discharge dampeners should be mounted *in* the discharge pipe.)

A suction stabilizer should be mounted in the suction line, as shown in Fig. 11.5. If the cylinder has an extra suction connection on the opposite end, a dampener there would be somewhat effective in absorbing velocity pulses, but no free gas would be separated from the incoming liquid. Also, current suction stabilizers contain baffles or other devices which are more effective in shielding the suction line from velocity variations when placed directly in the suction line. All liquid actually flows through the stabilizer prior to entering the pump.

11.4. PIPING NATURAL FREQUENCIES

To avoid excessive vibration of piping, supports of the pipe must be located close enough together, and be sufficiently rigid, so that the natural frequency of the pipe exceeds the exciting frequency from the pump. It must be remembered that the exciting frequency from many of the popular power pumps is *twice* the number of plungers times the crankshaft rotative speed.

11.5. REMEDIES FOR INSUFFICIENT NPSHA

If, in the design of the suction system, it is found that NPSHA is less than NPSHR, a number of remedies are available to the designer. Following is a list of options:

To increase NPSHA:

1. Increase the diameter of the suction pipe.
2. Reduce the length of the suction pipe by a more direct route, or by moving the pump closer to the suction vessel.
3. Install a suction bottle, stabilizer, or dampener adjacent to the pump liquid end. A fabricated bottle has often been used successfully at pressures below about 50 psig, but maintenance of a liquid level is required. A section of rubber hose in the suction line, adjacent to the pump, will often reduce acceleration head.
4. Elevate the suction vessel or raise the level of the liquid in the suction vessel.
5. Subcool the liquid (after it leaves the vessel).

To reduce the NPSH required by the pump, reduce the speed of the pump, or select a larger pump, running slower. At the lower speed, it may be possible to operate the pump with light suction valve springs, or none at all.

If the above steps are insufficient, impractical, or impossible, a booster pump must be utilized. A booster for a power pump is normally a centrifugal, although direct-acting recips and rotaries are sometimes used. The NPSH required by the booster must be less than that available from the system. The head of the booster should exceed the power pump NPSHR, suction line losses, and acceleration head by at least 20%. The booster should be installed adjacent to the suction vessel, and a pulsation bottle or stabilizer should be installed adjacent to the power pump to protect the booster from pulsating flow.

11.6. FOUNDATION AND BASEPLATE

Continuous-duty reciprocating pumps are often mounted on solid concrete foundations. Although many are mounted on baseplates which are bolted and grouted to the foundations, some users have found the base an unnecessary expense, and bolt the pump directly to the concrete.

Since most power pumps are designed with a plunger well which collects all liquid leakage from the stuffing box, and can be piped to a drain, the only function served by the base is to tie the driver and pump together. When the driver is mounted over, and supported by the power end of the pump, the base serves no function other than to collect liquid which drains from the liquid end upon disassembly (if the base is so constructed).

Even when the driver is mounted on the same plane as the pump, a baseplate is unnecessary because each may be bolted directly to the foundation, sometimes resting on sole plates embedded in the concrete. With a belt drive, the motor is mounted on an adjustable base. The cost of the baseplate and the expense of grouting are thus eliminated.

For portability, a pump and driver are often unitized on a skid-type base, as seen in Fig. 11.7, having its ends tapered and rounded so that it can be readily winched onto the bed of a truck. It is desirable that such a skid extend beyond the extremities of all mounted equipment so that the equipment will not be damaged during loading, transit, or unloading.

11.7. ROUTING OF ELECTRICAL CONDUIT

Electrical conduit may be routed to a power pump for an electric motor, a low-oil-pressure switch, an electric heater(s), and other auxiliary equipment. It is not unusual to find such conduit installed so that it blocks, or interferes with, routine maintenance and/or pump operation. It is important that the conduit be routed in such a manner as to provide adequate access to the pump for operation and maintenance.

FIG. 11.7. A duplex piston power pump and motor driver mounted on a skid-type base. (*Courtesy of GASO Pumps, Inc., a Goulds Pumps Company*)

11.8. MOUNTING OF DIRECT-ACTING PUMPS

When direct-acting pumps are installed, the drive end must be free to float axially on the base or foundation. If firmly bolted down, the drive end is unable to move to allow for expansion of liquid or drive cylinders upon warming to operating temperature. Also during operation, the yoke (cradle), which unites the liquid and drive ends, is alternately in tension and compression. If both ends of the pump are tied down, the base will absorb some of the force from the drive end, causing the yoke to bow, thereby causing misalignment between the drive and liquid cylinders. This is particularly true for pumps with tie-rod construction. The drive cylinder tie-down bolts must therefore be smaller than the holes in the drive cylinder foot, and must not restrict the axial movement of the cylinder.

Chapter 12
MAINTENANCE

12.1. GENERAL

Although this chapter is devoted to maintenance, maintenance information is also contained in other chapters. The chapters on stuffing boxes, packing, and auxiliary equipment are of particular value to maintenance personnel.

12.2. STORAGE OF PUMP

Maintenance of the pump and its auxiliary equipment should begin the day the unit arrives at the plant. If the equipment is to be stored for even a few weeks prior to operation, it should be checked to see that it is properly prepared for storage. If the unit is to be stored outdoors, a greater degree of protection is required, especially if it is a humid environment (such as a coastal area). Exposure to the elements and the environment of a plant under construction can cause more damage to the internals of machinery than the normal wear of operation. Unless the equipment manufacturer recommends an alternate storage procedure, the following may be used as a guide.

12.2.1. Power End of a Power Pump

A power pump power end should be opened and inspected, any dirt or moisture cleaned out, and the internals sprayed with oil or some other preservative. The preservative should be selected for the anticipated period of storage. As an alternative to spraying, a vapor-space-inhibiting oil may be poured into the power end. The power end should then be totally sealed. Sealing the power end requires removing the breather and any other auxiliary equipment which would allow the power end to breathe, and plugging all openings. All auxiliary equipment which has been removed should be tagged and stored for reinstallation prior to start-up. Every few weeks the crankshaft should be rotated by hand about $1\frac{1}{3}$ revolutions.

12.2.2. Gear Unit

A separate or bolt-on gear unit should be filled to the proper level with a vapor-space-inhibiting oil and sealed, or as otherwise instructed by the gear manufacturer.

12.2.3 Drive End of a Direct-Acting Pump

In a similar fashion to the power end, a direct-acting pump drive end should be dried, sealed, and preserved. Some users have completely filled drive ends with oil for long-term storage, although this requires installation of packing in the stuffing boxes which has resulted in corrosion (pitting) of the rods.

12.2.4. Liquid End

The liquid end should be opened, dried, and preserved. Preservation may take the form of a spray, a vapor-space-inhibiting oil, or a packet of deliquescent to absorb moisture. The problem with the deliquescent material is that it may inadvertently be left in the pump and cause damage to the pump or downstream equipment during start-up.

Packing and plungers should be removed, and the stuffing boxes and cylinder openings sealed. Packing should not be left in an idle pump for more than a few days. Otherwise the plunger (or rod) and stuffing box sealing surfaces may become corroded. Some packing materials contain small amounts of elements which tend to corrode metals, but even if the packing is totally inert, moisture trapped in the packing may corrode one or both surfaces. If the pump is equipped with pistons instead of plungers, the pistons and rods will be left in the pump, and the box sealed with a material which will minimize corrosion to the rod.

12.2.5. Mechanical Lubricator

To minimize moisture accumulation and corrosion, a mechanical lubricator should be filled with oil and covered with a plastic sheet.

12.3. PREPARATION FOR START-UP

12.3.1. Coupling Alignment

If the pump is driven through one or more flexible couplings, the couplings must be properly aligned to avoid excessive load on driver and pump bearings, and to avoid vibration. Alignment of a flexible coupling is actually an alignment of the axes of the two shafts. The degree of misalignment permitted is a function of the type of coupling and the operating speed. A coupling manufacturer will often indicate an allowable misalignment for his product which exceeds that which is satisfactory for the driver or driven equipment. An "all metal" coupling operating at 1800 rpm should normally be aligned to within 0.003 in. total indicator runout. Couplings with elastomeric elements may allow larger misalignment.

A coupling need not be disassembled for shaft alignment. As shown in Fig. 12.1, dial indicators may be used to indicate alignment with the coupling completely assembled.

12.3.2. V-Belts

Although not as critical as for a coupling, good alignment of a belt drive is required to obtain proper operation and life. The two sheaves must line up so that the individual belts enter and exit the grooves straight on. The driver must also be adjusted so that its shaft is parallel to the pump shaft. As discussed in Chapter 10, all belts in a set should contain the same match number. Proper belt tension is discussed in the section on v-belts found later in this chapter. New belts should be tensioned $\frac{1}{3}$ tighter than a set that has run-in.

12.3.3. Alignment Between Plunger and Stuffing Box

Nothing is more critical to the life of packing and plungers than the alignment between the plunger and stuffing box. If the plunger runs off center in the box, the plunger may rub on stuffing box bushings and become scored. If the clearance between the plunger and bushings is large enough to avoid contact (which in itself may lead to short packing life), an off-center plunger will bear heavily on one side of the packing, causing excess friction, and may not be contacted heavily enough on the other side to contain leakage. If short packing life is being experienced, alignment between the plunger and box should be the first item checked. No packing will survive in a stuffing box with a misaligned plunger.

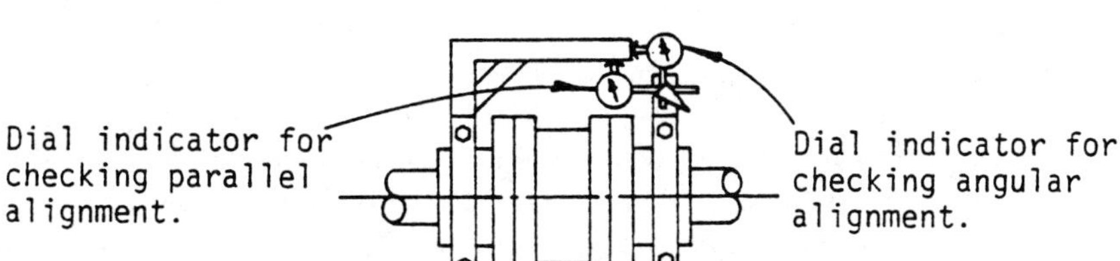

FIG. 12.1. The coupling may be left assembled when alignment is checked.

To contain packing in the stuffing box, the clearance between box bushings and the plunger must be small. Higher discharge pressures require smaller clearances. At 10,000 psi, the diametral clearance must be held to about 0.004 in. This tight clearance requires a close alignment between the stuffing box and plunger. Higher discharge pressures therefore require closer plunger alignment.

The stuffing box normally aligns to the fluid cylinder, and the fluid cylinder to the power frame. The plunger aligns to the crosshead extension, which aligns to the crosshead, which, in turn, aligns to the power frame (in the crosshead bore). If any one of these points of alignment is off, the plunger will not align properly with the stuffing box. A misalignment at any of these points can be caused by a design error, by a machining error, or by an assembly error. The assembly error can occur at the factory or in the field.

On some large horizontal power pumps, the fluid end is supported independently from the power end. With such units, the fluid end must be aligned with the power end in the field.

Easy Ways to Check Plunger Alignment. A plunger or box bushing which has rubbed heavily on one side is evidence of misalignment. Alignment can also be checked prior to operation by a number of methods. With the packing, gland, and all box bushings except the throat removed from the box, the plunger should rest easily in the bottom of the throat bushing. With all parts installed, the plunger should couple to the crosshead extension without being pushed or pulled to one side.

A Complex Way to Check Plunger Alignment. Figures 12.2 and 12.3 show a more complex, but very accurate, method to check fluid cylinder-to-plunger alignment using a dial indicator. By lifting the plunger from its lowest position, the running elevation of the plunger is determined. When the close-clearance alignment bushing slips into the cylinder bore, the plunger should be at its running elevation. The fluid end pictured here is supported independently of the power end so that adjustment in cylinder height is made by shimming the liquid end. For liquid ends which are doweled to the power end, such field adjustment is not possible.

Some Causes of Plunger Misalignment. With pumps that have the liquid end doweled to the power end, misalignment can have numerous causes. Improper machining of the fluid cylinder, stuffing box, box bushings, plunger, crosshead extension, crosshead, or power frame can cause plunger misalignment. Improper assembly of the box to the cylinder, the extension to the crosshead, or the plunger to the extension can cause

FIG. 12.2. A dial indicator may be used to determine the proper running position of the plunger. (*Courtesy Union Pump Co.*)

FIG. 12.3. A close-clearance bushing should slide into the stuffing box aligning bore when the plunger is at its running position. (*Courtesy Union Pump Co.*)

misalignment. A metal filing, paint, dirt, or a burr at any of these mating junctions can cause misalignment. Even an uneven tightening of the stuffing box bolting can cock the box slightly, and cause misalignment.

12.3.4. Alignment Between Piston Rod and Stuffing Box

For piston pumps, a round rod can be used to check alignment. The rod must be the same diameter as the packing cross-section (equal to the radial gap between the piston rod and box bore). With gland, packing, and bushings removed, the rod is inserted into the packing space in the stuffing box. If it is loose on one side of the piston rod, but cannot be rolled around to the other side, the piston rod is not centered in the box. This check should be performed with the piston at both extremes and the center of its stroke.

To maintain alignment during operation, the drive cylinder should be free to float on its base. The drive cylinder should not be clamped down. (See Section 11.8 for a more complete discussion.)

12.3.5. Adjustment of Drive for Mechanical Lubricator

If the lubricator is driven by a belt or chain, the drive should be checked for proper alignment and tension. If the lubricator is driven by a reciprocating linkage, the linkage should be adjusted so that the ratchet clicks through about two teeth when the power pump makes one revolution. The knuckles should be checked, as the pump is rolled over by hand, to see that they do not bind at any point. The knuckles should be lubricated with oil or grease.

12.3.6. Cleaning Pump Internals

Prior to the initial start-up, the pump and auxiliary equipment should be cleaned. If a pump is started with rust or dirt in the power end, its life may be significantly reduced. If the pump was properly prepared for storage, cleaning the unit will be a simple task, but if moisture or dirt has invaded the unit, the job will be more difficult.

Cleaning Power End Internals. The power end should be drained, opened, and inspected. Any dirt should be wiped out. Any rust on any of the bearing surfaces, such as crosshead bores, should be thoroughly removed. The power end should then be reassembled and filled to the proper level with the oil recommended by the manufacturer. Any auxiliary equipment removed for storage should be reinstalled.

Cleaning Gas End Internals. The gas end of a direct-acting pump should be drained, opened, and internally inspected. Any dirt or rust should be removed. Oil should be sprayed or wiped onto the cylinder bore and valve to provide initial lubrication.

Cleaning Liquid End Internals. Although less critical than the drive end, the liquid end may benefit from a cleaning and check of the valves. More important is a cleaning and inspection of all stuffing box components prior to packing installation.

Cleaning the Lubricator Sump. If the pump is equipped with a mechanical lubricator, it is important that the top be removed and the sump wiped out. These units always seem to accumulate water during a period of storage.

12.3.7. Priming Lubricator Lines

After reassembly, the lubricator should be filled with the proper lubricant. (For most services, steam cylinder oil is best.) Prior to installing stuffing box packing, the lubricator should be operated by hand to pump oil through the lubricator lines and into the stuffing box. This procedure primes the lubricator, fills the lines, and flushes most of the solids from the lines. If the stuffing boxes are equipped with check valves, disconnecting the oil lines just above the valves for the initial flushing will minimize the solids carried into the valves. The oil lines should then be reconnected, and priming should continue, until oil is seen flowing into each box. The oil should be wiped from the box to remove any solids flushed from the lines or valves.

12.3.8. Packing

After the lubricator lines have been flushed, and the oil wiped from the box bore, the packing may be installed. Removal of the plunger and all bushings from the box will enhance inspection and cleaning of the box and installation of the packing. For a more complete discussion of packing installation, the reader is referred to the liquid end section found later in this chapter. For a complete discussion of packing, see Chapter 5.

12.3.9. Lubricants

Before start-up, maintenance personnel should confirm that all lubricants required by the pumping unit have been provided in the proper grade and quantity. The power end of a power pump should be filled to the proper level with the type of oil recommended by the pump manufacturer, any gear unit should also be so provided, any packing lubricator should be filled with the correct oil for optimum packing life (usually steam cylinder or rock drill oil), and any lubricator provided for the gas end of a direct-acting pump should be filled with the proper oil. If driven by steam, the d-a pump will require a steam cylinder oil. If driven by cool air or gas, an industrial oil is usually adequate.

Horizontal power pump power ends which lubricate by the "lift-gravity" or similar system will be provided with important prelubrication to the bearings if a pint or so of oil is poured into the trough over the crossheads prior to initial start-up. If the breather is located over this trough, the oil may be poured through the breather connection.

12.3.10. Rotation of Pump and Driver

The pump should be checked to determine the proper direction of rotation. A rotation arrow will normally be found somewhere on the power end. Horizontal pumps normally rotate such that the top of the crankshaft moves toward the fluid end. If a power pump is operated backwards, the liquid end will still function satisfactorily. Pumping action therefore cannot be used as a criterium for confirming proper rotation. A power end running backwards may be noisy, run hot, and require excess power. Bearings may be damaged because of insufficient lubrication.

Motors of 1800 rpm and lower are usually furnished with cooling fans that function satisfactorily when rotated either direction. Therefore, if the motor is initially found to be rotating incorrectly, rotation can be reversed simply by switching two of the leads inside the conduit box. If the motor housing contains a rotation arrow, the fan is probably unidirectional. If opposite rotation is necessary to satisfy the pump, the motor manufacturer should be consulted.

12.3.11. Precharge of Bladder-Type Dampeners

The bladders of bladder-type dampeners must be precharged with a dry gas. Nitrogen is the normal choice. The pressure of the precharge is normally between 60% and 80% of the line pressure. For example, if the pump discharge pressure is 1000 psig, then the precharge pressure is normally between 600 and 800 psig. Most systems seem to run smoothest with about 80% precharge pressure. Some bladder-type dampeners, though, cannot tolerate a precharge pressure above 500 or 600 psig, so the manufacturer's instructions must be followed.

Precharging of the bladder must be performed with no liquid in the dampener. When the dampener is in service, the nitrogen is compressed by the liquid to line pressure. Therefore, a gage installed in the charging connection will read line pressure regardless of the amount of gas in the bladder. It will not indicate precharge pressure. To determine precharge pressure, the pump must be stopped, and the pressure in the line reduced to a pressure below precharge pressure. A gage in the charging connection will then indicate the actual amount of precharge in the bladder.

12.4. LIQUID END MAINTENANCE

Normally, more maintenance is required by the liquid end than the drive end of a reciprocating pump because of the shorter lives of packing and plungers, and sometimes valves and seats.

12.4.1. Packing

Packing is the component which normally requires the most maintenance on a reciprocating pump. Although the author is aware of two installations where continuous-duty power pumps have routinely experienced *two years* (17,000 hours) packing life, a more typical value is about three months (2500 hours).

In addition to the maintenance information contained in the chapters on packing and stuffing box designs, the following is offered. Different packing styles and stuffing box designs require different packing installation procedures. With all styles of packing, though, the author has found that soaking each ring in *clean* oil prior to installation enhances break-in and extends packing life. If the plunger can be removed from the stuffing box, the job of removing old packing and bushings is considerably eased; the box bore can be wiped and inspected; and installation of the new packing is easier.

Installing the Packing. Noncrushable style packing is easy to install, although care must be exercised to install each ring in the correct order and orientation. The gland is left a little loose until the plunger is reinstalled and fastened to the extension. The gland is then tightened and kept tight at all times.

The more versatile square braided packing requires a bit more skill. Because the bottom ring of packing (adjacent to the throat bushing) is affected so little by gland adjustment, this ring should be firmly seated during installation, using a rod with a flat end, or a stack of box bushings. Each successive ring should be oriented so that the split is off-set from the split in the adjacent ring by 90 to 120°, and should be seated with progressively less force.

Setting the Packing. After the box components, including the plunger, have all been installed, it is advisable, before exposing the liquid end to pumpage, to "set" braided packing. The packing is set by tightening the gland tightly with the gland wrench. If allowed to sit a few minutes under this load, the packing will flow, as it conforms to the plunger and stuffing box bore. The gland should then be tightened again. This process should be repeated until the packing stops flowing. The gland should then be completely loosened, and the packing allowed to expand for 10 to 15 minutes. The gland should then be drawn up only finger tight—no wrench—and should not be adjusted further prior to start-up. At this time, the block valves may be opened, and pumpage allowed into the pump.

Monitoring the Packing. During the first few hours of operation of new packing, each stuffing box should be monitored for excessive temperature. It is normal for some boxes to run warmer than others, under apparently identical conditions. Box temperature may exceed pumping temperature by as much as 50°F. Only if this exceeds the temperature rating of the packing are steps necessary to reduce box temperature.

12.4.2. Plungers

To extend its life, a plunger is usually coated with a hard material, but even so, the average life of a plunger is about six months. If a plunger is scored or undersized, new packing will wear out quickly. Optimum packing life is therefore obtained with a new, smooth plunger surface. The plunger surface should be ground or lapped to a finish of 8 micro-inch rms.

Whether a plunger is discarded or reconditioned is an economic determination. Lower cost prevails. Many small coated plungers cost more to properly prepare for a new coating than to make from scratch. If reconditioning costs less, it must be ascertained that the reconditioned part is equivalent to a new plunger.

12.4.3. Stuffing Box Bushings

The stuffing box bushings used for most packing must have square, flat faces, and have minimal clearance over the plunger and in the stuffing box bore. To properly support a packing ring, the bushing should not be chamfered or rounded at the corners. Due to the action of the plunger, packing, and pumpage, a box bushing wears, its average life being about six months. Stuffing box bushings should therefore be replaced at about every second change of the packing.

12.4.4. Stuffing Boxes

Although maintenance personnel will always change packing when it leaks, and will usually change plungers and bushings when necessary, the stuffing box is often not considered a wearing part, but it is. As packing flexes in the bore of the box, each time the pressure cycles from suction to discharge, the bore of the box is worn. Lip-type packing rings will wear grooves in the box bore faster than will braided packing. Maintenance personnel will often puzzle over the fact that packing life was satisfactory when their pump was new, but over a period of years has dropped to one or two weeks. They don't realize the importance of the box bore in determining packing life—or that the bore has been worn by the packing. The average life of a stuffing box is about two years. If the box is replaceable, it should be replaced when worn. If the box is integral with the liquid cylinder, the bore may need to be repaired by welding or sleeving.

12.4.5. Valves, Seats, and Springs

The check valves in the liquid ends of reciprocating pumps are provided in a variety of styles and materials and therefore require various maintenance procedures. The life of a valve and seat is a function of the nature of the pumpage, the presence of gas, operating pressure, and other factors. The life of a typical valve and seat is about two years in continuous duty, but they may require resurfacing about every six months. If valves are allowed to leak too long, severe "wire-drawing" of the seating surfaces may render them "nonrepairable" and require their replacement. The lives of these components are therefore extended by resurfacing the sealing surfaces prior to the occurrence of significant leakage. If leakage of the valves is suspected, because of *continuous* low capacity or *continuous* rough operation, the unit should be disassembled and inspected. (Intermittant low capacity and/or rough operation is usually caused by gas being ingested by the liquid end.)

When a valve and/or seat is ground or lapped, material is removed from the component. The result is lower spring force, higher valve lift, and weaker parts. The reduction in spring force and the increase in valve lift may cause the pump to run rougher. The weaker parts may lead to breakage of the parts. The manufacturer's recommendation should therefore be followed for minimum part thickness. An increase in valve lift of about 25% is often acceptable.

Wing-Guided Valves. All-metal wing-guided valves should be lapped in place using first a coarse, then fine, valve grinding compound. A grey, "frosted"-appearance, ring should be visible as a complete circle on both the valve and seat. After a valve has been lapped to a seat, it should not be changed to another seat without relapping. Each valve should be numbered with a vibrating pencil or equivalent, and a matching number stamped into the cylinder block near the valve cover so that during maintenance, each valve is returned to its proper seat.

Disc Valves. All-metal disc valves cannot readily be lapped in place, and must therefore be resurfaced in a grinding and/or lapping machine. The disc and seat are removed from the pump and independently resurfaced. For optimum sealing, the surface finish on both the disc and seat should be about 8 micro-inch rms.

Disc valves which utilize a plastic or rubber-faced disc should have the disc replaced when worn. The metal seats should be resurfaced as described in the above paragraph.

Ball Valves. When a ball becomes grooved or wire-drawn, it must be replaced or resurfaced. If the mating seat has become grooved or washed, it must be resurfaced or replaced. Special equipment may be available to resurface the seat in the pump. Otherwise, the seat must be removed from the pump and resurfaced in a machine shop.

Valve Springs. The valve spring must be strong enough to seat the valve before the piston or plunger has gained much velocity on the return stroke. Failure to do so will result in the valve being slammed onto its seat, sending a shock wave into the suction and/or discharge line, and causing premature wear of the seating surfaces. Proper stiffness is more important for the discharge spring than the suction spring. If the discharge spring is "stiffer" than the suction spring, it is important that these two parts not be interchanged during maintenance.

Because these spring undergo millions of compression cycles in just a few days of operation, it is not unusual that some will lose their original height. Therefore, during maintenance of the valves, the height of each spring should be measured. If shorter than recommended by the manufacturer, or worn in any way, the spring should be replaced.

12.4.6. Liquid End Gaskets

Spiral-Wound. One popular type of gasket used on the liquid end of reciprocating pumps is of the spiral-wound construction, as used in many radially-split process-type centrifugal pumps, and between raised-face pipe flanges. This gasket consists of a spiral of metal (usually stainless steel), with intermediate layers of asbestos, PTFE, or an asbestos substitute. Although normally provided with a fixed-depth groove to establish a fixed amount of compression, they are sometimes provided in grooves with no limit on compression. Although the inherent "spring" in these gaskets make them more suitable than some others for reuse, good maintenance practice dictates that new gaskets be used upon reassembly.

Jacketed. The jacketed gasket, as shown in Fig. 12.4, consists of a soft steel cover wrapped around a soft core such as asbestos or PTFE. The gasket can be "single-jacketed," in which the jacket covers just one side plus the edges of the core, or "double-jacketed," in which an additional jacket covers the other side of the core. Fig. 12.4 shows a double-jacketed gasket.

When installing a jacketed gasket, the smooth side (the wrap-around jacket) should be placed against the cylinder (toward the pressure), as indicated in Fig. 12.5.

The jacketed gasket is used in a groove that allows unlimited compression of the gasket. The soft metal jacket yields as it conforms to the irregularities of the sealing surfaces. Because of this compression and deformation, this gasket seldom provides a satisfactory seal if reused.

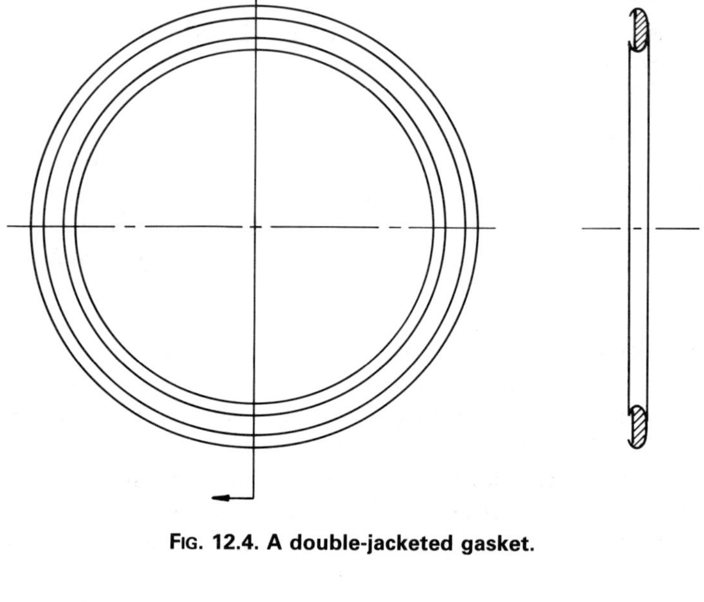

FIG. 12.4. A double-jacketed gasket.

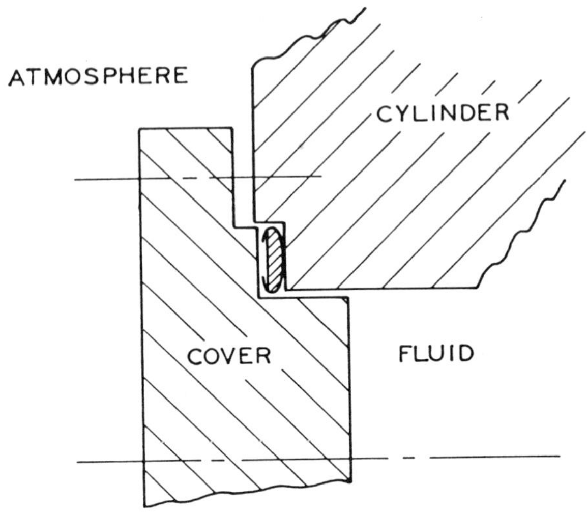

Fig. 12.5. A jacketed gasket should be installed with the smooth side toward the pressure.

"O" Rings. When suitable for the temperature, and compatible with the pumpage, "O" rings are sometimes used as liquid end gaskets. "O" rings impose lower loads on cover bolting, and are usually inexpensive and easy to replace. When used in an end-face groove, though, they require more accurate torquing of the fasteners. If the bolting is under-torqued, the "O" ring will extrude. If an "O" ring has not deformed or deteriorated during operation, it may be reused.

12.4.7. Cylinder Tie Studs

Conventional horizontal power pumps, and some vertical units, place the cylinder tie studs in tension each time the pump is on the discharge stroke. A power pump running 300 rpm imposed over *one million* stress cycles on each stud in only *three days* of continuous operation. If any fastener is cycled through a large range of tensile stress, it is more subject to a fatigue failure than one which experiences a small change in stress, even if the latter fastener is exposed to a *higher* stress.[1] During a plant visit, the author observed an installation of a power pump in which the liquid end was "breathing" at the junction with the power end. Because the user failed to tighten the nuts, as recommended, the studs experienced a fatigue failure in just a few weeks of operation, whereupon the liquid end separated from the power end. Had the nuts been properly torqued, the failure would have been averted. It is therefore essential that tie studs subject to such cyclic stresses be fully torqued to the recommended value, and periodically checked to assure that the torque is maintained.

12.5. POWER END

12.5.1. Inserts

Inserts (crank pin bearings) are the bearings at the large end of the connecting rod. These bearings usually require minimal maintenance unless the oil becomes contaminated with water, solids, or a solvent. Some power pump connecting rod assemblies are made with shims between the rod and cap so that the clearance of the bearing over the crankpin is adjustable. Other power pumps are made with "precision" bearings, i.e. no adjustment, so when the bearing wears, it must be replaced. The material of these bearings is typically a babbitt bearing surface backed by bronze or steel.

The clearance of the bearing over the crankpin can be measured by assembling the bearing and connecting rod on the pin with a length of plastic gaging material (obtainable at an automotive supply house) in the clearance. Upon tightening the connecting rod bolts, the plastic material will be flattened. Upon disassembly, the *width* of the plastic is measured with a special gage, supplied with the plastic, to determine the clearance.

If the crank pin has become damaged, two remedies are available to repair the unit. If undersize inserts are available, the crankpin can be ground undersize to the proper diameter to match the bearings. Otherwise the crankpin must be ground undersize and then restored to its original diameter with chrome plating, a flame-sprayed material, or the equivalent. To obtain a shaft with a strength equal to the original, it is important that the coating and grinding be performed so as not to create any thermal or mechanical stresses in the shaft, and so as to obtain the required surface finish and radii at the junctions of the pin with the cheeks of the shaft.

12.5.2. Crosshead Pins and Pin Bearings

Crosshead pins (wrist pins) are typically made of hardened and ground steel. In some smaller pumps, the pin presses into the rod, and rotates in the crosshead, using the crosshead itself as the bearing. Larger pumps typically have the pin locked into the crosshead with set screws or snap rings, and the connecting rod is fitted with a pressed-in bronze bushing which oscillates with the rod, on the pin. When installing a replacement bushing, it must be bored, reamed, or otherwise sized for the proper clearance over the pin.

Accelerated wear of the pin bearing may be an indication of high suction pressure or diluted oil. Accelerated wear of the crosshead pin may be an indication of solids in the oil. (The solids embed themselves in the softer bearing and cut the harder pin.)

12.5.3. Oil Seals

A power pump power end contains oil seals on the input rotating shaft (crankshaft or pinion shaft) and on the reciprocating crosshead extensions. Some pumps utilize small stuffing boxes with adjustable glands at these points. Others utilize rubber lip seals. Maintenance is seldom required for the seal on the rotating input shaft. The reciprocating seals require more attention. The crosshead extension seals will frequently fail on a new pump shortly after start-up. Such a premature failure is usually the result of a rough surface on the extension. Such extensions have been found to be rusty, partly painted, wrapped in tape, or coated with a heavy rust preservative (which sometimes contains embedded solids). Seals, and particularly lip seals, cannot seal against such surfaces, and usually will be damaged when they try. Therefore, prior to initial start-up, each extension should be cleaned and polished its entire length. Lip seals should be packed with grease, and packing should be oiled or greased to provide lubrication during the critical start-up period.

12.5.4. Main Bearings

The main bearings in power pumps are usually either sleeve-type or roller-type. Sleeve bearings are similar in construction, and require maintenance similar to the inserts as described above. Roller bearings are furnished as straight rollers or tapered rollers. Tapered rollers can be either single or double. The only one of these three types which is adjustable is the single roller when mounted on the end of a shaft. By adding or removing shims from the end cap, the end play in the shaft can be increased or decreased. Too much end play will cause the pump to knock. Too little end play will result in hot bearings and reduced bearing life. End play should be set in accordance with the manufacturer's recommendations. When setting the end play, it is important to maintain the proper axial position of the shaft. This can be checked by observing the gap on each side of each connecting rod inside each crosshead.

Failure of a roller bearing is normally caused by a fatigue failure of the surface of a race or roller. When any part of a roller bearing fails, the entire bearing should be replaced—both cup and cone.

12.5.5. Power End Gaskets

Power end gaskets are usually a soft paper, and can be reused only if undamaged during disassembly. These gaskets are often cut or torn during disassembly, so a spare set should be on hand prior to maintenance.

12.6. TYPICAL LIFE OF 'EXPENDABLES'—POWER PUMPS

Users will occasionally request an estimate of the life of wearing parts ("expendables") on power pumps. This is a difficult assignment because of the variety of services these pumps perform, the variety of conditions under which they operate, and variations in the quality of maintenance.

The life of liquid end components depends in part on speed, pressure, pumpage temperature, pumpage corrosiveness and abrasiveness, stuffing box-to-plunger alignment, stuffing box bushing clearances, packing lubricant, NPSHA, and the degree of shock conditions in the piping system.

The life of power end components depends in part on speed, load, suction pressure, ambient air temperature and cleanliness, oil quality, NPSHA, and the degree of shock conditions in the piping system.

In spite of all of these variables, an attempt is made below to list the approximate average life of expendables in a typical power pump operating continuously in a typical system.

Liquid end:

Packing	3 months
Plungers	6 months
Stuffing box bushings	6 months
Stuffing boxes (the bore)	2 years
Valves*	2 years
Seats*	2 years
Valve springs	1 year
Gaskets	Replace when removed

Power end:

Oil seals on crosshead stubs	6 months
Oil seals on rotating shafts	2 years
Connecting rod bearings (inserts)	2 years
Crosshead pin bearings	2 years
Crosshead pins	4 years
Crosshead extensions	3 years
Main bearings	5 years

* Metal valves may require relapping or regrinding every 6 months.

Pinion bearings (when applicable) ________________ 5 years
Gaskets ______ (unless undamaged) ______ Replace when removed

12.7. V-BELTS

To operate properly, v-belts must be tensioned properly. Too little tension will result in excessive slipping, resulting in inefficiency, heat, and a reduction in belt life. Too much tension will reduce the lives of the belts and the bearings in the pump and driver, and may add to the bending moment of the crankshaft (possibly contributing to a crankshaft failure). Proper tension is slightly above the point where slip is eliminated. This can best be determined by measuring the speed ratio (with a tachometer) at no load (discharge pressure = suction pressure) and full load. The change in speed ratio should not exceed 1%. During the first 36 hours of operation, the tension may require adjustment to compensate for the initial stretch and wearing-in (seating of the belts in the grooves).

Spare belts should be stored in a cool, dry area.

12.7.1. Match Numbers

As stated in Chapter 10, belts are often marked with a number to indicate a slight variation from the average length produced. This number is in addition to the belt size, which is indicated by a number such as "3V710." These "match numbers" are usually "46," "48," "50," "52," or "54." All belts in a drive should contain the same number so that the load will be shared equally among the belts.

12.7.2. Installation

To properly install a set of belts, the driver should be moved toward the pump until each belt can easily be installed in each groove by hand. Prying or rolling a belt onto the sheave may damage the belt. All slack should be

Table 12.1. Proper Tensioning Forces for V-Belts.

Cross Section	Small Pulley Diameter (in.)	Force (lbs)	
		New Belts	Old Belts
3V	2.65—3.65	4.4—6.7	3.3—5
3V	4.12—6.90	6.7—10	5—7.5
5V	7.10—10.9	17—27	13—20
5V	11.8—16.0	20—31	15—23

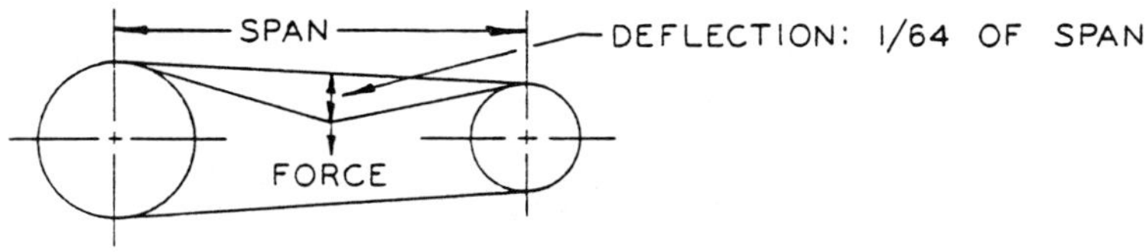

FIG. 12.6. Proper belt tension is obtained when each belt deflects $\frac{1}{64}$ of the span when pushed laterally with the force indicated in Table 12.1.

moved to one side of the drive, then the driver should be moved away from the pump until the tension in each belt is in accordance with Table 12.1. The driver should be checked to see that its shaft is parallel to the pump shaft, and the alignment of the two sheaves should be given a final check. The belts should flow into and out of each groove without any lateral distortion.

12.7.3. Belt Tension

Belts should be tensioned in accordance with Table 12.1. The force indicated in the table is applied laterally to each belt as indicated in Fig. 12.6. The resultant deflection should be $\frac{1}{64}$ of the span of the drive. If the deflections are not uniform among the belts in a drive, the cause should be identified and eliminated.

REFERENCE

1. *Helpful Hints,* 6th ed. (Bulletin DC-6), Port Chester, NY: Russell, Birdsall & Ward Bolt and Nut Co., 1971.

Chapter 13
OPERATION

13.1. UNLOADING TO START

In the discussion of system design in Chapter 11, it is recommended that a bypass line be installed to allow the pump to be started in an unloaded condition. Numerous benefits result from such a procedure, the primary one being that all pumping chambers have an opportunity to become primed. Each pumping chamber of a reciprocating pump is an independent pumping entity, operating in *parallel* with each other chamber. A triplex pump has three chambers, a quintuplex—five. It is possible for only one of these chambers to be primed, while all others remained vapor-locked. The writer has witnessed both triplex and quintuplex pumps in operation, in plants, with only one pumping chamber primed. The results were reduced capacity, noisy (pounding) operation, high vibration, and wildly-swinging gage readings.

A reciprocating pump *tends* to be self-priming, but with the large clearance volume common in most recips, it is impossible for the plunger to produce much discharge pressure when the chamber is filled with gas. A pump will normally be full of air after maintenance, and some chambers gain air that leaks through packing during idle periods. Often, upon start-up, gas is drawn into the pump from the suction line. For these reasons, it is necessary that discharge pressure be kept low during the first half-minute or so of operation. This gives each pumping chamber an opportunity to clear itself of gas and be operating fully primed when exposed to discharge pressure. With most systems, it is possible to keep discharge pressure low only by having a bypass line. This line should not connect into the pump suction line, because the gases would again be ingested by the pump.

Other benefits accrue from starting a power pump against negligible discharge pressure. As shown in Chapter 6, starting torque will be only about 25% of full load torque, allowing the use of a motor with normal starting torque, and reducing the time of inrush current. Couplings, v-belts, gears, or chains will not be overloaded. Also, the power end of the pump has an opportunity to establish full lubricating films on all sliding surfaces, and the plungers have an opportunity to become wetted by the lubricant and/or the pumpage.

To start a pump with a bypass system, the suction, discharge, and bypass valves are fully opened. The pump is then started and operated for 30–60 seconds, or until the power end oil temperature reaches the minimum recommended by the manufacturer. The bypass valve is then slowly closed, and the pump is on stream.

If made absolutely necessary by process requirements, such as possible emergency conditions, that a power pump be started against system pressure, it is recommended that:

1. The driver starting torque capability exceed running torque by at least 50%.
2. Drive train components be suitable for the driver starting torque.
3. The pump not be allowed to sit idle more than 10 hours—or provisions be made for prelubricating power end bearings.
4. Pumping chambers be kept primed.

13.2. GAS SYMPTOMS

If, during normal operation, the pump suddenly begins to run roughly, or the capacity drops suddenly, it is probable that gas has been ingested by one or more pumping chambers. With the above suggested system it is necessary only to open the by-pass valve until the pump clears itself. If a pump continues to be plagued by gas, the source of the gas should be identified and eliminated.

13.3 MONITORING THE PUMP

During the initial startup, it is important that the pump be monitored for abnormal operating conditions.

13.3.1. Discharge Pressure

The discharge gage is an indicator of pump operation, as well as system design. Some fluctuation of the reading is inevitable with a reciprocating pump, but the degree of fluctuation, and its frequency, indicate the adequacy of the system and smoothness of pump operation. With a triplex pump, the gage should fluctuate with a frequency of six times pump rpm. If the fluctuation is one times rpm, something's wrong. One or two pumping chambers may not be primed, or one or more valves are not functioning correctly.

13.3.2. Stuffing Box Temperature

The first few minutes of operation are critical to the longevity of packing. The temperature of the stuffing box is usually a good indication of conditions inside the box. The temperature of the box should normally not exceed the temperature of the pumpage by more than 50°F. After break-in, box temperature will usually exceed pumpage temperature by no more than 25°F.

If high box temperature is caused by the gland being too tight, the gland should be loosened slightly.

13.3.3. Oil Temperature

Power end oil temperature is an indicator of the condition of the power end and the load on the power end. If bearings or bearing mating surfaces have deteriorated, bearings will run hot. If suction or discharge pressure is in excess of rated, bearings will run hot. Other causes of a hot power end are excess ambient temperature, blockage of air flow, exposure to direct rays of the sun, an adjacent piece of hot equipment, incorrect oil, insufficient oil, contaminated oil, and reverse rotation. If designed to be operated with an oil cooler, the cooler may be fouled, or the coolant may be of inadequate flow or too hot.

To prevent excessive carburization of the oil, the temperature should not be allowed to exceed 180°F. Close monitoring is recommended if the temperature exceeds 160°F.

13.3.4. Sound

Many power pumps are noisy, particularly at higher speeds. The sound from the pump, though, should be regular. If clicking of the valves can be heard, it should be a regular "clickity-clickity-clickity." Any surging or irregular sound may indicate a "vapor-locked" condition or gas moving through the pump.

A power pump with inadequate suction conditions or faulty valve parts (such as a weak or broken spring) will sometimes emit a knocking sound from the power end which sounds like a loose rod or a broken gear tooth. Although the shock is originating in the liquid end, the sound gets transmitted through the plunger and crosshead into the power end. A knocking or slapping sound can also be caused by reverse rotation.

If the flow from a power pump is extremely irregular (such as caused by vapor-lock or a stuck valve) some pulsation dampeners may slam as the bladder (or equivalent) tries to compensate for the resulting velocity fluctuation.

13.3.5. Driver Load

An ammeter indicating driver load can be used to add knowledge of pump operation. The reading will indicate if the pump is requiring more or less power than rated. A uniform reading (except with simplex and duplex pumps) indicates a uniform sharing of load among plungers. A significant pulse equal to pump rpm indicates improper operation.

13.4 BREAK-IN

A break-in may be required by a new power pump or one in which new parts have been installed.

13.4.1. Power End Break-In

Parts of the power end which may require break-in are crossheads (and the crosshead bores in the power frame or liners), sleeve main bearings, sleeve pinion bearings, sleeve-type crankpin bearings, sleeve-type crosshead pin bearings, and tapered roller bearings. As a rule, sliding-contact surfaces require a break-in, rolling-contact surfaces do not. Why then the tapered roller bearings? At the ends of the rollers, on the cone, is a flange which absorbs a load axially from each roller. That is a sliding surface.

The degree of break-in required by a power end is a function of bearing loads, pump speed, number of sliding bearing surfaces, and the smoothness and perpendicularity of these surfaces as provided by the pump manufacturer. Some pumps exibit a stable or dropping power end temperature after one or two hours of operation. Others require eight to ten hours to reach a stable or dropping temperature. The ideal break-in procedure for a power pump is to operate the unit at an incrementally-increasing load, waiting until power end temperature stabilizes before proceeding to the next level. For example, the unit could be operated first at no-load, then 25% load, then 50%, then 75%, and then finally 100% load. If the manufacturer has performed such a break-in prior to shipment, some of these steps may be unnecessary. If it is not possible to step the discharge pressure on the pump, a less desirable procedure is to run the unit at full load for periods of time, with intervening shut-downs for cool-down. The manufacturer's recommendations should be used as a guide.

Because any temperature above 180°F causes a rapid deterioration of most lubricating oils, this is normally the cut-off point for power end operation. If, during a break-in, oil temperature begins to approach 180°F, it may be necessary to reduce discharge pressure, or shut the pump down until it cools.

13.4.2. Packing Break-In

The only component of a liquid end which may require break-in is packing. Types of packing which may require break-in include hard braided rings and v-rings of PTFE. The procedure is exactly the same as for a power end, with the first choice being incrementally-increasing discharge pressure. Again, temperature is the characteristic to monitor, but a desired maximum cannot as readily be established—being a function of numerous factors, including the temperature of the pumpage. Box temperature will rise to a maximum, level off, then drop as the packing seats in. It is normal for some boxes to run warmer than others, under apparently identical conditions. Box temperature may exceed pumpage temperature by as much as 50°F. Only if this exceeds the temperature rating of the packing are steps necessary to reduce box temperature.

Appendix A

Trouble-Shooting Table—Power Pumps (Written Primarily for Single-Acting Plunger Pumps)

Symptom	Possible Cause	Remedy
Low capacity and Pump pounds or vibrates	Insufficient NPSHA	Provide more NPSH
	Excessive acceleration head in suction line	Install suction stabilizer
	All pumping chambers not primed	Prime all chambers
	Excessive gas in liquid caused by:	
	Entrained bubbles	Separate gas
	Gas pocket in suction line	Vent suction line
	Air leaks into suction line	Seal all leaks
	Air leaks through stuffing box	Provide sealing liquid
	Suction pipe inlet insufficiently submerged (vortexing)	Increase submergence or stop vortex
	Relief valve discharge piped to suction line	Pipe to suction vessel
	By-pass line piped back to suction line	Pipe to suction vessel
	Broken or weak valve springs	Replace springs
	Valve stuck open	Clean and deburr valve
Low capacity (only)	Speed too low (belts slipping?)	Increase speed
	Pump valves leaking	Relap or replace
	Piston rings and/or liner worn	Replace
	Relief valve leaking	Repair
	Cylinder washed out around valve seat	Repair or replace
	Undersize plungers or pistons	Install correct size
Pump pounds or knocks (only)	Insufficient NPSHA	Provide more NPSH
	Excessive acceleration head in suction line	Install suction stabilizer
	Pulsation dampener improperly charged	Charge to proper pressure
	Loose piston or rod	Tighten
	Low oil level in power end	Fill to proper level
	Gear tooth cracked or broken	Replace gear
	Pump running backwards	Correct rotation
	Excessive clearance in connecting rod bearings	Adjust or replace
	Worn crosshead pin and/or bearing	Replace worn parts

(continued)

Symptom	Possible Cause	Remedy
	Excessive end play in main bearings	Adjust end play
	Plunger not tight in crosshead extension	Tighten
	Valve seat loose in cylinder	Repair cylinder and replace seat
	Loose sheaves or bushings	Tighten
Piping vibrates	(Same as causes of pump pounding above plus the following)	
	Excessive pressure variation in discharge line	Install discharge pulsation dampener
	Piping inadequately supported	Install supports at proper locations
	Too many short-radius elbows and tees	Minimize turns—use long-radius els
Pump requires too much power	Discharge pressure too high	Reduce discharge pressure
	Speed too high	Reduce speed
	Packing too tight	Loosen gland
	Pump running backwards	Correct rotation
	Plungers or pistons too large	Change to smaller diameters
	Misaligned coupling	Align coupling
	Bearings too tight	Increase clearance or end-play
	Low voltage to motor	Increase voltage
	Belts too tight	Loosen belts
Short packing life	Incorrect packing	Install correct packing
	Inadequate packing lubrication	Improve lubrication
	Misalignment of plunger (or rod) with stuffing box	Improve alignment
	Worn plunger (or rod)	Replace worn part
	Worn box bushings	Replace bushings
	Worn stuffing box bore	Repair or replace box
	Packing gland too tight	Loosen gland
	Packing gland too loose	Tighten gland
	Too much packing in box	Reduce number of rings
	Pump operated only partially primed (packing run dry)	Fully prime before imposing pressure
	Entrained gas	Eliminate gas
	Insufficient NPSH (cavitation)	Provide more NPSH
	Packing installed incorrectly	Install correctly
	Icing in packing or on plunger	Increase lubrication and/or injection
	Abrasives in pumpage	Strain or filter inlet

Symptom	Possible Cause	Remedy
Excessive "wear" of liquid end parts	Corrosion	Eliminate corrodant or upgrade materials
	Cavitation (inadequate NPSHA)	Provide more NPSH
	Abrasion	Reduce abrasives
Cracked liquid cylinder	Discharge pressure too high	Reduce discharge pressure
	Hydraulic shock caused by cavitation or gas	Provide more NPSH or eliminate gas
	Discharge valve seat loose in cylinder (causing excess internal pressure)	Replace faulty parts
	Defect in material or manufacture	Repair or replace
Hot power end (over 180°F) and/or Short life of power end parts	Pump running backwards	Correct rotation
	Contaminated oil	Fill with clean oil Eliminate contaminant
	Incorrect oil	Fill with correct oil
	Oil level too high	Lower oil level
	Oil level too low	Raise oil level
	Bearings too tight	Increase clearance or end-play
	Misaligned coupling	Align coupling
	Belts too tight	Loosen belts
	Suction pressure too high	Reduce suction pressure or reduce plunger diameter
	Discharge pressure too high	Reduce discharge pressure or reduce plunger diameter
	Discharge valve stuck open	Clean and deburr valve
	Speed too low	Increase speed
	Speed too high	Reduce speed
	Insufficient cooling	Provide adequate cooling
	Pump exposed to heat source	Shield pump from heat
Cylinder stud failure	Nut not adequately torqued	Torque properly
	Discharge pressure too high	Reduce discharge pressure
	Plunger diameter too large	Install smaller plungers
	Excessive piping loads on cylinder	Reduce pipe loads
Connecting rod twisted	Plunger tightened too tightly in crosshead extension	Replace rod Torque properly

(*continued*)

Symptom	Possible Cause	Remedy
Broken crankshaft	Discharge pressure too high	Reduce discharge pressure or plunger diameter
	Suction pressure too high	Reduce suction pressure or plunger diameter
	Pump pounding	(See "pump pounds")
	Plunger diameter too large	Install smaller plungers
	Defect in material or manufacture	Replace defective parts

APPENDIX B

Trouble-Shooting Table—Direct-Acting Pumps

Symptom	Possible Cause	Remedy
Low capacity and Pump pounds	Insufficient NPSHA	Provide more NPSH
	Excessive acceleration head in suction line	Install suction stabilizer
	All pumping chambers not primed	Prime all chambers
	Excessive gas in liquid caused by:	
	Entrained bubbles	Separate gas
	Gas pocket in suction line	Vent suction line
	Air leaks into suction line	Seal all leaks
	Air leaks through stuffing box	Provide sealing liquid
	Suction pipe inlet insufficiently submerged (vortexing)	Increase submergence or stop vortex
	Valve stuck open	Clean and deburr valve
Low capacity (only)	Speed too low	Increase speed by increasing gas pressure differential
	Liquid end valves leaking	Relap or replace
	Piston rings and/or liner worn	Replace worn parts
	Relief valve leaking	Repair valve
	Cylinder washed out around valve seat	Repair or replace
	Undersize plungers or pistons	Install correct size
Pump pounds or knocks (only)	Insufficient NPSHA	Provide more NPSH
	Excessive acceleration head in suction line	Install suction stabilizer
	Loose piston or rod	Tighten
	Valve seat loose in cylinder	Repair cylinder and replace seat
	Piston hitting head or end of cylinder	Decrease length of stroke
	Pump reversing too soon	Increase length of stroke
Piping vibrates	(Same as causes of pump pounding above plus the following)	
	Excessive pressure variation in discharge line	Install discharge pulsation dampener
	Piping inadequately supported	Install supports at proper locations
	Too many short-radius elbows and tees	Minimize turns—use long-radius els

(continued)

Symptom	Possible Cause	Remedy
Short packing life	Incorrect packing	Install correct packing
	Inadequate packing lubrication	Improve lubrication
	Misalignment of plunger (or rod) with stuffing box	Improve alignment
	Worn plunger (or rod)	Replace worn part
	Worn box bushings	Replace bushings
	Worn stuffing box bore	Repair or replace box
	Packing gland too tight	Loosen gland
	Packing gland too loose	Tighten gland
	Pump operated only partially primed (packing run dry)	Fully prime before imposing pressure
	Entrained gas	Eliminate gas
	Insufficient NPSH (cavitation)	Provide more NPSH
	Packing installed incorrectly	Install correctly
	Abrasives in pumpage	Strain or filter inlet
	Icing in packing or on plunger	Increase lubrication and/or injection
Excessive "wear" of liquid end parts	Corrosion	Eliminate corrodant or upgrade materials
	Cavitation (inadequate NPSHA)	Provide more NPSH
	Abrasion	Reduce abrasives
Cracked liquid cylinder	Discharge pressure too high	Reduce discharge pressure
	Hydraulic shock caused by cavitation or gas	Provide more NPSH or eliminate gas
	Discharge valve seat loose in cylinder (causing excess internal pressure)	Replace faulty parts
	Ram pressure produced against closed discharge valve	Install relief valve
	Defect in material or manufacture	Repair or replace
Cylinder stud failure	Nut not torqued high enough	Torque properly
Pump jerks through stroke or Pump stalls near end of stroke	Liquid end misaligned with gas end caused by:	
	Improper assembly	Realign
	Feet of both ends of pump tied down	Loosen gas end tie-down nuts (let float)
	Excess pipe loads	Reduce pipe loads
	Packing not adequately lubricated	Improve lubrication
	Drive piston rings not adequately lubricated	Improve lubrication
	Packing gland(s) too tight	Loosen gland(s)
	Excess water in drive end	Drain water

Symptom	Possible Cause	Remedy
	Gas valve set improperly (on simplex units only)	Reset valve
	Gas valve(s) worn	Repair seating surfaces
	Auxiliary gas line not connected (on simplex units only)	Connect auxiliary line
Pump fails to start	Low drive gas pressure	Increase pressure
	High liquid pressure	Reduce pressure
	Water in drive cylinder	Drain water
	Auxiliary gas line not connected (on simplex units only)	Connect auxiliary line
Pump jumps though stroke, hesitates, then jumps back	Auxiliary gas line not connected (on simplex units only)	Connect auxiliary line
Pump stalls when speed is reduced	Auxiliary gas line connected below throttling valve (SX units only)	Connect auxiliary line above throttling valve
	(also see "Pump jerks through stroke")	

APPENDIX C

Symbols for Quantities

Quantity Symbol	Quantity	Symbol Description
a	acceleration	lower case "a"
A	area	upper case "A"
C	constant or clearance volume	upper case "C"
d	diameter or differentiation	lower case "d"
D	displacement	upper case "D"
E	expansion	upper case "E"
F	force	upper case "F"
g	acceleration of gravity	lower case "g"
h	fluid energy (head)	lower case "h"
K	constant	upper case "K"
L	length or liquid	upper case "L"
m	mass	lower case "m"
n	pure number	lower case "n"
N	speed of rotation	upper case "N"
NPSH	net positive suction pressure	upper case "NPSH"
NPSHA	NPSH available from system	upper case "NPSHA"
NPSHR	NPSH required by pump	upper case "NPSHR"
p	pressure	lower case "p"
PTFE	polytetrafluoroethylene (equivalent to DuPont's Teflon®)	upper case "PTFE"
P	power	upper case "P"
Q	capacity (flow rate) (volume per unit time)	upper case "Q"
S	speed (velocity)	upper case "S"
SG	specific gravity	upper case "SG"
t	time	lower case "t"
T	temperature	upper case "T"
v	velocity or specific volume	lower case "v"
V	volume	upper case "V"
w	specific weight	lower case "w"
W	weight	upper case "W"
Δ	differential	upper case delta
η	efficiency	lower case eta
θ	angle	lower case theta
ω	angular rotative speed	lower case omega
Subscripts		
a	absolute or acceleration	lower case "a"
acc	acceleration	lower case "acc"
atm	atmospheric	lower case "atm"

QUANTITY SYMBOL	QUANTITY	SYMBOL DESCRIPTION
c	chambers, pumping	lower case "c"
d	discharge	lower case "d"
dr	drive end	lower case "dr"
e	exhaust	lower case "e"
f	friction	lower case "f"
g	gage or gas	lower case "g"
i	inlet or input	lower case "i"
L	liquid or liquid end	upper case "L"
m	mechanical	lower case "m"
max	maximum	lower case "max"
o	output	lower case "o"
p	plunger or piston or pipe	lower case "p"
r	rod, piston	lower case "r"
s	steam or stroke or stagnation or suction	lower case "s"
t	tank	lower case "t"
v	volumetric or valve or vapor	lower case "v"
vel	velocity	lower case "vel"
z	elevation	lower case "z"

APPENDIX D
Symbols for Units and Metric Conversion Factors

Quantity (1)	Typical Application	U.S.-Customary Unit	Abbrev. or Unit Symbol	Metric Unit	Unit Symbol	Conversion Factors: U.S. to Metric	Conversion Factors: Metric to U.S.
Acceleration		foot per second-squared	ft/s^2	meter per second-squared	m/s^2	0.3048*	3.281
Area	flow area	square inch	in^2	square millimeter	mm^2	645.2	0.001550
Capacity	flow rate of fluid	gallon per minute	gpm	cubic meter per hour	m^3/h	0.2271	4.403
		gallon per minute	gpm	liter per minute	L/min (3)	3.785	0.2642
Density (mass)	density of fluid	pound per cubic foot	lb/ft^3	kilogram per cubic meter	kg/m^3	16.02	0.06243
Energy, work		foot pound-force	ft-lbf	joule [N−m] (2)	J	1.356	0.7376
Force		pound-force	lbf	newton	N	4.448	0.2248
Fluid energy	head of fluid	foot	ft or ′	meter	m	0.3048*	3.281
Heat (energy)		British thermal unit	Btu (4)	kilojoule	kJ	1.055	0.9478
Heat flow rate (power)		British thermal unit per minute	Btu/min (4)	watt [J/s] (2)	W	17.58	0.05687
Length	dimension	foot	ft or ′	meter	m	0.3048*	3.281
	dimension	inch	in or ″	meter	m	0.0254*	39.37
	coating thickness	mil		micrometer	μm	25.4*	0.03937
	surface finish	microinch	μin	micrometer	μm	0.0254*	39.37

Mass	component mass	pound	lb or #	kilogram	kg	0.4536	2.205
Pressure	fluid pressure	pound-force per sq. in.	psi or lbf/in^2	kilopascal $[kN/m^2]$ (2)	kPa	6.895	0.1450
Power	mechanical power	horsepower [550 ft-lbf/s] (2)	hp	kilowatt [kJ/s] (2)	kW	0.7457	1.341
Specific weight (force)	fluid	pound-force per cubic foot	lbf/ft^3	kilonewton per cubic meter	kN/m^3	0.1571	6.366
Rotational frequency	shaft speed	revolutions per minute	rpm or r/m	revolutions per minute	r/min	1*	1*
Temperature	fluid temperature	degrees Fahrenheit	°F	degrees Celsius	°C	$(T_F - 32) \times 5/9$*	$(T_c \times 1.8) + 32$*
Velocity (speed)	fluid velocity	foot per minute	ft/m or fpm	meter per second	m/s	0.00508*	196.9
		foot per second	ft/s or fps	meter per second	m/s	0.3048*	3.281
Volume	fluid volume	cubic inch	in^3	milliliter $[cm^3]$ (2)	mL (3)	16.39	0.06102
		cubic foot	ft^3	cubic meter	m^3	0.02832	35.31
		gallon	gal	cubic meter	m^3	0.003785	264.2
		gallon	gal	liter	L (3)	3.785	0.2642

* Exact

[1] As used in ISO-31 and ISO-1000, "quantity" means "measurable property."

[2] Symbols in brackets, [ ], are explanatory only.

[3] The U.S. Department of Commerce, National Bureau of Standards, has established the capital letter "L" as the preferred unit symbol for liter for the U.S.

[4] All factors in this table are based on the Btu-International Table

APPENDIX E
SAVING ENERGY WITH POWER PUMPS
(A Case Study)

by Barry W. Brown*

In recent decades, applications for reciprocating power pumps have been reserved for high discharge pressure, constant flow conditions. In certain applications demanding variable flow at discharge pressures rendering centrifugal pumps impractical, variable speed slip type couplings (eddy-current, hydraulic) have been used. Since the efficiency of these drives is limited to the ratio of output speed to input speed (times 100), the power consumption remains essentially constant throughout the capacity range. The inefficiency of the drive system results in excess power consumption and cost whenever the pump is operated at less than maximum speed.

For many years, the solution to the high pressure variable flow application has been to use a constant speed multi-stage centrifugal pump and throttle the discharge to maintain the required discharge pressure. The advantages of this system are a generally lower first cost for the pump and drive and a simple control system.

The disadvantages of the constant speed centrifugal pump arise from its basic operating characteristics. Two things happen as you throttle the discharge: 1) the efficiency of the pump decreases and 2) the pump develops a higher discharge pressure that is then throttled to meet the requirements of the system head which are lower at the reduced flow. The net result is again excess power consumption and cost.

Figure E.1 graphically presents throttling of a centrifugal pump as it affects efficiency and power consumption.

When there was an abundance of cheap power, the inefficiency of the centrifugal pump could be justified on the basis of lower first cost and control simplicity; however, in recent years, because of the general increase in energy cost, the evaluation has swung in favor of the more efficient reciprocating unit.

From an energy conservation standpoint, reciprocating plunger type power pumps have three distinct advantages: 1) they have a higher effi-

* Barry W. Brown, LCDR USN (Retired). Mr. Brown is a District Manager for Union Pump Company, officed in San Francisco, California.

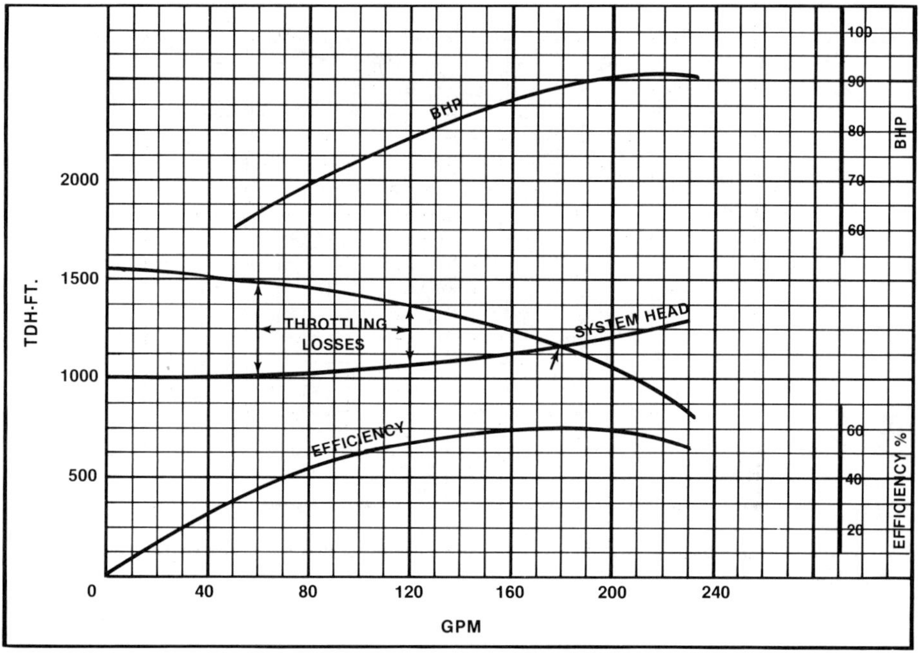

FIG. E.1. Performance curve for a typical centrifugal pump illustrating efficiency variation with load, and throttling losses.

ciency (typically 90%) than centrifugal pumps for similar hydraulic conditions, 2) the efficiency remains constant throughout the range of flow, and 3) the pump discharge pressure adjusts automatically to the system pressure. The problem now becomes one of marrying this pump to an energy efficient variable speed drive.

The major motor control manufacturers now offer variable frequency drives which, when coupled with conventional induction motors, produce variable speed drive systems with high efficiencies. These controls operate by taking standard 60 hertz power, rectifying it to DC, and then electronically converting it back to variable frequency AC. The concept is not new; however, due primarily to quantum jumps in electronic technology, this product has become an affordable, reliable means to provide variable speed control.

The logical question now is: "How do system efficiencies compare?" Figures E.2 and E.3 provide the answers. Figure E.2 shows motor efficiency, pump efficiency, and wire-to-water efficiency of the constant-speed centrifugal pump with throttling control. This was developed from the actual pump curve in Figure E.1.

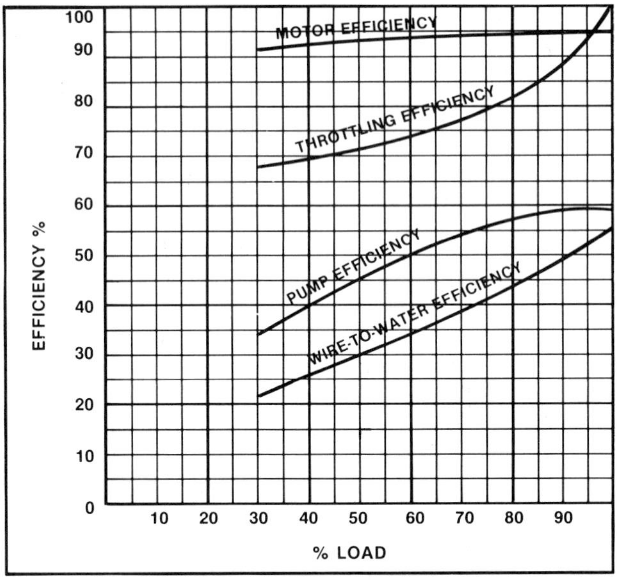

FIG. E.2. Motor, pump, and wire-to-water efficiencies for the centrifugal pump of Fig. E.1.

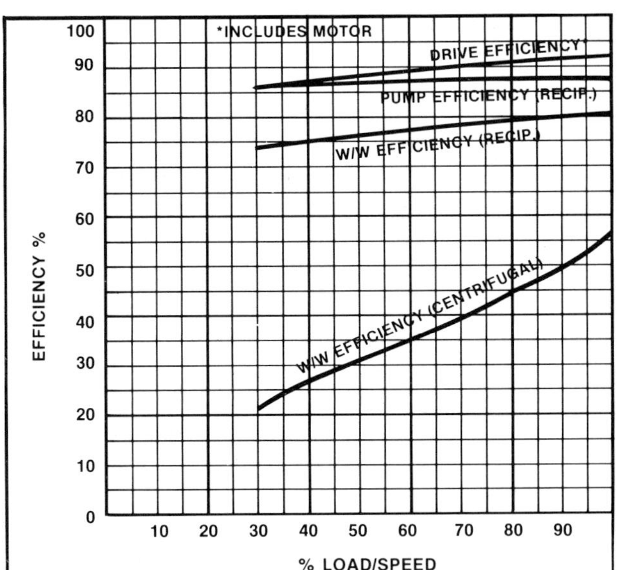

FIG. E.3. Drive, pump, and wire-to-water efficiencies for a triplex plunger pump with variable frequency, variable speed drive.

Table E.1. Energy Consumption Evaluation (Multi-Stage Centrifugal Pump vs Power Pump).

Assumptions

1. Total operating hours per year—8000.
2. Full load operating hours—4000.
3. ⅔ load operating hours—2500.
4. ⅓ load operating hours—1500.
5. Energy cost—0.075 $/kW–h—no power factor or peak load penalties.
6. Drive motors are high efficiency type.

Centrifugal	**Full Load**	**⅔ Load**	**⅓ Load**
RPM	3550	3550	3550
GPM	180	120	60
Pump head (ft)	1150	1370	1470
System head (ft)	1150	1050	1020
Efficiency (W–W)	56%	38%	23%
Power input (hp)	93.3	84.0	67.0
Power input (kW)	69.5	62.6	49.9

Reciprocating	**Full Load**	**⅔ Load**	**⅓ Load**
RPM	275	183	91.5
GPM	180	120	60
Disch. pressure (psig)	500	454	442
Efficiency (W–W)	81%	77%	74%
Power input (hp)	64.8	41.2	20.9
Power input (kW)	48.3	30.7	15.6

Savings	**Full Load**	**⅔ Load**	**⅓ Load**
kW savings	21.2	31.9	34.0
Hours	4000	2500	1500
kW–h savings	84,800	79,750	51,000
Unit power cost ($)	0.075	0.075	0.075
Energy cost savings	$ 6360	$ 5981	$ 3825
Total Energy Cost Savings			$16,166

First Cost Comparison

Centrifugal

Typical multi-stage centrifugal pump with 100 hp motor and combination starter $34,000

Reciprocating

Typical triplex power pump with 75 hp motor and variable frequency speed control $51,000

Difference in first cost. $17,000
First year energy savings $16,166
Payout period . . . . 13 months

Figure E.3 shows drive efficiency (including motor), pump efficiency, and wire-to-water efficiency for an actual triplex plunger pump equipped with variable frequency speed control. For comparison, the wire-to-water efficiency for the centrifugal pump has been added.

The improvement in efficiency is quite dramatic. Table E.1 illustrates the effect of this improved efficiency on actual energy costs. The energy costs shown are typical for a small industrial plant with a 3000 kW total electrical load. The savings would have been even larger if the system resistance had been more friction than static. This evaluation also does not account for the fact that pumps are usually selected for a head larger than required by the system. With a centrifugal pump, this additional head must be throttled, and is a continuous additional loss.

While the intent of this evaluation is to illustrate the energy cost savings for a variable flow system, it should also be pointed out that the energy cost savings and payout period are no less dramatic for a constant flow system.

Referring to the "SAVINGS" tabulation in Table E.1, we see that if the pumps were to operate at full load for the full 8,000 hours, the annual energy cost savings with the power pump would be \$12,720. Since we no longer require variable speed control, the first cost of the complete reciprocating pump system would be reduced \$16,000 to a new total of \$35,000—only \$1,000 more than the centrifugal pump. The *complete* cost of the power pump system could be recovered in a three-year period on energy cost savings alone.

The energy cost savings are impressive. Actual installations on board ocean-going ships, which use the power pumps primarily for in-port boiler-feed service, have demonstrated that such dramatic savings can be obtained.

INDEX